Protein Engineering
by Semisynthesis

Protein Engineering
by Semisynthesis

Carmichael J. A. Wallace, D.Phil.

CRC Press
Boca Raton   London   New York   Washington, D.C.

Library of Congress Cataloging-in-Publication Data

Protein engineering by semisynthesis / edited by Carmichael J.A.
    Wallace.
            p.   cm.
    Includes bibliographical references and index.
    ISBN 0-8493-4727-4 (alk. paper)
    1. Protein engineering. 2. Proteins—Chemical modification.
I. Wallace, Carmichael J. A.
RP248.65.P76P7435 1999
660.6′3—dc21   99-16241
                                                                    CIP

Preface

Protein engineering is in an exponential growth phase. The advent of reliable methods to clone, mutate, and then overexpress plasmid-borne genes has meant that virtually every functional member of a cell's expressed protein complement is amenable to structural manipulation. The impending completion of multiple genome sequence projects means that we will soon have an index of all expressible genes, to most of which we will be able to assign an identifiable product as the practice of proteomics grows, through the powerful combination of two-dimensional electrophoresis and mass spectrometry.

Once components of a proteome are identified and assigned in a genome, homology considerations, and isolation and characterization, should indicate function, but the ultimate and crucial question—how do they work?—is, of course, the province of protein structural biology and engineering. Providing the answers will be a massive undertaking, in comparison to which the current human genome project may seem trivial. Yet the acceleration of progress seen in the 30 years since I commenced the study of biochemistry gives me confidence that, within my lifetime, a substantially complete biochemical description of a complex eukaryote (H. sapiens) will emerge.

Most of that protein engineering will be by the techniques of molecular biology developed in the last 15 years. But the drive to understand protein function in terms of its structure significantly predates those developments. As soon as three dimensional structures of sufficient resolution to match to sequences appeared, obvious target residues for engineering experiments were revealed. The only conceivable routes to their replacement by other side-chains were chemical ones. Chemical modification experiments were already common, and many workers were drawn to total synthesis of peptides and, occasionally, proteins. A small group of pioneers, aware of the limitations of both of the above, developed protein semisynthesis. This had the simple goal of preparing homogeneous defined analogs of proteins using a limited number of chemical steps on synthetic intermediates derived from the native protein.

This book is a record of their achievement and, given the small number of researchers in the field, it is a substantial one. The introductory chapter sets out the general principles that have evolved, but the individual stories of the attainment of those goals with the five most intensively studied models are detailed in the later chapters.

However, the book is not intended as a historical document but as a resource for the active protein engineer. Although most analog generation will be by genetic methods, difficult cases involving more drastic modification, or complex experiments that require site-specific markers and labels, or non-coded amino acids, remain the forte of semisynthesis.

My goal, then, is to provide the protein engineer who aspires to do such experiments with a user's guide. Initially, semisynthesis of different proteins inspired, often, very different approaches, but lately there has been a convergence of thinking and a concentration on generalized methods—and concerted efforts to simplify the technology. In Chapter 3, I present the results of this endeavor with respect to protein fragment religation, undoubtedly the most demanding aspect of semisynthesis and still the focus of intensive research effort. These stereoselective and chemoselective methods have, I believe, made semisynthesis accessible to any experienced protein engineer, particularly since such methods are often synergistic hybrids of genetic and chemical approaches. I would welcome the opportunity to offer practical advice to any interested reader via e-mail, cwallace@is.dal.ca.

Contributors

Seetharama A. Acharya
Albert Einstein College of Medicine
Bronx, New York

J. Gwynfor Davies
St. Georges Medical Hospital
London, United Kingdom

Marilynn Scott Doscher
Wayne State University School of
  Medicine
Detroit, Michigan

Parimala Nacharaju
Albert Einstein College of Medicine
Bronx, New York

Keith Rose
University Medical Center
Geneva, Switzerland

Harald Tschesche
University of Bielefeld
Bielefeld, Germany

Carmichael J.A. Wallace
Dalhousie University
Halifax, Nova Scotia, Canada

Herbert R. Wenzel
University of Bielefeld
Bielefeld, Germany

Acknowledgments

Naturally, I want to thank a number of people and organizations who have provided support and encouragement to me in my involvement in the field. Foremost among them is Robin Offord, my erstwhile supervisor, and the source of inspiration for many of the ideas contained within this book. Indirectly, there are Lord David Phillips, for the vision to back Robin in tackling unknown territory, and Robert Chambers, for supporting and encouraging my move to Dalhousie. My colleagues and coworkers over the past 25 years, both within the Offord group and since have, of course, been indispensable and would be, conventionally, too numerous to name; but nevertheless I will. To all of these I owe a debt of gratitude; they are listed in roughly chronological order.

Anthony Rees	Derek Saunders	Keith Rose
Edward Wawrzynczak	Charles Bradshaw	Monique Rychner
Barbara Battistolo	Amanda Proudfoot	Blaise Corthesy
Angela Brigley	Kevin Leighton	Geoffrey Stone
Lori Campbell	Douglas Craig	Anthony Woods
Jonathan Parrish	Ken Niguma	Anne Rich
Hugh Gillis	Nicole Chan	Fiona Robinson
Ian Dawe	Christian Blouin	Jingyu Mu
Karen Black	Robert Boudreau	Timothy Shutt
	Janet Hankins	

Many outside collaborators have been invaluable, too, both in common projects or simply wide-ranging discussions. In particular, I should thank Geoffrey Moore, Giampietro Corradin, Gianfranco Borin, Maurice Schmidt, Bruno Amati, Yvonne Paterson, Hans Bosshard, Stephen Kent, Ian Clark-Lewis, Michael Smith, Guy Guillemette, Martin Caffrey, Yoshikazu Tanaka, David McIntosh, Esa Tuominen, Paavo Kinnunen, Grant Mauk, and Gary Brayer.

Financial support for my research from the Medical Research Councils of Great Britain, Canada, and South Africa, the Natural Science and Engineering Research Council of Canada, the Swiss National Scientific Research Foundation, the Dalhousie Medical Research Foundation, and the Boehinger Ingelheim Foundation is gratefully acknowledged.

About the Author

Carmichael Wallace was born in the UK and received his secondary education at Wolverhampton Grammar School, only a short distance from where Charles Mac-Munn discovered cytochrome c in 1884. He studied biochemistry as an undergraduate at Wadham College, Oxford from 1968 to 1972, having the good fortune to have Prof. R.J.P. Williams as tutor. Unaware at the time of the coincidence described in the first sentence, he chose cytochrome c as the subject of graduate work. D.Phil. studies in the Laboratory of Molecular Biophysics in Oxford, completed in 1976, were supervised by Robin Offord. Prof. Offord had established a protein chemistry laboratory that was seminal in the development of semisynthesis, but most of the unit, directed by Prof. Sir David (later Lord) Phillips, was devoted to protein crystallography. The intellectually stimulating atmosphere of this Institute, which trained many of the present generation of prominent crystallographers, instilled a continuing passion for the understanding of protein structural biology.

When Robin Offord left Oxford in 1980 to take up the position of Head of Medical Biochemistry at the University of Geneva, Margaret Thatcher had commenced the decimation of the British University system, so Dr. Wallace accepted an offer to join him there as part of a team that made substantial contributions to the development of semisynthesis over the next seven years. In 1987, he left to set up his own laboratory, as Associate Professor at Dalhousie University in Nova Scotia. Research has continued on the development of semisynthetic methods and into the structure-function relationships of cytochrome c; in addition, molecular biological approaches to protein engineering of cytochrome c have become part of his Laboratory's arsenal. In 1993, he was promoted to Professor and has been chair of the Natural Sciences and Engineering Research Council of Canada's Cell Biology Grant Selection Committee, a Councillor of the Canadian Society of Biochemistry, Molecular and Cell Biology, and an MRC of South Africa Distinguished Visiting Scientist during his sabbatical year at the University of Cape Town. Carmichael Wallace was married in 1974 to Lorna Llewellyn; they have three children, only the eldest of whom has so far commenced the study of biochemistry!

Table of Contents

1 Protein Semisynthesis

Carmichael J.A. Wallace

CONTENTS

1.1 INTRODUCTION

Semisynthesis is a means of engineering protein structure that is based on chemical manipulation of the protein itself. Now that we have become familiar with protein engineering methodologies that depend on mutagenesis of the gene that encodes the protein of interest, the obvious question to be answered is that of why bother with the more direct, but more complex, operation at the level of the protein?

The answer lies in the limitations of the genetic approach. The enormous advantages of using living organism as bioreactors, of fidelity and potentially unlimited productive capacity, are slightly offset by their intolerance of toxic or grossly deformed structures, and their incapacity to appropriately insert non-coded amino acids or specifically locate labels for biophysical techniques. However, the growth in familiarity with, and use of, mutagenic protein engineering has served to increase the ambitions and expectations of the engineers so that the interest in chemical methods that substitute for or complement the genetic approach will, in all probability, increase as a consequence. The purpose of this volume, then, is to acquaint the reader with both the philosophy and methodology of protein semisynthesis and to provide some practical examples of both the achievements of the method in illuminating protein structure-function relationships and how to set about optimizing the engineering of any interesting protein.

0-8493-4727-0/00/$0.00+$.50
© 2000 by CRC Press LLC

In principle, we could prepare any structural variant of a natural sequence by total synthesis. The de novo approach is normally undertaken via the sequential addition of amino acids on an insoluble support. Solid-phase peptide synthesis is in common use[1] but generally, for technical reasons, for the preparation of structures smaller that the average protein.

Such structures can be, and are, stitched together to make protein-sized assemblages, but an argument for semisynthesis first articulated by Offord[2] is that the desired synthetic product most often differs from the natural protein at only a single amino acid residue, and it therefore makes sense to use the naturally occurring protein as a ready-made source of the bulk of the final sequence.

Clearly, this limits semisynthesis to those proteins of relative natural abundance, and the subjects of later chapters reflect that limitation. However, another contribution of the emergence of genetic manipulation is the ability to produce in bulk, in heterologous hosts, proteins that were of low abundance. Now, therefore, almost any protein should be a potential target for semisynthetic engineering.

Total synthesis and semisynthesis are not rival techniques; they have different applications, and there is genuine synergy between them. The non-natural portions of most semisynthetic proteins are generated by solid-phase peptide synthesis and, as described in Ch. 3, the selective methods of fragment condensation in total synthesis take as their inspiration the earlier adoption of such approaches in semisynthesis.

The crucial role of proteins in cellular organization, and the staggering variety of function they manifest, is a consequence of their extreme structural variability. This means that synthetic schemes are inevitably chemically complex and may necessarily be unique to each protein. Thus, the development of the fully versatile scheme represents the investment of a great deal of time and effort. An idealized scheme is shown in Figure 1.1, and real examples are shown elsewhere in this book (e.g., Chapter 8, Figure 8.8). Not surprisingly, an imperative in the development of semisynthetic methods has been simplicity. Most of the successful and productive routes are so because they represent the exploitation of "short-cuts" that have vastly simplified the idealized fragment-condensation route of Figure 1.1. Offord[3] has categorized several types of protein semisynthesis, but all the other forms represent simplification of the fragment-condensation archetype.

We therefore will first examine the mechanics of the full-blown approach before dealing with the variants. The individual chapters that follow provide numerous examples of all types, and a great variety of solutions to the problems that great structural variation throws up, although a unifying theme will be the search for simplicity. But to make semisynthesis accessible to every aspirant protein engineer, an absolute requirement is a generalized simplified approach that can be adopted as a starting point for any protein. The reader can assess the progress we have made toward that goal.

The concept of semisynthesis did not arise fully formed, but, like most such ideas, evolved from work in a number of protein chemistry laboratories, and initially from those concerned with the study of ribonuclease A. The observation by Anfinsen[4] that the active conformation of this protein could be recovered after complete denaturation and disulphide bond reduction led to the formation of the "thermodynamic

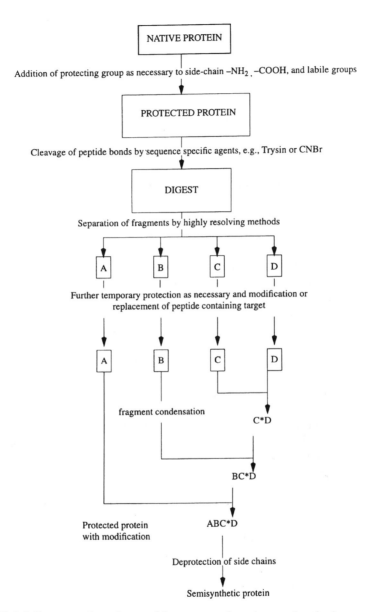

FIGURE 1.1 Representative scheme of fragment-condensation semisynthesis summarizing the steps outlined in this chapter.

hypothesis" that won him the Nobel Prize. It states that the native tertiary fold is that in which ΔG of the whole system is at a minimum and hence is driven by the optimization of the interatomic interactions that can be derived from, and so determined by, the sequence. This inspired the confidence that chemical synthesis could yield a functional protein; the proof was provided by the development of the ribonuclease S system (Chapter 5).

Thus, the first semisynthesis—a functional association between a synthetic peptide (residues 1–13) and a fragment (21–124) of the natural protein—then called a *partial synthesis,* was achieved in 1963 by Hoffman's group.[5] That same year, Anfinsen's group[6] achieved the same type of semisynthesis by carboxypeptidase A truncation of the natural peptide by 5 residues to give fully active 1–15:21–124.

Semisynthesis continued with this system but did not expand into other proteins until the late 1960s. Offord described a scheme for fragment-condensation semi-synthesis in 1969[2,7], and the first model for this type of synthesis was provided by Burton and Lande[8] in 1970, in which they ligated a synthetic tetrapeptide to the N-terminus of natural porcine ß-melanotropin (8 residues) to prepare an analog of the human 12-residue hormone. Other early targets were insulin[9–10] and soybean trypsin inhibitor.[11] There followed, in the 1970s, a rapid expansion in the development and use of semisynthesis, as the strategy was adopted by a number of laboratories for structure-function studies, often of a protein in which they had an established interest. In some cases, more than one group took up the challenge of an individual protein (e.g., cytochrome *c*) and developed unique solutions to the problems it posed. This developmental phase is largely described in the individual chapters that follow and was crowned in 1977 by the gathering of most of those involved in the field in Italy for an "International Meeting on Protein Semisynthesis," organized by Offord and Di Bello. The proceedings volume of this meeting[12] provided a valuable com-pilation of most of the research then current and was followed in 1980 by Offord's monograph,[13] which contains a comprehensive assembly of strategies and protocols. This book will not therefore attempt to repeat the endeavor, but it is appropriate now to summarize the principles, using examples of proteins that are not covered in succeeding chapters. For further technical detail, the reader is referred to that monograph, the later chapters of this volume, and the references therein.

Since those two major works, progress in the field reflects two major trends. First was the realization, in the early 1980s, of the hope that enzymes could be exploited as primary tools of resynthesis as well as fragmentation in fragment-condensation synthetic schemes. The second was the first application of site-directed mutagenesis to proteins under semisynthetic scrutiny in the mid-80s. This has led some investigators to switch entirely to the genetic technology, while others have exploited the unique capabilities of semisynthesis in new niche roles and worked on techniques to expand those capabilities. These developments have been recorded in a number of reviews published during that period[14–20] and are considered in detail in the rest of this book.

1.2   THE SEMISYNTHETIC STRATEGY

The objective of most semisynthesis is the creation of a point mutation at a single site in a protein. This may be to create a clinical or industrial tool, but most such analogs will be for investigation of the basis of the relationship between protein structure and function and will be informed mutations. The information that defines the targets, then, will include at least alignments of sequence from multiple species, and probably a crystal structure of the protein. In fact, it was the availability of this data, as much as of plenty of the protein, that determined the early choice of subjects for semisynthesis. Indeed it was probably the desire to make use of this type of structural data that drove the development of all protein engineering technology. Most semisynthesis groups had their origins in structural biology laboratories.

The basic scheme of fragment condensation semisynthesis as set out in Figure 1.1 includes a number of principal steps. Each of these phases will require several distinct chemical and physical processes, and the potential for variation within steps is great. Before all else, the natural protein, extracted in pure form from its characteristic tissue or organism, a heterologous host or a chemical supplier's bottle, will probably be *chemically modified* as a preparatory measure for a later step. Then *fragmentation* will normally ensue. The exception is in the case of *stepwise semisynthesis* when subsequent chemistry is performed with the whole protein. Fragmentation must be followed by the *separation* from one another and *purification* of the fragments. The fragment containing the sequence of interest is then subjected to procedures that will enable the desired *residue substitution*. At this stage, temporary chemical modification (*protection*) may also be used to facilitate *religation*. Depending on the number of fragments involved, this step can vary vastly in degree of complexity, and it will probably be followed by a final *deprotection* phase to remove any temporary chemical masking employed. Religation via peptide bonds is avoided in two variants on the scheme. In *non-covalent semisynthesis* or *disulphide-bonded semisynthesis* the final product has the global structure of the holo-protein, but the fragments are linked together by either non-covalent or disulphide, rather than peptide, bonds.

1.2.1 FRAGMENTATION

Statistically, target residues will be remote from either terminus and access to them will require fragmentation of the protein. The means chosen will perhaps reflect the proximity of the target to a cleavage site, but mostly it will be determined by the balance between good access and a limited, manageable number of fragments for reassembly. Early schemes were sometimes overambitious and required too many religation steps. The subsequent limiting of ambition and ingenious approaches to access have meant that the norm in many semisyntheses are methods that yield just two fragments.

Given the size of most proteins, this might at first glance seem a difficult objective, but two successful approaches have evolved to achieve it. The most specific and efficient means of protein fragmentation, of course, is the use of proteolytic enzymes. Their deployment as a primary tool of protein sequencing had led to detailed knowledge of their properties. Yet, despite the often narrow specificity of these enzymes, digests would normally provide large numbers of peptides. The important breakthrough was the discovery that digestion could be limited by harnessing the protective properties of the native conformation. Limited proteolysis by conformational restriction is one of the pillars on which protein semisynthesis is built, and it will be discussed in detail in some of the subsequent chapters. Normally, complete digestion requires prior denaturation of the substrate, or extended digest times, or both. Thus, it was found that, in many cases, short exposure to proteases of the protective, folded, conformation would restrict cleavage to just one or a few of many potentially vulnerable bonds.

The second set of tactics consists of chemical ones. These require cleavage to be targeted to rare sites, using highly specific reagents. An example is the use of cyanogen bromide, which is absolutely specific for methionine residues, which have an average frequency of occurrence in a sample of 200 proteins of just 1.7%.[21] In conformity to this average, Lysozyme (129 residues) has two methionines at positions 12 and 105 (Figure 1.2), providing in principle an ideal system for structural engineering at either end of the molecule.[22-24] The figure shows that methionine residues undergo conversion to homoserine in the course of this cleavage reaction. This is at once a blessing (see Chapter 3) and a curse: the religation of these three fragments to give [Hse[12], Hse[105]] lysozyme results in an inactive product, no doubt due to the transformation.[24] Although this potential has not been further developed in Lysozyme, in another protein, cytochrome c (Chapter 7) that also contains two methionine residues, cyanogen bromide fragments have been the starting point for the semisynthesis of well over 100 different analogs.

Tryptophan occurs with a frequency of 1.1%, and specific reagents are available for this residue too.[25] Myoglobin has two tryptophans, both near the N-terminus at positions 8 and 13, so that cleavage gives two small and one large fragment. Again, chemical cleavage agents specific for tryptophan inevitably result in side-chain modification and, since this is substantially more drastic than the methionine-to-

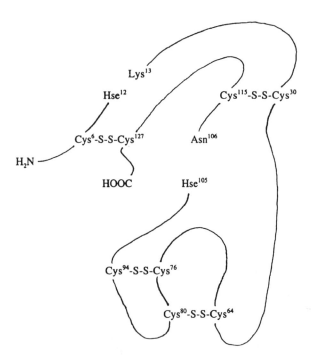

FIGURE 1.2 Sketch of the structure of hen egg-white lysozyme, much simplified, showing the locations of the disulfide bridges in relation to the CNBr cleavage sites.

homoserine transformation, the small peptides are not used further in semisynthesis, but, being small, can readily be replaced by synthetic analogs in a fragment condensation semisynthesis.[26]

Although, in general, enzymes are of lower specificity, that can be enhanced by chemical means. The most common example is the use of chemical modification (temporary protection) of lysine residues so that proteolysis by trypsin is restricted to peptide bonds C-terminal to arginine residues. An excellent example is provided in work on phospholipase A2.[27] The structure of the pancreatic zymogen, prophospholipase A2 is shown in Figure 1.3. Amidination with methyl acetimidate (elsewhere termed acetimidation or acetimidylation) blocks all nine lysine ε-amino groups, leaving five arginines as potential sites of sensitivity to trypsin. One of these, at position 7, is the site normally cleaved in the transformation of zymogen to active enzyme in the gut by trypsin and is very labile. In the active enzyme cleavage occurs at only one of the four arginines, residue 13, so the chemical restriction imposed by acetimidylation may be complemented by conformational protection as discussed above. However, a third factor may come into play here. The other three arginines (50, 59, and 106, Figure 1.3) are adjacent either to a disulphide linked cysteine or an acidic residue. Similar apparent nearest-neighbor

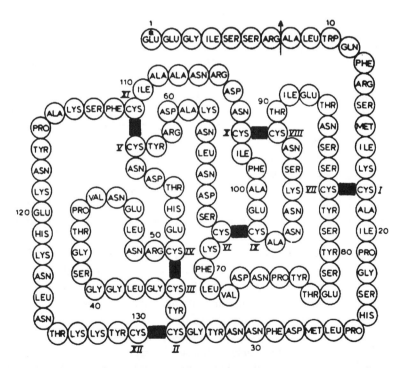

FIGURE 1.3 Primary structure of the zymogen form of porcine pancreatic phospholipase A₂, showing the zymogen activation site at Arg 7, and the other arginine residues, potential cleavage sites for trypsin in N^ε-acetimidyl PLA₂ at positions 13, 50, 59, and 106. Residue 1 is pyroglutamic acid. Reprinted with permission from Slotboom, A.J., and deHaas G.H., *Biochemistry,* 14, 5394, 1975. Copyright 1975 American Chemical Society.

effects have been noted in cytochrome c (28; N. Chan and C.J.A. Wallace, unpublished data). Unfortunately the small fragment, residues 8–13, did not complement the large fragment, so non-covalent semisynthesis was not possible in this example.

1.2.2 FUNCTIONAL GROUP PROTECTION

The temporary masking of side-chain functional groups has purposes other than restriction of proteolysis. It is important to ensure that side-chain carboxyl and amino groups do not participate in any peptide bond-forming chemistry, and the integrity of labile functional groups is not threatened by side reactions during such procedures. Similarly, both classes of groups should be protected against unwanted modification during procedures to manipulate peptide sequence. To discriminate between the α- and ω-COOH and α- and ϵ-NH$_2$ groups of the fragments for condensation or manipulation, the simplest course is to apply that protection before fragmentation, although there may be good reason and appropriate techniques to do so after cleavage. Modification of specifically side-chain functional groups can be applied at either stage.

The trend toward conformationally directed and chemoselective methods of fragment condensation (Chapter 3) means that current strategies employ minimal (or no) protection, but in the early days of semisynthesis, this was an important point of focus, and protection remains crucial when chemical procedures to manipulate peptide sequence by, for example, sequential degradation are employed. Amino protection is often applied, but prior carboxyl group protection is much rarer. The reasons for this are manifold. One is practical: there are fewer useful and convenient blocking agents. Sequential degradations from C-termini almost inevitably employ enzymes with inherent specificity for α-COOH. In general, when coupling peptides, it is the carboxyl group that is activated, and the procedure itself can have the power of discrimination. Examples are discussed in Chapter 3. The carboxyl component in many of the semisyntheses we will encounter is small (as little as one amino acid) and is prepared synthetically, so that any necessary side-chain protection is built into the sequence during its preparation. Protection of other functional groups is now rarely encountered, since the need for them was often predicated on the harsh and less discriminating chemistry for -NH$_2$ or -COOH protection, or fragment condensation.

Therefore, no detailed account of reagents and methodologies is presented in this volume. Offord provides many worked examples of protection, both prior to and post-cleavage, in his monograph[13], and the chapters on individual proteins herein will demonstrate many different solutions to problems requiring protection. However, the reader may find useful a compilation (Table 1.1) of the amino and carboxyl protecting groups that have found favor in semisynthesis.

Offord[13] also describes the logistics of protection. If more than one type of group is to be masked, then compatibility of blocking methods is an important consideration, although in semisynthetic schemes the obsession with orthogonality often seen with total synthesis is usually unnecessary. The most extensive protection pattern ever employed was that devised for the Offord group's ambitious attempt at lysozyme semisynthesis. They were working with 11 tryptic peptides, and prior protection with

TABLE 1.1
Protecting Groups for Side-Chain Functions

Name	Structure	Properties	Deprotection	Ref.
Amino group protection				
Acetimidyl	NH ‖ CH$_3$-C-	Retains + charges, water solubility	High [NH$_3$]	29
t-Butyloxycarbonyl	O ‖ (CH$_3$)$_3$C-O-C-	Uncharged, lowers H$_2$O solubility	Strong acid	30
Citraconyl	O ‖ HOOC-CH=C(CH$_3$)-C-	Changes + to − charge, maintains solubility at neutral pH	pH 2	31
Maleyl	O ‖ HOOC-CH=CH-C-		pH 3.5	30
Methylsulphonyl-ethyloxycarboxyl-	O ‖ CH$_3$-SO$_2$-(CH$_2$)$_2$-O-C-	Uncharged, somewhat reduces water solubility at alkaline pH	Strong base	32
Cbz-Met*	O ‖ Cbz·Met·C-		CNBr	33
Phthaloyl*		Uncharged and quite nonpolar, lowering aqueous and increasing organic solubility	Hydrazine	34
Carbobenzoxy	CH$_2$·O·CO		Very strong acid	2
-dimethyl*	(CH$_3$-)$_2$	Retains charge, solubility	Not removed	35
-Guanidinyl*	NH$_2$-C- ‖ NH	Retains charge, solubility	Not removed	35
Carboxyl group protection				
methyl	CH$_3$-O-		Strong base	36
benzyl	CH$_2$·O·	All eliminate charge and decrease polarity to varying degree	Very strong acid	30
anisyl	CH$_3$-O- CH$_2$O-		Strong acid	30

*These groups are less useful and infrequently encountered.

the maleyl group was necessary both to restrict trypsin action to arginine residues and to mask ε-amino groups during subsequent chemical coupling of the fragments. Prior to that, disulphide bridges were reduced to permit resolution of the fragments after proteolysis, and the highly reactive cysteine sulfhydryls were blocked by reaction with sodium sulphite, giving the S-sulphonate derivative. After fragmentation, peptide α-amino groups so revealed were masked by either the t-butyloxycarbonyl (Boc) or benzyloxycarbonyl (Cbz) groups. Then, the sole histidine residue was blocked by alkylation with either the Cbz- or Boc-trifluoroethylamine group. Protection by the formyl group of the other potentially labile side-chain, tryptophan, was investigated, but deemed superfluous. Finally, carboxyl groups were converted to benzyl or anisyl (p-methoxybenzyl) esters by use of the corresponding diazoalkane.[30] To distinguish between α- and ω-COOH groups, advantage was taken of the esterolytic activities of the protease that generated the fragment; trypsin. Only benzyl or anisyl esters of the terminal carboxyl group are substrates for this activity. Offord[37] has suggested an equally ingenious approach to discrimination between α- and ω- carboxyls in CNBr fragments. Since these are released with C-terminal homoserine lactone, the α-carboxyl is already differentially protected, and esterification of side chains can be undertaken with groups that will resist the mild alkaline conditions necessary to hydrolyse the lactone and free the C-terminus for peptide bond formation.

Other approaches to the use of enzymatically labile groups as protecting agents for both side-chain and terminal functions in peptides have been reviewed by Glass.[38]

1.2.3   FRAGMENT ISOLATION

Limiting the number of fragments generated by the proteolysis system chosen not only makes the task of reassembly easier, it also simplifies the separation of the fragments from one another. Using a protocol that generates just two or three fragments, it is likely that the molecular weights will be so disparate that excellent resolution on appropriate gel-exclusion media will be obtained. This has therefore proved to be the first or only separation method for the majority of the proteins subject to semisynthesis. Where complete resolution has not been achieved by sizing methods only, ion-exchange chromatography has generally been the method of choice. Numerous examples are provide in subsequent chapters and the references therein.

These methods were developed for, and are often dependent on, aqueous buffers. It seemed reasonable to adopt the techniques that had worked so well for protein purification. Thus, there was a need in peptide handling in semisynthesis to maintain water solubility, and so an additional role of the protecting groups discussed above, and hence a consideration in the choice, has been the retention of polarity (see Table 1.1).

If that is achieved, then these methods will also be useful for purification and isolation of the products of peptide structural modification, and, of course, of fragment religation in the process of reassembly of the semisynthetic protein. Occasionally, a peptide will be derived from a nonpolar sector of the protein primary sequence, or polarity will be reduced by the need to protect additional protecting groups after fragmentation. In such cases chromatography using organic solvents or cosolvents may be required. These methods, including reversed-phase HPLC and partition chromatography on lipophilic gel-filtration media, can be extremely powerful and peptides differing by the presence or absence of a single methyl protecting group can be satisfactorily resolved.[36]

It is obvious that complete resolution of proteolytic fragments requires the reduction of any disulphide bridges that link them. However, retention of those bonds during an initial chromatographic step, then a second stage after reduction, may prove a facile means to optimize separation. However, I know of no scheme that has adopted this tactic. Another important consideration is that many of the

fragments generated in limited proteolyis schemes are capable of mutual complementation to give non-covalent complexes that are, in fact, stable to chromatography.
This is particularly true of the fragment systems that have been chosen as the basis
for semisynthesis, and so these are generally resolved in denaturing media. Gel
filtration in weak acid media is common, and ion-exchange buffers that contain high
proportions of urea have been very useful to a number of groups, giving sharp peaks
and very reproducible runs.[39]

1.2.4 FRAGMENT SEQUENCE MODIFICATION

In the vast majority of semisyntheses, only one of the fragments generated by
proteolysis will have its sequence changed. In fact, change would normally be
confined to a single residue, since the functional consequences of a unique modification will be simplest to interpret. However, there are times (see Chapter 8 for a
number of examples) when two or three residues in the same peptide have been
replaced.

The ways in which sequence change can be achieved are many, and indeed many
ingenious tactics have evolved in the development of semisynthesis. The simplest
is the replacement of the entire fragment by its homolog from the protein of another
species. Such interspecific hybrids—chimerae—have been often prepared with
cytochrome c peptides, and more recently with hemoglobin fragments using entirely
naturally derived peptides. In other cases, for example phospholipase A_2, the same
sort of chimera of natural sequences has been achieved by combination of a natural
and synthetic peptide.

Most of these studies have come to the perhaps intuitively obvious conclusion
that these hybrids rarely differ functionally from the parent protein. The variation
introduced this way is confined to naturally evolutionary variable residues, though
if there is some functional distinction between the proteins from the two parent
species, then the chimeric construct can aid in the location of that structural element
responsible for the difference. However, even this analysis will be complicated if
the two species differ at several residue positions.

Normally, though, the target of modification will be a residue that interspecific
sequence comparisons has shown to be conserved, and thus of mechanistic consequence, and so some structural manipulation of a peptide sequence is required. That
could be as simple as treatment with one of the reagents used for whole-protein
chemical modification. That technique of protein engineering is severely limited by
the usually indiscriminable nature of the reactions utilized. However some reagents
can be directed to a single type of side-chain, and if that residue occurs only once
in a proteolytic fragment, uniquely specific modification is achieved. This approach
to analog generations was adopted early in the development of semisynthesis when
in 1965 chemically modified S-peptide was incorporated in the Ribonuclease S
complex.[40] It has since been adopted for the covalent semisynthesis of cytochrome c
(Chapter 8).

The range of reagents used for the covalent modification of proteins is very
wide, but all aspects and very many examples of their use have been comprehensively
described by Lundblad and Noyes.[41] Some of the residues likely to be useful are

cysteine, tryptophan, histidine, arginine, methionine and tyrosine; and the latter three have been targeted. Relatively specific reagents are available for each, and either the natural frequency of these residues or the fragmentation method employed is likely to result in peptides with single representations of these side-chains. Another strategy in which the sequence of natural fragments is modified before reincorporation is one of sequential degradation and reassembly, either at the N- or C-terminus of the peptide. Although aminopeptidases exist that might give controllable and limited removal, little exploration of this potential has occurred. The reason is the ease and convenience of the Edman degradation (Figure 1.4). Provided the ε-amino groups are blocked as discussed above, reaction of phenylisothiocyanate is confined to the α-amino group. Strong acid causes the cyclization and spontaneous peptide bond cleavage of the N-terminal amino acid adduct, and leaves the α-amino group of the subsequent amino acid free, either for a further cycle of the Edman reaction, or a peptide bond-forming reaction that will introduce the semisynthetic replacement. Examples of its use will be found in many of the chapters that follow.

The exopeptidases that degrade peptides from the C-terminus are both more specific than aminopeptidases and more used, given a lack of a useful chemical degradation method. Carboxypeptidase B activity is naturally limited to lysine and arginine residues, although we have found a useful ability of the enzyme to remove ε-acetimidyl lysine.[36] The specificity of carboxypeptidase A is much broader and so can "run" along a sequence of bulky and nonpolar residues. However, charged and small polar residues are essentially untouched so that the enzyme can be stopped in its tracks, and clean one- or two-residue truncations can be achieved.[36] Sometimes, truncated species, such as the ribonuclease S peptide shortened by five residues with carboxypeptidase A or cytochrome *c* peptides shortened at either N-or C- terminus (Chapter 8), can be intermediate participants in non-covalent semisynthesis, but more frequently the residues removed will be reinstated in the sequence.

In this context, peptide bond formation is usually via chemical activation of the carboxyl group participant in the dehydration. A number of activated species have been tried, and some examples are included in the discussion of stepwise semisynthesis below, and detailed in Offord's monograph.[13] Perhaps the most popular reagents for the extension of truncated proteolytic fragments have been the Boc amino acid N-hydroxysuccinimide esters. These are quite safe and stable, easy to make, and widely available from suppliers, thanks to their popularly in many areas of research and manufacturing. Although not water soluble, they are used in polar nonaqueous solvents (DMF, DMSO) in which most minimally protected peptides are soluble. Reaction readily goes to completion and alternative N-protection can be employed if the removal conditions for the Boc group are too harsh. The combination of Edman degradation and subsequent elongation with Boc amino acid NOHS esters has been successfully employed to replace five amino acids in some natural peptides and there is no reason to suppose that could not be further extended. However, considerations of time and economy might motivate an alternative route.

The usefulness and effectiveness of this tactic, though, has another important consequence. If sequential degradation has been undertaken at the C-terminus of a fragment, then residue replacement is easier to achieve by addition of the missing residues at the N-terminus of the adjacent peptide via NOHS esters, prior to fragment

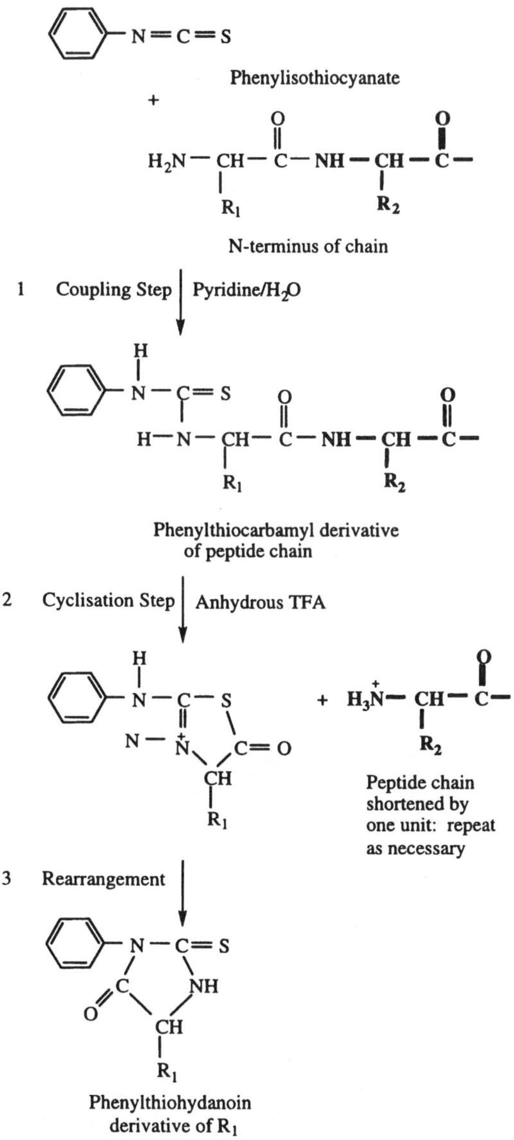

FIGURE 1.4 Reaction scheme of the Edman degradation.

religation. It is also much more efficient to adopt this approach, since the activated carboxyl species is of necessity used in excess in any fragment elongation reaction.

An example that embraces all the above aspects is presented in the semisynthesis of cytochrome *c* from its three CNBr fragments,[36] and can be summarized as in Figure 1.5. Although the replacement of methionine by homoserine implicit in the CNBr fragmentation method is tolerable at position 65, it is not at 80. Hse can be specifically removed, however, by CpA. (If desired, residue 79 can be excised with

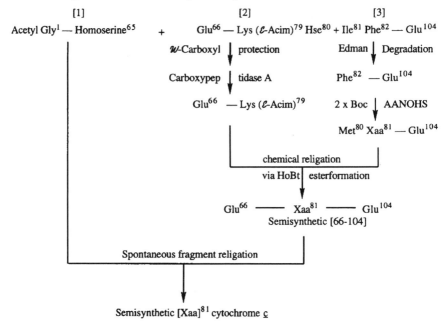

Three fragments derived from -NH2 protected cytochrome c

FIGURE 1.5 Semisynthetic route to a residue 81 analog of horse cytochrome c based on initial CNBr fragmentation. Typical chemical and enzymatic manipulations of the smaller fragments to truncate and extend them precede a classical chemical coupling. The product fragment is then condensed with heme-containing fragment 1–65 by the stereoselective spontaneous religation method.

CpB). One cycle of degradation and resynthesis of residue 81 (which could be extended) is followed by addition of methionine to the N-terminus of that peptide. Fragment condensation follows.

However, the replacement of single residues at the C-terminus of peptides is not unknown. The context in which this usually occurs is in those semisyntheses where reversed proteolysis is a primary tool (Chapter 3), and the most elegant example is provided by the use of carboxypeptidase Y, where degradation and resynthesis occur in a single-pot operation. Extensive development work in this area was done by the Carlsberg Research Group (e.g., Ref. 42) and has been reviewed.[43,44]

CpY, unlike A or B, is a serine protease and hence a covalent acyl-enzyme intermediate is initially formed. In the presence of sufficient amino acid ester or amide, aminolysis of the acyl-enzyme is favored over hydrolysis and a transpeptidation is achieved. More complex, multistep processes with the same goals have been elaborated with other enzymes, and are discussed in the same reviews,[43,44] or in the chapters on insulin, protease inhibitors, or cytochrome c that follow.

The replacement of an entire proteolytic fragment by a totally synthetic substitute is perhaps the most simple and direct approach, and has eventually found favor with almost all research groups using semisynthesis. Whether this is achieved by solid-phase or solution methods, in one piece or piecemeal, depends on the problem to be solved. This is not the place for a extensive discussion of peptide synthesis, or the roles for which one technique might be more appropriate that another. Consult Bodanszky's monograph for a comprehensive account of the entire field[45] or Kent's review on solid-phase peptide synthesis.[1] The strategy is applicable to extremes. The entire 39-residue fragment (66–104) of cytochrome c has been made in many variants by solid-phase methods to replace the natural CNBr fragment.[46] In phospholipase A_2, CNBr treatment removes just an N-terminal octapeptide, and this was restored in variant forms by first coupling on a single amino acid, then a 7-residue fragment made by the solid-phase method.[47] Restoration of the N-terminal 14-amino acid fragment of myoglobin removed by BPNS-Skatole cleavage at Trp has been achieved by the successive addition of two solid-phase peptides, of 8 and 5 residues, after initial coupling of a single amino acid at position 14.[26] The adoption of this latter tactic for both PLA$_2$ and Mb seems to be driven by a need to avoid sensitive amino acids at the C-terminus of a fragment to be activated for coupling. In Mb the subsequent addition of the remaining sequence as two peptides was necessitated by the very low solubility of the entire 1–13 fragment in the organic medium required by the coupling method.

Although solid-phase methods have dominated in this context, solution syntheses have been used by the Tesser group to make substantial segments of the cytochrome c sequence (Chapter 8). Usually, short stretches are prepared by stepwise methods, then, after purification, linked by successive fragment condensations as illustrated in Figure 1.6.

1.2.5 Fragment Condensation

Figure 1.6 showed the type of fragment condensation typically used in the total synthesis of small peptides. The earliest approaches to ligation of the larger fragments required in a protein semisynthesis used the same logic and methods. Thus, as discussed above, even a minimal protection scheme required long-term blocking of ε-NH$_2$, ω-COOH and vulnerable side-chain groups, and temporary masking of α-NH2 functions. The problem is the formation of a covalent (amide) bond between two groups with inherently low reactivity, carboxylic acid and amine, by dehydration. The obvious complication of distinguishing between terminal and side-chain groups on peptides has been considered above. To achieve useful reaction rates, one of the participating groups must be activated: usually, and in semisynthesis exclusively, the carboxyl group. The activation takes the form of a dehydration to form an unstable bond readily susceptible to aminolysis. Such bonds are inevitably even more labile to hydrolysis and reaction media must be kept rigorously anhydrous. This fact can lead to solubility problems if protecting groups are chosen to facilitate handling in aqueous media. Mass action considerations require high concentrations of reacting peptides in solution, and often substantial excesses of the activated component, which can be uneconomical. The solvents employed, or the high reactivity of the agents needed for activation, may be damaging to the peptide, and in the process of coupling, side-reactions, or racemisation of the activated C-terminal amino acid, may occur.

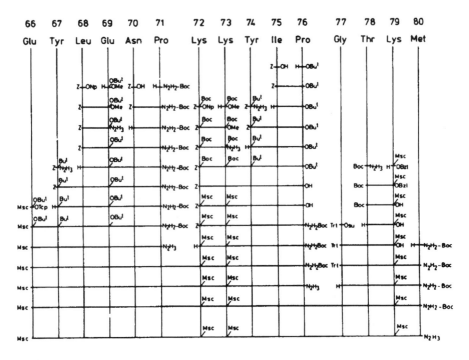

FIGURE 1.6 Classical "solution synthesis" approach to the creation of a totally synthetic fragment for later incorporation into a semisynthetic analog of cytochrome *c*. Taken from Ref. 12 with permission.

With this litany of tribulation, it is sometimes amazing that multiple-fragment condensation semisyntheses were attempted. One of the most challenging undertaken was that of lysozyme,[48] in which seven tryptic peptides were spliced to give two large segments, representing the bulk of the protein sequence, from residue 1 to 45 and residue 62–113. This study served as a trial of a number of protocols used in total synthesis, to evaluate their efficiency and the level of unwanted products. The results led to the selection of a methodology that has moderate efficiency and causes no racemisation: this has since proved applicable to fragment condensations of other proteins, including insulin and cytochrome *c*. Dicyclohexylcarbodiimide is the dehydrating agent used to form an ester between α-COOH and 1-hydroxybenzotriazole in DMF solution. After an empirically determined optimum time, the preformed ester is mixed with the amino component in basic conditions. The resulting aminolysis of the active ester can give yields of up to 40% of coupled product with equimolar amounts of the two fragments.

An alternative is the use of the azide of the α-carboxyl group as the activated species. This tactic has been used in fragment condensation semisynthesis of lysozyme,[48] myoglobin,[49] and cytochrome *c*,[50] most frequently when activating totally synthetic fragments. One reason for this choice is that the hydrazide group, the azide precursor, can be built into the peptide during synthesis (Figure 1.6). With natural peptides, the carboxyl group must be esterified, then subjected to hydrazinolysis.[48] If the peptide is prepared by solid-phase synthesis, the protocol is even more elegant: the finished peptide can be removed from the supporting resin by hydrazinolysis, directly yielding the intermediate.

Offord[37] suggested a related tactic. CNBr fragments have C-terminal homoserine; under the low pH digest conditions, its favored form is the lactone or internal ester. This ring structure is unstable and can be opened for subsequent coupling reactions, at higher pH. Alternatively, incubation with hydrazine can directly yield the hydrazide intermediate for an azide coupling.

A great deal of ingenuity and hard work has gone into making these types of methods appropriate for a role in semisynthesis, but that has not prevented a gradual drift from condensation methods that rely on strong activation and mass action to more subtle and directed techniques that can be collectively termed *entropic activation*. The movement was fueled by developments that were both serendipitous and the consequence of applied logic.

It had been appreciated since 1937 that proteolytic enzymes could be exploited as a tool in peptide synthesis if the reaction conditions could be so manipulated as to shift the equilibrium in favor of peptide bond formation. The enzyme could both accelerate attainment of equilibrium and impose an absolute specificity of reacting groups.

Since the majority of limited fragmentation protocols employed for semisynthesis relied on proteolytic enzymes, it was clear that the most facile means of fragment condensation would be a direct reversal of this process. It was the work of Laskowski's group with protease inhibitors, far more recently, that provided the initial clues as to how this could be done for sizable peptides and made reverse proteolysis a reality.

Kinetic and thermodynamic limitations to reformation of the peptide bond are massively reduced by two factors: there is a single site of susceptibility to protease, and the fragments are held together, and the cut site termii in proximity, by the retention of the covalent and non-covalent forces that normally determine conformation. Since the evolutionary pressure on protease inhibitors is to long-term blockage of the enzyme active site, it is no surprise that this class of protein fulfills these criteria. While we normally consider proteolysis to be irreversible, with trypsin the K_{HYD} for soybean trypsin inhibitor (STI) was only 4.0.[51] The recognition of this fact was followed by its exploitation in earlier semisyntheses of both STI and trypsin-kallikrein inhibitor.[52] The extension of enzyme-catalysed peptide bond synthesis to semisynthesis, in general, was a consequence of a chance observation. When incubated with trypsin in 60% glycerol, the K_{HYD} for STI shifted to 0.5.[51] The reason for this change was determined to be the increase of the pK of the carboxyl group in the presence of the organic cosolvent, and in some cosolvent systems this is sufficient to decrease K_{HYD} (increase K_{SYN}) 80-fold. Details of the application of this principle are provided in Chapter 3, but an immediate consequence of the discovery was its testing on the majority of the two-fragment non-covalent complex systems then known, using the enzyme that generated them as catalyst, but at far higher concentration than in the hydrolytic process. Since such systems usually fulfilled the criteria discussed above, very often these trials were successful and so provided a basis for subsequent semisyntheses.[53,54]

At about the same time, another important serendipitous discovery occurred. The value of CNBr as a limited cleavage agent has been discussed. In bovine pancreatic trypsin inhibitor (BPTI) the two fragments generated by the CNBr reaction remain linked by disulphide bonds and, presumably, internal non-covalent interactions. It was, however, noted that the gap in the nicked inhibitor disappeared on standing in non-denaturing conditions.[55] This startling spontaneous reformation of a peptide bond was revealed to be due to aminolysis of the homoserine lactone ring by the adjacent amino group, a reaction that is normally very slow but is accelerated by the same conformational factors operating in the reverse proteolysis systems; maintenance of the global tertiary structure, and consequent proximity of the cut ends. In this case, the activation of the -COOH group stems from internal ester formation; in the others, the ester linkage in the acyl-enzyme intermediate.

This phenomenon also proved to have some generality and has been used in a number of semisyntheses, notably of cytochrome *c*, and the principle underlying both spontaneous and enzyme-catalysed resyntheses has been developed in the devising of general methods to simplify fragment condensation semisynthesis (Chapter 3).

1.2.6 THE WORK-UP

The *entropic activation* type of fragment condensation occurs under physiologically compatible conditions and avoids protection/deprotection cycles. Thus, exposure to conditions that can cause chemical damage or physical denaturation of the product is avoided. Nevertheless, some consideration must be give to the finishing touches. Whatever the method, couplings rarely go to completion, so the desired product must be resolved from the intermediates as well as reagents. In principle, any of the separation methods used for fragment isolation and purification described above could be used, and most have been.

If any form of chemical protection has been used, it normally must be removed, since with the rare exception (e.g., acetimidyl protection of lysine NH_2), chemical modification of a residue side chain will alter structure or function of the product protein. Techniques for the removal of all the blocking groups discussed above have been well established and are extensively documented by Offord,[13] as well as the dangers to the product integrity posed by them. These problems of compatibility of protecting methods will, of course, have been accommodated for in the initial selection of *orthogonal* schemes of protection but may still require some precautions, or the presence of scavenging agents, during the deprotection steps. In some cases, remedial actions may be necessary, either for covalent or non-covalent (denaturation) damage to the product. An example of the former would be the partial oxidation of methionine side-chains to the sulfoxide, which is easily reversed by treatment with thiols.

Denaturation of protein structure is a common consequence of either the chemical handling in the final stages, or of the conditions, frequently deliberately denaturing, of the separation methods employed. After removal from these conditions to a physiological environment, many semisynthetic products will spontaneously fold to the native confirmation. In some cases, though, intervention may be necessary, and the simplest tactic is the dissolution of the denatured material in a mildly denaturing medium, such as 8M urea, followed by dialysis or buffer exchange to remove the chaotropic agent. Such methods have also proved to be of great utility in the renaturation of proteins from inclusion bodies in heterologous expression systems.[56]

The spontaneous refolding of polypeptide chains to the native conformation also aids in the formation of the correct disulfide bonds, if any, in the final product. The other forces that determine protein stability and structure will bring into proximity those sulphydryl groups that need to be linked. For the practical purposes of a semisynthesis involving a protein with natural -S-S- bridges, it is usually sufficient that the oxidation reaction be prevented prior to correct folding by the use of reductants, and then initiated by the deliberate introduction of O_2.

Only when the desired product is not that which would naturally fold are there likely to be problems: the case in point is insulin. This is biosynthesised as proinsulin and the disulphide bonds are formed prior to proteolytic processing to the hormone form. Thus, forming the correct bridges between the two separated chains of insulin is a difficult and inefficient task. Nevertheless, protocols have been evolved to maximize the yield, although it is significant that most semisynthetic work on insulin has been done with the chains kept joined.

1.3   VARIANTS ON THE BASIC METHOD

Described above is the basic progression of a fragment condensation semisynthesis, but I have alluded to the fact that some of the steps may be omitted, in variants on the scheme, to achieve a simpler semisynthesis. Indeed, the first semisyntheses reported were examples of *non-covalent semisynthesis*, in which synthetic S-peptides were complexed with naturally derived S-protein to give viable ribonuclease S analogs.

This approach is enormously appealing, since it avoids the most troublesome element of the fragment-condensation methodology, the religation. It is also potentially widely available, because so many proteins are know to exhibit the propensity to form functional non-covalent complexes from limited numbers of large fragments that this may be a universal property of proteins. This interesting possibility remains to be fully explored.

Thus, in addition to ribonuclease, extensive non-covalent semisynthesis, in which both fragment religation and the preparatory protection and subsequent deprotection are avoided, has been undertaken with cytochrome *c* (Chapter 8), and to a lesser extent with other proteins. However, unlike ribonuclease S, cytochrome *c* complexes rarely exhibit full functionality and, although analogs of complexes here proved useful, this limitation has led to a preference for fully covalent structures.

The complexes themselves can prove useful. In cytochrome *c,* such a large range is known that the difference in functionality can be correlated with the position of the missing peptide bond and conclusions on structure-function relationships in the protein drawn from their comparison.[57]

A way to covalently link fragments without making peptide bonds is through the formation of disulphide bridges. Since this is a "chemoselective" method of fragment ligation (see Chapter 3), protection and deprotection protocols are once again avoided. Insulin, as a consequence of its proteolytic generation from proinsulin, naturally occurs as a pair of fragments linked by two -S-S- bonds and has served as a model for this type of semisynthesis, but in the case of interleukin 2 (IL2), disulphide ligation has substituted for peptide bond formation in the preparation of semisynthetic analogs. The strategy of Ciardelli's group was to prepare a synthetic peptide and analogs thereof that corresponded to the C-terminal helix, residues 100–133, of this 4-helix bundle protein. These were then joined to the remainder of the molecule via oxidative linkage of cysteines 58 and 105.[58,59] The large fragment was generated as such by expression of a recombinant gene in *E. coli*, thus overcoming an obvious limitation to semisynthesis of this scarce protein.

Insulin can also be used to exemplify the most popular prevarication from the fragment condensation method, that of stepwise semisynthesis. By focusing on residues at either terminus of the intact protein, both cleavage and fragment religation are avoided, and only that protection necessary to avoid collateral damage during stepwise degradation and elongation is required. The operations become essentially those described in Section 1.2.4 above.

Besides insulin, at least three proteins, phospholipase A2, myoglobin and ferredoxin have been tackled this way, though mostly at the N-terminus, given the simplicity of use of the Edman degradation and chain extension with protected amino acid active esters. Ferredoxin is a small protein with a conserved aromatic residue at position 2, making it an ideal candidate for this tactic and techniques. Tyr^2 was replaced by either leucine or glycine (60), and although Leu^2 ferredoxin was fully enzymatically active, the glycine-containing analog of the apoprotein could not reassociate with iron and sulfide ion, leading to the conclusion that a bulky hydrophobic residue, but not necessarily an aromatic one, was required to maintain the hydrophobicity of the pocket containing the iron-sulfur cluster.

Such an approach is not so simple with insulin: again a consequence of its two-chain structure. Thus, a number of strategies to confine reaction to one or other N-terminus have evolved and are described in detail in Chapter 4. Once differentiation is achieved, the manipulations are analogous, and residues up to five positions from the terminus have been replaced.[61]

Insulin is, however, the only protein where C-terminal stepwise semisynthesis has been undertaken, despite the complication of two termini. A number of schemes are described in Chapter 4, and one became the basis for a commercial route to the transformation of porcine to human insulin.

1.4   AFTERWORD

The strategies of protein engineering using protein chemistry, collectively described here as *semisynthesis,* have evolved with the relatively limited set of proteins described in this volume, often along very different lines. The advantage of this rather fragmented development is that a very wide range of tactics and techniques have been explored and have resulted in a large repertoire at the disposal of anyone proposing semisynthesis on a novel subject. But implicit in this process is the suspicion that new tactics must be evolved for every novel protein and that years of preliminary development work must be expected before any payoff.

I think that suspicion would have been well justified 10 or 15 years ago, but what the following chapters should reveal to the reader is a convergence on a philosophy and set of principles and practice that should make the process accessible to any newcomer.

As originally elaborated, semisynthetic schemes were often lengthy, cumbersome, and very labor intensive. This rapidly created the imperative to simplify the process, and the discovery of reverse proteolysis and the adoption of this, of genetic manipulation of precursors, and the evolution of stereoselective and chemoselective methods, has created an unprecedented efficiency of operations and drastically reduced the number of them that are required to create a protein analog.

Whether we need to create protein analogs this way is the other pertinent question an interested bystander might pose. The unique capabilities of chemical methods have been discussed above and elsewhere and need not be repeated here, and the ability of semisynthesis to extend chemical protein engineering into a size range unattainable by repetitive stepwise synthesis is freely acknowledged by the leading proponents of solid-phase peptide synthesis. Assuredly, most protein engineering goals can be achieved by genetic methods with naturally encoded amino acids, but some of the most imaginative structure-function experiments will demand semisynthesis. Some have characterized the method as a dinosaur, but even the dinosaurs have found a modern-day niche role as birds.

REFERENCES

1. Kent, S.B.H., Solid-Phase Peptide Synthesis, *Annu. Rev. Biochem.,* 57, 957, 1988.
2. Offord, R.E., Protection of peptides of biological origin for use as intermediates in the chemical synthesis of proteins, *Nature* (London), 221, 37, 1969.
3. Offord, R.E., Protein semisynthesis in theory and practice, in *Semisynthetic peptides and proteins,* Offord, R.E. and Di Bello, C., Eds, Academic Press, London, 1978, 3.
4. Aninsen, C.B., Principles that govern the folding of protein chains, *Science,* 181, 223, 1973.
5. Hoffman, K., Finn, F.M., Haas, W. Smithers, M.J., Wolman, Y., and Yanaihara, N. Studies on Polypeptides, XXVI. Partial Synthesis of an enzyme possessing high RNase activity. *J. Am. Chem. Soc.* 85, 833, 1963.
6. Potts, J.T., Young, D.M., and Anfinsen, C.B., Reconstitution of fully active RNase S by carboxypeptidase-degraded RNase S-peptide. *J. Biol. Chem.* 238, 2593, 1963.
7. Offord, R.E., Some experiences in protein and peptide semisynthesis, in *Peptides 1972,* Hanson, H. and Jakubke, H.D., eds., North-Holland, Amsterdam 1973, 52.
8. Burton, J. and Lande, S., Semisynthetic polypeptides. Transformation of native porcine ß-Melanotropin into the Lysine-10 analog of the human hormone. *J. Amer. Chem. Soc.* 92, 3746, 1970.
9. Borras, F. and Offord, R.E., Protected intermediate for the preparation of semisynthetic insulins, *Nature* (London) 227, 716, 1970.
10. Brandenburg, D., Des -Phe -insulin, ein kristallines analogon des rinderinsulins, Hoppe-Seyler's Z. *Physiol. Chem.* 350, 741, 1969.

11. Sealock, R.W and Laskowski, M., Jr., Enzymatic replacement of the arginyl by a lysyl residue in the reactive site of soybean trypsin inhibitor, *Biochemistry*, 8, 3703, 1969.

12. Offord, R.E. and Di Bello, C., *Semisynthetic Peptides and Proteins.* Academic Press, London, 1978.

13. Offord, R.E., *Semisynthetic Proteins.* John Wiley and Sons, Chichester, UK., 1980.

14. Sheppard, R.C., Partial synthesis of peptides and proteins, in *The Peptides,* Volume 2, Gross, E. and Meienhofer, J., Eds., Academic Press, New York, 1980, 44.

15. Chaiken, I.M., Semisynthetic peptides and proteins, *CRC Crit. Rev. Biochem.,* 11, 255, 1981.

16. Offord, R.E., Protein engineering by chemical means?, *Protein Eng.,* 1, 151, 1987.

17. Medvedkin, V.N., Semisynthesis of proteins and peptides, *Bioorg. Khim.,* 15, 581, 1989.

18. Tesser, G.I., *Analogues of Some Small Protein by Semisynthesis,* Kontakte (Darmstadt), 1989, 29.

19. Offord, R.E., Chemical approaches to protein engineering, in *Protein Design and Development of New Therapeutics and Vaccines,* Hook, J.B., and Poste, G. eds., Plenum, New York, 1990, 253.

20. Wallace, C.J.A., Peptide ligation and semisynthesis, *Curr. Opin. Biotechnol.,* 6, 403, 1995.

21. Lehninger, A.L., Nelson, D.L. and Cox, M.M., *Principles of Biochemistry,* 2nd Edn., Worth, New York, 1993, 113.

22. Galpin, I.J. and Hoyland, D.A., Semisynthesis I—CNBr digestion of hen egg-white lysozyme, *Tetrahedron* 41, 895, 1985.

23. Galpin, I.J. and Hoyland, D.A., Semisynthesis II—formation of a covalently-linked 1-12/106-129 fragment of hen egg-white lysozyme, *Tetrahedron* 41, 901, 1985.

24. Galpin, I.J. and Hoyland, D.A., Semisynthesis III—the homoserine 12, 105 analogue of hen egg-white lysozyme, *Tetrahedron* 41, 907, 1985.

25. Fontana, A. And Gross, E., *Fragmentation of Polypeptides by Chemical Means, in Practical Protein Chemistry—a Handbook,* A. Darbre, Ed. John Wiley and Sons, New York, 1986, 67.

26. Simmerman, H.K.B., Wang, C.C., Horwitz, E.M. Berzotsky, J.A. and Gurd, F.R.N., Semisynthesis of sperm whale myoglobin by fragment condensation, *Proc. Natl. Acad. Sci. USA,* 79, 7739, 1982.

27. Slotboom, A.J. and de Haas, G.H., Specific transformations at the N-terminal region of phospholipase A_2, *Biochemistry*, 14, 5394, 1975.

28. Harris, D.E. and Offord, R.E., A functioning complex between tryptic fragments of cytochrome *c*, *Biochem. J.,* 161, 21, 1977.

29. Wallace, C.J.A. and Harris, D.E., The preparation of fully N-ε-acetimidylated cytochrome *c*, *Biochem. J.,* 217, 589, 1984.

30. Rees, A.R. and Offord, R.E., The preparation of protected fragments of lysozyme for semisynthesis, *Biochem. J.,* 159, 467, 1976.

31. Proudfoot, A.E.I., Offord, R.E., Rose, K., Schmidt, M. And Wallace, C.J.A., A case of spurious product formation during attempted resynthesis of proteins by reverse proteolysis, *Biochem., J.,* 221, 325, 1984.

32. Tesser, G.I. and Balvert-Geers, I.C., The methylsulfonylethyloxycarbonyl group: a new and versatile amino protective function, *Int. J. Protein Pept. Res.,* 7, 295, 1975.

33. Saunders, D.J. and Offord, R.E., A novel semisynthetic route to insulin analogues modified at the N-terminus of the A-chain, Hoppe-Seyler's Z. *Physiol. Chem.,* 358, 1469, 1977.

34. Borras, F., D.Phil. Thesis, University of Oxford, 1972, cited in reference 13 above.
35. Wallace, C.J.A. and Corthesy, B.E., Alkylamine derivatives of cytochrome c, Eur. J. Biochem., 170, 293, 1987.
36. Wallace, C.J.A. and Offord, R.E., The semisynthesis of fragments corresponding to residues 66-104 of horse-heart cytochrome c, Biochem. J., 179, 169, 1979.
37. Offord, R.E., The possible use of cyanogen bromide fragments in the semisynthesis of proteins and polypeptides, Biochem. J., 129, 499, 1972.
38. Glass, J.D., Enzymatic manipulation of protecting groups in peptide synthesis, in The Peptides, volume 9, Meienhofer, J. Ed., Academic Press, New York, 1987, 167.
39. Wallace, C.J.A. and Proudfoot, A.E.I., On the relationship between oxidation-reduction potential and biological activity in cytochrome c analogues, Biochem. J. 245, 773, 1987.
40. Finn, F.M. and Hofmann, K., Studies on polypeptides. XXXIII. Enzymic properties of partially synthetic ribonucleases. J. Amer. Chem. Soc., 87, 645, 1965.
41. Lundblad, R.L. and Noyes, C.M., Chemical Reagents for Protein Modification. CRC Press, Boca Raton, Fl., 1984.
42. Breddam, K., Widmer, F. And Johansen, J.T., Carboxypeptidase Y catalyzed transpeptidations and enzymatic peptide synthesis, Carlsberg Res. Commun. 45, 237, 1980.
43. Chaiken, I.M., Komoriya, A., Ohno, M. And Widner, F., Use of enzymes in peptide synthesis, Applied Biochem. Biotechnol., 7, 385, 1982.
44. Kullmann, W., Enzymatic Peptide Synthesis, CRC Press, Boca Raton, Fl., 1987.
45. Bodanszky, M., Principles of Peptide Synthesis, 2nd ed., Springer-Verlag, Berlin, 1993.
46. Wallace, C.J.A. and Clark-Lewis, I., Functional role of heme ligation in cytochrome c, J. Biol. Chem., 267, 3852, 1992.
47. van Scharrenburg, G.J.M., Puijk, W.C., Egmond, M.R., de Haas, G.H. and Slotboom, A.J., Semisynthesis of phosphoslipases A_2. Preparation and properties of arginine-6 bovine pancreatic phospholipase A_2, Biochemistry 20 1584, 1981.
48. Rees, A.R. and Offord, R.E., The semisynthesis of proteins of hen's-egg lysozyme by fragment condensation, Biochem. J., 159, 487, 1976.
49. Wang, C-C., DiMarchi, R.D. and Gurd, F.R.N., Replacement of the N-terminal tetradecapeptide sequence of sperm whale myoglobin, in Semisynthetic Peptides and Proteins, Offord, R.E. and Di Bello, C., Eds., Academic Press, London, 1978, 59.
50. Ten Kortenaar, P.B.W., Adams, P.J.H.M. and Tesser, G.I., Semisynthesis of horse heart cytochrome c analogues from two or three fragments, Proc. Natl. Acad. Sci. USA, 82, 8279 1985.
51. Homandberg, G.A., Mattis, J.A. and Laskowski, M., Jr., Synthesis of peptide bonds by proteinases. Addition of organic cosolvents shifts peptide bond equilibrium towards synthesis, Biochemistry, 17, 5220, 1978.
52. Jering, H. and Tschesche, H., Replacement of lysine by arginine, phenylalanine and tryptophan in the reactive site of the bovine trypsin-kallikrein inhibitor (Kunitz) and change of the inhibitory properties, Eur. J. Biochem., 61, 453, 1976.
53. Homandberg, G.A. and Laskowski, M., Jr., Enzymatic resynthesis of the hydrolysed peptide bond(s) in ribonuclease S, Biochemistry, 18, 586, 1979.
54. Homandberg, G.A. and Chairken, I.M., Trypsin-catalyzed conversion of staphylococcal nuclease-T fragment complexes to convalent forms, J. Biol. Chem. 255, 4903, 1980.
55. Dyckes, D.F., Creighton, T.E. and Sheppard, R.C., Spontaneous re-formation of a broken peptide chain, Nature (London) 247, 202, 1974.

56. Jaenicke, R. and Rudolph, R., Protein folding, in *Protein Structure: a Practical Approach,* Cerighton, T.E., Ed. IRL Press, Oxford, 1989, 191.

57. Wallace, C.J.A. and Proudfoot, A.E.I., On the relationship between oxidation-reduction potential and biological activity in cytochrome *c* analogues, *Biochem. J.* 245, 773, 1987

58. Ciardelli, T.L., Landgraf, B., Gadski, R., Strand, J., Cohen, F.E. and Smith, K.S., A design approach to the structural analysis of interleukin-2, *J. Mol. Recognition,* 1, 42, 1988.

59. Landgraf, B., Cohen, F.E., Smith, K.A., Gadski, R., and Ciardelli, T.L., Structural significance of the C-terminal amphiphilic helix of IL2, *J. Biol. Chem.* 264, 816, 1989.

60. Lode, E.T., Murray, C.L., Sweeney, W.V. and Rabinowitz, J.C., Synthesis and properties of *Costridium acidi-urici* [Leu^2]-ferredoxin: a function of the peptide chain and evidence against the direct role of the aromatic residues in electron transfer, *Proc. Natl. Acad. Sci. USA,* 71, 1361, 1974.

61. Saunders, D.J. and Offord, R.E., Semisynthetic analogues of insulin, *Biochem. J.,* 165, 479, 1977.

2   Methodological Advances in the Last Decade

Keith Rose

2.1   RECENT METHODOLOGICAL ADVANCES

Semisynthesis may be defined as the use of fragments of natural (biologically synthesized) proteins in the synthesis of new structures, so the total chemical synthesis of proteins normally has no place in an article on semisynthesis. However, the recent tendency of some of those involved in total synthesis to deprotect large fragments prior to combining them has brought to total synthesis the problems (and opportunities!) associated with semisynthesis. For this reason, a section covering the coupling of totally synthetic, unprotected fragments is included.

The last 15 years have seen an explosion of progress in the area of recombinant DNA technology. Consequently, the availability of relatively large amounts of proteins of fundamental and commercial importance, and the ease with which analogs may be made by site-directed mutagenesis, have had a major impact on the practice of protein semisynthesis. Indeed, having a recombinant-derived protein as starting material for a semisynthesis project is one of the most significant recent methodological advances, the advantages of which are discussed below.

Progress in the field of recombinant DNA technology has naturally gone hand in hand with improvements in protein handling and characterization. There have been improvements both at the scale of the small amounts of natural proteins needed for determination of amino acid sequence and biological activity, and at the scale of large amounts of recombinant-derived products that need to be refolded and purified. Those improvements in protein handling and characterization that have had a major effect on the practice of protein semisynthesis are discussed. In particular, improvements in the area of mass spectrometry have led to its becoming a most useful and widely applicable technique for the characterization of polypeptides and proteins.

Over the last 15 years also, the technique of reverse proteolysis has proven its utility, even on an industrial scale with the production of insulin of human sequence, first from porcine insulin and then from single chain precursors produced in yeast.[1] In spite of such successes, reverse proteolysis still tends to be regarded as a curiosity, or as a last resort, by many peptide chemists. For this reason, one of the objectives of the present chapter is to emphasise the usefulness of this particular technique. Reverse proteolysis may be exploited for the direct formation of peptide bonds (discussed in Chapters 3 through 8) or to attach reactive groups to the C-terminus (discussed in Chapters 3 and 4). From a methodological standpoint, considerable progress has been made recently in the areas of C-terminal amidation of recombinant polypeptides and of production of protein conjugates of defined structure, and these areas of application of reverse proteolysis are discussed in some detail at the end of the present chapter.

2.2 TOTAL CHEMICAL SYNTHESIS OF PROTEINS

The definition of *semisynthesis* given at the start of this chapter has been a useful one until now, and protein semisynthesis has been distinguished from total chemical synthesis by its very different methodology. Protein semisynthesis involves work with biologically synthesized proteins, which are generally soluble in water, easily damaged, highly multifunctional, and which, initially at least, lack differential protection. Total chemical synthesis, on the other hand, at least in its initial stages, involves work with highly protected species, organic solvents, and harsh reagents.

The field of total synthesis has made enormous strides over the past decade, with specialized research groups now capable of performing the routine synthesis of certain proteins of up to 70–140 residues in length,[2] which is about the limit attainable in a single run by the solid phase approach; beyond this limit, impurities having structures very close to the target sequence tend to dominate and are very difficult to remove entirely. Interesting results may be obtained even when the target sequence is in a minority (e.g., Clark-Lewis et al.[3]), although this clearly is not an ideal situation. When faced with longer or more difficult sequences, the classical approach is to prepare and purify protected fragments, which are then condensed together to form the target sequence. A recent example of this convergent approach is the synthesis of the 121-residue human midkine protein.[4] Unfortunately, protected fragments of considerable length are notoriously insoluble, and this lack of solubility is the principal problem encountered in the total chemical synthesis of proteins by fragment condensation.[4] Deprotection gives unprotected fragments, which are generally much more soluble in an appropriate solvent (often water) and are easier to purify, since the full range of protein purification techniques (including ion exchange chromatography and, sometimes, affinity chromatography) are then available.[5,6] The problem of condensation of unprotected synthetic fragments is less severe than for the semisynthetic approach, since the flexibility of a totally synthetic approach permits the building in of a distinction between terminal and side chain groups, where necessary. For example, by following techniques developed by semisynthetic chemists, totally synthetic analogs of the 104-residue protein cytochrome c are now accessible by combining a synthetic polypeptide comprising residues 1–65, made to terminate in homoserine lactone, with synthetic 66–104.[6,7] Here, the homoserine lactone serves as the unique active carboxyl function, and conformational assistance ensures that it encounters only the alpha-amino group of fragment 66–104.

More generally, through the use of a specially engineered enzyme (subtiligase), synthetic polypeptides esterified at the alpha-carboxyl group may be ligated to form totally synthetic proteins:[8] this is a very powerful technique.

When a peptide bond linking the fragments is not obligatory, unprotected synthetic polypeptides may be brought together through a variety of alternative chemistries. One instructive example is that of HIV protease, which exists as a homodimer of two 99-residue subunits. An analog of this protease has been synthesized by a site-specific alkylation reaction between two unprotected synthetic polypeptides.[5] Fragment H-(1-50)-$NHCH_2$-COSH was reacted with $BrCH_2CO$-[$Aba^{67,95}$](53-99)-OH to yield an analog of the 99-residue polypeptide chain, H-(1-50)-$NHCH_2CO$-SCH_2CO-[$Aba^{67,95}$](53-99)-OH. This analog, which was well characterized, formed a fully functional homodimer of 198 residues. Residues 51–52, both Gly in native HIV protease, were thus replaced by the thioester link -$NHCH_2CO$-SCH_2CO-, or more simply the NH of residue 52 was replaced by a sulfur atom. As the authors point out, the selective chemical ligation of large unprotected synthetic fragments offers several advantages: improved syntheses, the flexibility to modify both the structure of the fragments themselves and that of the backbone of the final polypeptide, and the possibility of synthesizing hybrid protein-nonprotein conjugates. However, while very elegant from a chemical and structural point of view, the thioester linkage is unstable[5] at neutral to alkaline pH, hydrolyzing with a half life of 2 h at pH 7.5. To be useful in a more general context, alternatives to the thioester linkage must be found. Some are already available, having been developed to ligate natural fragments in semisynthetic operations, and are discussed in the last section of this chapter.

The assembly of partially protected peptides on a solid support using enzymatic techniques is not yet developed sufficiently for the assembly of proteins (e.g., Slomczynska et al.[9]).

It is sometimes forgotten that protein semisynthesis was suggested by Offord[10] to exploit the availability of natural polypeptide fragments at a time when total chemical synthesis was much less highly developed than it is today. Now that large, unprotected, chemically synthesized fragments are being coupled by site-specific chemistry, it is almost time to abandon the term protein semisynthesis and bring all of the work under the heading of protein engineering (see the title of Offord's own review[11]).

2.3   RECOMBINANT DNA TECHNIQUES

When protein analogs containing coded changes are required, recombinant DNA mutagenesis techniques almost always offer the most convenient synthesis, but such methods are outside the scope of the present review. These techniques also offer facile incorporation of deuterium or ^{15}N throughout a protein molecule for NMR purposes, involving merely the use of appropriately enriched water or ammonium salts in the growth medium.

The introduction of non-coded structures through manipulation of the genetic code[12,13] is progressing. Species such as beta(o-nitrobenzyl)-Asp, iodo-Tyr, D-Phe, and even L-phenyllactic acid may be introduced. Interesting and elegant as these methods are, they are not widely practiced, and it is clear[12] that it will be some time before they can offer a convenient route to substantial (50 mg or more) amounts of protein analogs. Practically, then, if non-coded structures are to be introduced into a protein, the options remain total synthesis or semisynthesis.

Site-specific mutagenesis has been exploited to tailor precursors specifically for particular semisynthetic operations, either chemical or enzymatic. This tailoring is simply an extension of techniques used for cleavage of fusion polypeptides, where, for example, the target protein is preceded by a Met (cleavage site for BrCN) and internal Met residues are engineered out. In an analogous way, fragments designed for subsequent semisynthetic operations may be prepared by creating appropriate sites of chemical or enzymatic cleavage and eliminating unwanted ones.[14-17] These designed fragments may then be modified as required before being assembled by chemical or enzymatic techniques. The resulting combination of the advantages of recombinant techniques (freedom to introduce coded changes) and of semisynthesis (relative freedom to introduce non-coded changes) is very powerful. Examples of the use of the combination of recombinant DNA techniques and protein semisynthesis and are discussed in the last two sections of this chapter and in Chapters 3 and 8.

2.4   PROTEOLYTIC CLEAVAGE

The chemical and enzymatic techniques employed to prepare fragments suitable for amino acid sequencing (e.g., Allen[18]) are also useful for semisynthetic operations. Reagents and enzymes that are highly specific for a particular residue or sequence generally give high yields of fragments with little optimization, provided that the cleavage site is accessible, or can be made accessible, to the cleaving agent. This is not always so, however. For example, BrCN does not cleave efficiently when Met is followed by Ser, Thr, or Cys. Of course, if several potential sites of cleavage exist, there are problems even with a reagent of high specificity, and in such cases the ability to remove the extra sites by mutagenesis techniques is a great advantage,[15,16] as discussed above. On the other hand, when less specific cleavage reagents are employed, and especially relatively non-specific enzymes, to obtain particular fragments, this generally requires much optimization and, although high yields can sometimes be obtained in special cases due to conformational factors, success cannot be guaranteed.

As regards recent methodological advances, seven commercially available enzymes are worthy of mention.

1. Trypsin, which cleaves on the carboxyl side of Arg and Lys residues (except where these are followed by Pro), has been used from the very beginning of semisynthesis and is still widely used today. It is not widely recognised, however, that the porcine (hog) enzyme, available from the Sigma Chemical Co., is much more robust than the bovine enzyme and is active down to relatively low pH.[19,20] While these properties of the porcine enzyme are very useful for synthesis by reverse proteolysis,[19,21–23] they also offer flexibility in the preparation of fragments from substrates that require relatively low pH or the presence of organic solvents or urea for efficient digestion.[20] It may be useful to point out here that Davies et al.[21] contains a typographical error: the enzyme stock solution used for synthesis was 50 µg/µl, not 50 µg/ml as stated.

2. Another robust, commercially available enzyme is that isolated from strains of *Achromobacter* (*enzymogenes* for the enzyme from Calbiochem and Boehringer, *lyticus* for the enzyme from Wako Pure Chemical Co.) and which cleaves a polypeptide chain on the carboxyl side of Lys residues. Since this enzyme is also very useful for synthesis under conditions of reverse proteolysis, it is possible to follow the cleavage step, which liberates a desired fragment, directly with a synthetic step simply by adding nucleophile, lowering the pH, and adding a co-solvent if required.[24] Such an approach permits the preparation of well defined antibody conjugates[25] as discussed in detail toward the end of this chapter.

3. Clostripain, like trypsin, has been used for semisynthesis for many years. Its specificity for Arg residues can be useful, despite the fact that reducing conditions must be maintained. Renewed interest in the enzyme has been created by a better understanding of its use for reverse proteolysis with quite large protein fragments.[26]

4. Cleavage on the carboxyl side of Pro residues is possible using proline specific endopeptidase from *Flavobacterium meningosepticum*[27] available from Seikagaku Kogyo, Japan. Again, this is useful since the same enzyme may be used in a reverse proteolysis coupling reaction, as with insulin.[28] Disadvantages are slow cleavage at Ala, and the fact that larger proteins, even when denatured, are not cleaved.[27]

5. Cleavage on the amino side of Asp residues is possible with endoproteinase Asp-N from Boehringer, but the protease is very expensive (sold by the microgram). It also cleaves on the amino side of cysteic acid, but this property is more useful for preparing fragments for amino acid sequencing purposes than for semisynthesis.

6. Asparaginylendopeptidase, available from Pierce, cleaves on the carboxyl side of Asn residues. It is expensive and requires reducing conditions to be maintained.

7. Thermolysin, available from several sources including Sigma, is generally regarded as being too non-specific to be of much synthetic use. We found,[17]

however, that in the presence of organic co-solvent and a high concentration of a suitable nucleophile, thermolysin can be used to couple in good yield to the C-terminus of the B-chain of large insulin fragments which terminated -Ala, -Phe, or even -Arg. For this approach to be successful, and as expected from the specificity of thermolysin, the nucleophile should ideally resemble a free phenylalanyl (or similar hydrophobic) derivative.

New methods of peptide bond cleavage at selected positions are being developed all the time[29,30] but rarely become high-yield techniques of wide application.

2.5 SEPARATION, PURIFICATION, AND CHARACTERIZATION

2.5.1 SEPARATION AND PURIFICATION

The use of column packings having small particle size has greatly increased the resolution (performance) of liquid chromatographic separations and the operating pressure. The resulting technique, high-performance liquid chromatography (HPLC, where the P sometimes represents *pressure* and always, in the eyes of those who are concerned with the bottom line, represents *price*), may be applied to all chromatographic modes: reversed phase, size exclusion (gel filtration), ion exchange, hydrophobic interaction, chelate chromatography, chromatofocusing, and affinity chromatography. High-performance systems that avoid contact with metal (except titanium pump parts), and that operate at medium pressure, are most suitable for larger proteins and protein conjugates and are available from many instrument manufacturers; the best known are probably the FPLC and the newer, smaller-scale, SMART system from Pharmacia.

This is not the place to discuss in detail the theoretical and practical aspects of the HPLC of proteins and polypeptides, which are well described in textbooks.[31-33] Reference 32 is particularly useful, since it contains a good discussion of both soft-gel (open-column, conventional) chromatography and HPLC techniques for protein purification; of course, the same techniques may be successfully applied to the isolation of large fragments for semisynthesis. There has been a trend to move from soft gel separations to high-performance ones, with a consequent reduction in sample size, at the laboratory scale, normally imposed by the very high price of the media and the corresponding pumping equipment. This reduction in sample size has not adversely affected work in the field, since polypeptide characterization, which used to require very large amounts of material, is now very sensitive indeed, as we shall see below.

It is sometimes possible to profit from newer technology that offers high-speed separations, where these give sufficient resolution. For example, the flow-through chromatography technique (Poros Series columns, PerSeptive Biosystems, Cambridge, Mass.) leads to very flat plate height vs. reduced velocity curves and offers high throughput with good capacity and resolution. Various membrane-based chromatography devices (e.g., Mem-Sep, Millipore-Waters) offer a similar advantage, speed, while preserving a resolution which is often adequate.

The area of polypeptide and protein separation is evolving very rapidly, so the reader confronted with a difficult separation problem is also confronted with a great number of possible solutions. Reference 33 can be useful here, as it is regularly updated and exists in searchable CD-ROM form. Various peptide and protein interest groups exist on the Internet and can prove useful. Applications chemists at separation media manufacturers can sometimes be helpful, but the reader is strongly encouraged to ask the manufacturers for a demonstration column or device prior to purchase. My own group has been very agreeably surprised by the performance of some columns and equally disappointed by others; the fact that a system seems to work well for a given separation in the applications brochure does not mean that one's own separation will be equally well achieved.

High-performance displacement chromatography provides a means of increasing the capacity of a chromatographic column while maintaining resolution, a feature that facilitates the isolation of minor components. Use with recombinant growth hormone has been described.[34]

Capillary electrophoresis[33,35] is used to examine semisynthetic polypeptides (e.g., Bongers et al.[36]). Separating power is excellent, but the technique can be regarded only as preparative for the tiny amounts needed for amino acid sequence determination, mass spectrometry, or for some sensitive bioassays. It offers another dimension to reversed phase HPLC, should be considered when demonstrating homogeneity, and has the advantage that it may be coupled on-line to mass spectrometry.

Preparative isoelectric focusing can be very useful and is applicable up to about the gram scale with laboratory size units.

2.5.2 CHARACTERIZATION

Protein semisynthesis generally involves work with fragments of known amino acid sequence, so amino acid analysis and terminal group identification (by automated Edman degradation and carboxypeptidase digestion), or sometimes simply precise molecular weight determination, by mass spectrometry, is usually sufficient to characterize a fragment. All of these techniques now routinely consume only picomoles of material.

Electrophoresis on polyacrylamide gels is still a useful technique for monitoring cleavage and coupling reactions. The use of preprepared gels and well designed equipment (e.g. the Phast system from Pharmacia) has brought speed and convenience to what used to be a time-consuming and rather messy business. The resolution of gel electrophoresis is nonetheless rather limited compared to that of mass spectrometry, which has rapidly established itself as a key (if not the key) analytical procedure in the area of protein characterization.[33] More complete structural information is available by NMR spectroscopy, but this is not used for routine characterization as the equipment required is very expensive, and data interpretation is very time consuming.

Mass spectrometry has been used for over 30 years to assist the characterization of peptides and proteins. Originally reserved for specialists and for samples of low molecular weight, mass spectrometry has matured and become a user-friendly source of instant and precise information about molecules having relative molecular masses well in excess of 100 kDa. It should no longer be regarded as an expensive, complicated luxury requiring highly specialized personnel, for it is now an essential technique whose practice is well within the competence of a protein chemist. A recent masterly, yet accessible review of protein mass spectrometry is available.[37]

The would-be purchaser of mass spectrometric equipment is today faced with a baffling array of devices which operate on many different principles and which differ in price over at least an order of magnitude. To assist those protein chemists who are considering the purchase of a mass spectrometer, the practical performance of currently available commercial machines is critically considered here. Measurement by mass spectrometry consists of an ionization step followed by a mass analysis step, and we shall consider both of these aspects. Table 2.1 shows the characteristics of the principal ionization techniques used for proteins and large polypeptides.[37]

TABLE 2.1
Ionization Techniques Useful for Proteins and Large Polypeptides

Technique	Practical Mass Range for Proteins (kDa)	Practical LC	Typical Sens. (pmol)	Comments
FAB/LSIMS	ca. 15	yes	1–500	High mass samples need sector or ICR analyzer
TSP	ca. 4	yes	ca. 50	Limited mass range
PD	ca. 45	no	ca. 10	Can use cheap TOF analyzer
MALDI	>300	no	1	Can use cheap TOF analyzer
ES/ion Spray	ca. 150	yes	ca. 10	Multiply-charged ions can use cheap quadrupole analyzer

FAB, fast atom bombardment; LSIMS, liquid secondary ion mass spectrometry; LC, liquid chromatography; TSP, thermospray; PD, plasma desorption; MALDI, matrix assisted laser desorption ionization; TOF, time of flight; ES, electrospray ionization.

Desorption chemical ionization and field desorption are not considered, since they are relatively difficult techniques and of limited mass range. Table 2.2 shows the characteristics of the principal techniques of mass analysis.[37] Magnetic sector instruments (Table 2.2) generally incorporate an electric sector, too, to provide a double-focussing instrument. By coupling together a series of analyzers, it is possible to construct machines (e.g., so-called four-sector machines, or triple quadrupoles) that permit selection of a beam of ions of a given mass-to-charge ratio, excitation of these ions to promote fragmentation, followed by mass analysis of the fragments. This approach, known as MS/MS or tandem MS, a full description of which is outside the scope of this review, allows determination of the amino acid sequence and post-translational modifications to be made, but only for polypeptides of mass below 2500 or so. At present, no mass spectrometric technique produces a mass spectrum, running from low mass to the mass of a protein of 10–100 kDa, which has anything like enough information to permit deduction of the entire amino acid sequence; beyond a mass of a very few kDa, mass spectrometry essentially provides

TABLE 2.2

Mass Analyzers Considered from the Point of View of Analysis of Proteins and Polypeptides

Type	Mass Precision	Practical Resolution	Practical Mass Range (proteins) k Da	Comments
Magnetic Sector	Very good	High	Up to 20	Expensive; MS/MS good when part of a larger machine
Quadrupole	Good	Medium	<200*	Cheaper; MS/MS good when part of a larger machine
TOF	Good**	Medium**	>300	Cheaper
FTICR	Very good	Very high	>20*	Expensive but very versatile; essentially a research tool

FTICR, Fourier transform ion cyclotron resonance; *when multiply charged (m/z limitation 3000 or 4000); **when modern instrument used, e.g.,[39]—the best reflectron TOF analyzers equipped with delayed extraction offer resolution approaching that practically obtainable with magnetic sector instruments.

molecular weight information. This situation may change in the future if appropriate sample pre-treatment procedures can be found.

As an example of the utility of mass spectrometry, Figure 2.1 shows the ease with which the various steps (described by Rose et al.[23]) in the preparation of a semisynthetic insulin conjugate can be followed by this technique. Electrospray (ES) ionization with a quadrupole analyzer was used to characterize substrate, reagent, intermediates and product. Panel (a) shows the spectrum of the porcine insulin starting material. The signals at m/z 963.86 and 1156.44 are due to intact molecules associated with 6 and 5 protons, respectively. These values lead to an experimentally determined molecular weight of 5777.13, very close to the theoretical value of 5777.59. Panel (b) shows the spectrum of des-Ala^{B30}-insulin produced by cleavage of insulin with lysyl endopeptidase. The experimentally determined molecular weight is 5706.06, very close to the theoretical value of 5706.51. Loss of the B30 Ala residue is thus confirmed by the mass difference between intact insulin and this sample; loss of any other residue would give a quite different molecular weight. A small amount of undigested material is visible in the mass spectrum. In its ability to detect minor components, electrospray mass spectrometry is more rapid and precise that conventional characterization by end-group determination and amino acid analysis. Panel (c) shows the spectrum of des-Ala^{B30}-insulinyl-NHNHCONHNH$_2$, formed by the coupling of carbohydrazide specifically to the C-terminus of the B-chain by reverse proteolysis catalyzed by the same protease used to prepare the truncated insulin.[23] The experimentally determined molecular weight is 5778.33, very close to the theoretical value of 5778.58. A second component, responsible for signals at m/z 966.06 and 1156.67, is due to a species of molecular weight 5790, most probably the hydrazone formed with formaldehyde during storage

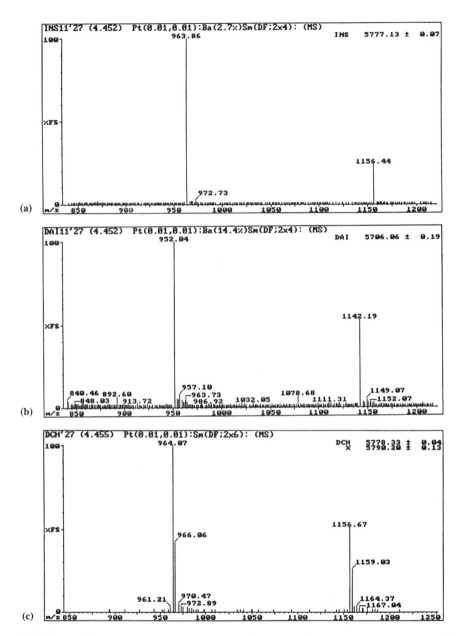

FIGURE 2.1 Electrospray ionization mass spectra of substrate, intermediates, reagent and semisynthetic product. The spectra were obtained in positive ion mode on a Trio 2000 instrument equipped with a 3000 m/z RF generator (VG Biotech, now Micromass, Altrincham, UK) operated at a flow rate of 2 μl/min with methanol, water, acetic acid (49.5:49.5:1 by vol.) as solvent: (a) porcine insulin starting material; (b) des-Ala^{B30}-insulin intermediate; (c) des-Ala^{B30}-insulin-NHNHCONHNH$_2$ intermediate; (d) HCO-C$_6$H$_4$-m-CH=NOCH$_2$CO-ferri-oxamine reagent; (e) des-Ala^{B30}-insulin-NHNHCONHN=CH-C$_6$H$_4$-m-CH=NOCH$_2$CO-ferri-oxamine product *(continued next page).*

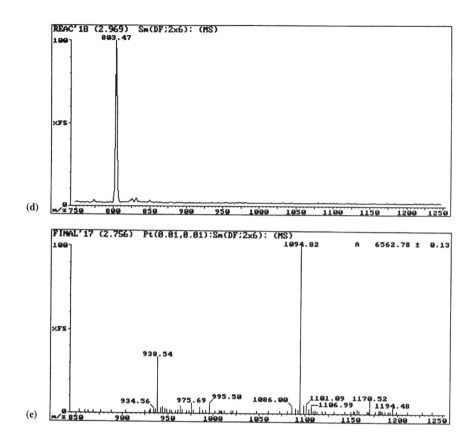

FIGURE 2.1 *(continued)*

in aqueous methanol containing acetic acid. Panel (d) shows the spectrum of the reagent HCO-C$_6$H$_4$-m-CH=NOCH$_2$CO-ferrioxamine. The signal at m/z 803.47, close to the predicted value of 803.69, is due to the protonated molecular ion. Panel (e) shows the spectrum of the conjugate formed between des-Ala^{B30}-insulinyl-NHNHCONHNH$_2$ and HCO-C$_6$H$_4$-m-CH=NOCH$_2$CO-ferrioxamine. The two components react cleanly and site-specifically, under very mild conditions, to give a conjugate, des-Ala^{B30}-insulinyl-NHNHCONHN=HC-C$_6$H$_4$-m-CH=NOCH$_2$CO-ferrioxamine, held together by a hydrazone bond.[23] The experimentally determined molecular weight is 6562.78, close enough to the theoretical value of 6563.25 to be sure that all has gone as planned.

It is sometimes possible to observe non-covalent association by mass spectrometry. A common feature of fast atom bombardment (FAB) and liquid secondary ion mass spectrometry (LSIMS) spectra (glycerol clusters are non-covalent, and one often observes "sample plus proton plus matrix", "sample plus proton plus trifluoroacetic—or other—acid", "sample plus proton plus sample," etc.), non-covalent association is also observed with the electrospray technique (association with phosphoric or sulfuric acid is well documented, and we have observed association between a protein and a detergent molecule) and with matrix assisted laser desorption ionization (MALDI). Although usually considered an artifact, the ability to observe non-covalent association can be regarded as an advantage.[37,38]

In recommending mass spectrometric equipment for the protein chemistry laboratory, it is assumed that the major thrust of research concerns semisynthetic and synthetic proteins. The characterization of tiny quantities of native proteins of completely unknown structure that may possess post-translational modifications generally requires much more expensive and sophisticated equipment than that recommended here. MALDI equipped with a high-resolution time-of-flight (TOF) analyzer,[39] and ES with a single quadrupole mass analyzer, are relatively inexpensive, robust, and easy to operate. An ES quadrupole instrument may be purchased with exchangeable sources so that reagents may, where necessary, be characterized by alternative ionization techniques such as electron impact or chemical ionization, and some models may be upgraded to a triple quadrupole (for MS/MS) if needs change. The extra expense and complexity associated with magnetic sector instrumentation is difficult to justify solely for semisynthetic applications unless work is exclusively with polypeptides below about 8 kDa, where the higher resolution of a sector machine may be exploited usefully.

The ability to characterize substrates, reagents, intermediates, and reaction products almost as soon as they are isolated (and even before, in mixtures) is so powerful that all groups in the field will need access to mass spectrometry if they are to remain competitive.

2.6   CARBOXYL-TERMINAL AMIDATION

Many medium-sized polypeptides are C-terminally amidated (e.g., vasoactive intestinal polypeptide, calcitonin gene related polypeptide, growth hormone releasing factor, amylin, etc.), and this post translational modification is generally required for biological activity. Such polypeptides may be produced on a laboratory scale by solid-phase peptide synthesis, but there is some reluctance to use that technique on a vast scale in view of the expense and of the problems of handling and disposing of huge amounts of toxic chemicals. Recombinant DNA techniques permit the facile preparation of polypeptides, but the primary product is not amidated. Use of pituitary amidating enzyme, either coexpressed with product or in a separate in vitro reaction and which converts -Xaa-Gly-OH into -Xaa-NH$_2$, has proved useful.[36,40] Some groups have turned to reverse proteolysis as a means of amidating specifically the C-terminus of a polypeptide produced (or to be produced) by recombinant techniques. While direct amidation using ammonia as nucleophile gives low yields, about 25–30% at equilibrium,[17,36] the use of amino acid amides as nucleophile can lead to high yields,[17,41,42] but this is not always the case.[36,43]

Although ammonia itself couples poorly, many compounds may be coupled through an amino group in nearly quantitative yield by enzyme-catalyzed synthesis. Thus, the coupling by reverse proteolysis of a suitably protected ammonia, which could subsequently be deprotected to liberate the desired amide, would be expected to increase yields over those obtained by direct amidation at the expense of an extra step (the deprotection). Experiments in this direction have been reported,[44] but the ideal combination of enzyme and nucleophile has yet to be found.

Along with the synthesis of human insulin,[1] C-terminal amidation provides good examples of the modern approach to the synthesis of polypeptides and proteins, where the strength of the recombinant DNA approach is used to provide large quantities of a suitable substrate from which the target structure is obtained by semisynthetic transformation.

2.7 SITE-SPECIFIC MODIFICATION AND PROTEIN CONJUGATES

Quite often, a single, unmodified polypeptide does not possess all of the properties (binding, enzymatic activity, radioactivity, fluorescence, cytotoxicity) required for a given application. Additional properties may be conferred by forming a conjugate with another molecule (e.g., polypeptide, reporter group, cytotoxic agent, etc.). Many useful conjugates (e.g., radiolabeled proteins, antibody-enzyme conjugates for ELISA) are prepared by methods that lead to product heterogeneity. However, with increasing development of clinical applications (of antibody-enzyme and antibody-toxin conjugates), there is a move toward producing homogeneous products.

This is not the place to discuss the general chemical conjugation literature, which has been reviewed elsewhere[45–47] and even has a dedicated journal, *Bioconjugate Chemistry,* published by the American Chemical Society. We shall concentrate here on techniques that are being used to modify proteins in a site-specific manner in order to yield products that are homogeneous. The advantages of a homogeneous product are not just theoretical. Such a product can be characterized (and so shown to be identical even if a procedure is changed on scale-up), and the interpretation of biological results is simplified. Random chemical conjugation has been considered unsuitable for large-scale production for in vivo application,[48] and the ill-defined nature of the reaction products are thought to limit the ability to assess accurately the results of animal and clinical trials of such engineered agents.[49]

Site-specific reaction, necessary if a homogeneous product is to be formed, can be attempted through a variety of approaches, and we shall consider these in turn.

2.7.1 REACTIVITY OF SIDE CHAINS

In some cases, it is possible to exploit the reactivity of a single residue (or sequence of residues) either naturally present or introduced into an analog by recombinant DNA techniques. Typically, Cys is used (which possesses a thiol group for alkylation or disulfide bond formation), either as a residue introduced specially for the purpose[50] or naturally present,[51] but other residues such as Trp and Met may be exploited if present only once in the primary structure and not necessary for function. This side-chain modification approach has been extended to include a particular sequence of residues (for example, a sequence that permits site-specific phosphorylation, or metal binding), and it is easy to imagine others.

2.7.2 INSERTION INTO A DISULFIDE BOND

Insertion into a disulphide bond has been shown to be possible.[52] It is little used and deserves more attention.

2.7.3 THE AMINO-TERMINUS

Specific acylation of an N-terminal amino group while sparing side-chains is said to be possible using iodoacetic anhydride for relatively small peptides[53] or isothio-cyanates with substrates the size of antibodies.[54] While isothiocyanates, under certain conditions, definitely favor alpha over epsilon amino groups, in our hands the discrimination is not absolute, even with a protein of 20 kDa (R.M.L. Jones and K. Rose, unpublished results).

2.7.4 CARBOHYDRATE GROUPS OF GLYCOPROTEINS

Oxidation of the carbohydrate residues of a glycoprotein, either chemically with periodate or enzymically with galactose oxidase, produces aldehyde groups. These aldehyde groups may then be exploited in a subsequent reaction (reductive alkylation or hydrazone formation) to form a conjugate (reviewed by O'Shannessy and Quarles[55]). An improvement involves replacing the reductive alkylation step by simple incubation with an aminooxyacetyl-ligand. Conjugation then occurs under exceedingly mild conditions through oxime formation, which provides a stable conjugate without the risk, inherent in the reductive alkylation approach, of stabi-lizing by reduction unwanted inter- or intra-molecular links.[56]

2.7.5 OXIDATION OF N-TERMINAL SER AND THR

The extreme susceptibility of 1-amino-2-hydroxy compounds to oxidation by peri-odate has been exploited for the formation of well-defined protein conjugates.[57] When Ser (or Thr) occupies the N-terminal position (but not elsewhere, see Fig. 2.2), it may be oxidized under exceedingly mild conditions to yield an alpha-N-glyoxylyl-polypeptide. This resulting aldehyde group may then be exploited for conjugation specifically to the thus-modified N-terminus, as shown in Fig. 2.2. Although Fig. 2.2 shows hydrazone formation as the ligation step, oxime[58] and thiazolidine[59,60] formation are also much used. Internal residues of delta-hydroxylysine would also be oxidized,[57] but such residues are rarely found in proteins. It is possible that sialic acid groups, which are present on some glycoproteins and particularly susceptible to periodate,[61] may compete with oxidation of N-terminal Ser or Thr to some extent, although this has not yet been carefully tested. N-terminal oxidation has been used to attach a variety of groups to the N-terminus of polypeptides, including attachment of a biotinyl hydrazide to murine interleukin-1-alpha with full retention of receptor binding and biological activity[57] and attachment of a fluorescent probe to interleukin-8 to visualize and quantify receptor expression.[58] Oxidation of N-terminal Ser or Thr has been used as part of a strategy for "cassette-type" engineering of granulocyte colony stimulating factor,[16,62] to link beta-lactamase to an antibody fragment[63] and to link a carboxypeptidase to an antibody fragment.[64] In view of the mildness of the

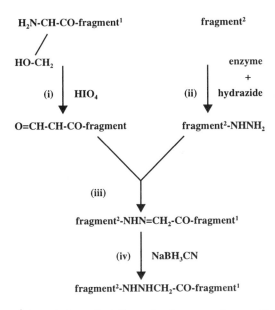

FIGURE 2.2 Oxidation of N-terminal Ser and subsequent site-specific conjugation. (i) A fragment possessing N-terminal Ser (or Thr) is oxidized by very mild periodate treatment to create an N-terminal aldehyde group. (ii) The fragment to be ligated to this one is modified at its C-terminus by reverse proteolysis, to attach a hydrazide group ($-NHNH_2$ itself, or $-NHNHCONHNH_2$, or similar group). (iii) Mixing these two fragments then allows site-specific ligation by formation of a hydrazone bond, which may be stabilized if required by mild reduction with cyanoborohydride (iv).

reaction conditions (fully aqueous solutions, pH 4.5–8.3, room temperature, micromolar concentrations) and the fact that site-specific reaction can be achieved, coupling through oxidized N-terminal Ser or Thr deserves to be used more widely.

2.7.6 ENZYMIC MODIFICATION OF THE C-TERMINUS

We shall consider here the application of reverse proteolysis to the formation of well defined protein conjugates through modification at the carboxyl terminus. Enzymes may be used, when an appropriate nucleophile is available, to attach directly such a nucleophile to the alpha-carboxyl group of the polypeptide or protein. This approach was used to attach a biotin group.[65,66]

A more general approach[17,23] involves the enzyme-catalyzed coupling of a small, relatively hydrophilic linker molecule in a first step. This linker, which possesses a reactivity not found in proteins (for example that of an aldehyde or hydrazide group) becomes, in a second step, the site of a spontaneous chemical coupling reaction (hydrazone formation, for example). This technique permits the preparation of conjugates that are chemically well defined and has recently been shown to be applicable even to very large antibody fragments.[25,64] By using such techniques to attach to a protein an oligomer of defined chemical structure and itself capable of reacting specifically with a given number of ligand molecules, it is possible to prepare conjugates of defined structure that carry several ligand (e.g., drug) molecules attached to a single site.[67]

The general approach involving the ligation of two species, each possessing a chemical group that reacts each with the other but not with itself, has been likened to the press-stud fitting[68,69] or to the dovetail joint,[5] and is often referred to as *chemoselective ligation*. It is discussed in more detail in the following chapter.

Finally, although out of order, I must draw attention to a most recent article concerning renaturation of proteins through slow dialysis.[70] While the technique described is similar to established procedures of oxidative refolding of small recombinant proteins,[33] the modified procedure (which uses a high-pressure pump) leads to high yields of renatured proteins of high complexity (antibodies). The availability of this renaturation technique should offer further flexibility in the choice of semisynthetic options when faced with a complex multi-subunit protein, since it may no longer be necessary in such cases to always try to preserve the disulfide bonds intact.

ACKNOWLEDGMENTS

I thank Mr. Pierre-Olivier Regamey for obtaining the mass spectra shown, and the Fonds National Suisse de la Recherche Scientifique and the Sandoz Stiftung for the purchase of our mass spectrometric equipment and for its continued support of our developments in the area of protein chemistry.

Note: review supplied in 1993; additional references added in 1996.

REFERENCES

1. Markussen, J., *Human insulin by tryptic transpeptidations of porcine insulin and biosynthetic precursors,* MTP Press, Lancaster, England, 1987, 170 pp.
2. Brown, A.R., Covington, M., Newton, R.C., Ramage, R. and Welch, P., The total chemical synthesis of monocyte chemotactic protein-1 (MCP-1), *J. Peptide Sci.,* 2, 40-46, 1996.
3. Clark-Lewis, I., Aebersold, R., Ziltener, H., Schrader, J.W., Hood, L.E. and Kent, S.B.H., Automated chemical synthesis of a protein growth factor for hemopoietic cells, interleukin-3, *Science,* 231, 134-139, 1986.
4. Inui, T., Bodi, J., Kubo, S., Nishio, H., Kimura, T., Kojima, S., Maruta, H., Muramatsu, T. and Sakakibara, S., Solution synthesis of human midkine, a novel heparin-binding neurotrophic factor consisting of 121 amino acid residues with five disulphide bonds, *J. Peptide Sci.,* 2, 28-39, 1996.
5. Schnolzer, M. and Kent, S.B.H., Constructing proteins by dovetailing unprotected synthetic peptides: backbone-engineered HIV protease, *Science,* 256, 221-225, 1992.
6. Vita, C., Gozzini, L., Di Bello, C., Total synthesis of horse heart apocytochrome *c* by conformation-assisted condensation of two chemically synthesized fragments, *Eur. J. Biochem.,* 204, 631-640, 1992.
7. Di Bello, C., Vita, C., Gozzini, L., Total synthesis of horse heart cytochrome *c,* *Biochem. Biophys. Res. Commun.,* 183, 258-264, 1992.
8. Jackson, D.Y., Burnier, J., Quan, C., Stanley, M., Tom, J. and Wells, J.A., A designed peptide ligase for total synthesis of ribonuclease A with unnatural catalytic residues, *Science,* 266, 243-247, 1994.

9. Slomczynska, U., Albericio, F., Cardenas, F. and Giralt, E., Studies on the enzymatic coupling of peptide segments on the solid support, *Biomed. Biochim. Acta,* 50, 10/11, S67-73, 1991.

10. Offord, R.E., Protection of peptides of biological origin for use as intermediates in the chemical synthesis of proteins, *Nature,* 221, 37-40, 1969.

11. Offord, R.E., Chemical approaches to protein engineering, in *Protein design and the development of new therapeutics and vaccines,* Hook, J.B. and Poste, G., eds., Plenum, New York, 1990, pp 253-282.

12. Bain, J.D., Switzer, C., Chamberlin, A.R. and Benner, S.A., Ribosome-mediated incorporation of a non-standard amino acid into a peptide through expansion of the genetic code, *Nature,* 356, 537-539, 1992.

13. Ellman, J., Mendel, D., Anthony-Cahill, S., Noren, C.J. and Schultz, P.G., Biosynthetic method for introducing unnatural amino acids site-specifically into proteins, *Meth. Enzymol.,* 202, 301-336, 1991.

14. Camble, R., Edge, M.D. and Moore, V.E., Properties of interferon alpha2 analogues produced from synthetic genes, *Proc. 9th Amer. Pep. Symp.,* 1985, pp 375-384.

15. Wallace, C.J.A., Guillemette, J.G., Hibiya, Y. and Smith, M., Enhancing protein engineering capabilities by combining mutagenesis and semisynthesis, *J. Biol. Chem.,* 266, 21355-21357, 1991.

16. Gaertner, H.F., Rose, K., Cotton, R., Timms, D., Camble, R. and Offord, R.E., Construction of protein analogues by site-specific condensation of unprotected fragments, *Bioconj. Chem.,* 3, 262-268, 1992.

17. Rose, K., Fisch, I., Vilaseca, L.A., Meunier, A., Werlen, R., Pochon, S., Jones, R.M.L., Dufour, B., Rossitto, I., Regamey, P.-O. and Offord, R.E., Site-specific modification of natural and biosynthetic polypeptides by reverse proteolysis, in *Innovations and perspectives in solid phase synthesis and related technologies,* Epton, E., ed., Intercept (Andover, UK), 1992, pp 129-134.

18. G. Allen, Sequencing of proteins and peptides, Laboratory techniques in biochemistry and molecular biology, vol. 9, revised 2nd ed., Burdon, R.H. and van Knippenberg, P.H., eds., Elsevier, Amsterdam, 1989.

19. Yagisawa, S., Studies on protein semisynthesis. I. Formation of esters, hydrazides, and substituted hydrazides of peptides by the reverse reaction of trypsin, *J. Biochem. (Tokyo),* 89, 491-501, 1981.

20. Proudfoot, A.E.I., Davies, J.G., Turcatti, G. and Wingfield, P.T., Human interleukin-5 expressed in *Escherichia coli:* assignment of the disulfide bridges of the purified unglycosylated protein, *FEBS Lett.,* 283, 61-64, 1991.

21. Davies, J.G., Rose, K., Bradshaw, C.G. and Offord, R.E., Enzymatic semisynthesis of insulin specifically labelled with tritium at position B-30, *Protein Engineering,* 1, 407-411, 1987.

22. Rose, K., Herrero, C., Proudfoot, A.E.I., Offord, R.E. and Wallace, C.J.A, Enzyme-assisted semisynthesis of polypeptide active esters and their use, *Biochem. J.,* 249, 83-88, 1988.

23. Rose, K., Vilaseca, L.A., Werlen, R., Meunier, A., Fisch, I., Jones, R.M.L. and Offord, R.E., Preparation of well-defined protein conjugates using enzyme-assisted reverse proteolysis, *Bioconj. Chem.,* 2, 154-159, 1991.

24. Morihara, K., Oka, T., Tsuzuki, T., Tochino, Y. and Kanaya, T., *Achromobacter* protease I-catalyzed conversion of porcine insulin into human insulin, *Biochem. Biophys. Res. Commun.,* 92, 396-402, 1980.

25. Fisch, I., Künzi, G., Rose, K. and Offord, R.E., Site-specific modification of a fragment of a chimeric monoclonal antibody using reverse proteolysis, *Bioconj. Chem.*, 3, 147-153, 1992.

26. Yagisawa, S., Watanabe, S., Takaoka, T. and Azuma, H., High-efficiency transpeptidation catalyzed by clostripain and electrostatic effects in substrate specificity, *Biochem. J.*, 266, 771-775, 1990.

27. Yoshimoto, T., Walter, R. and Tsuru, D., Proline-specific endopeptidase from *Flavobacterium*, *J. Biol. Chem.*, 255, 4786-4792, 1980.

28. Seikagaku Kogyo Co. Ltd., Manufacture of human insulin, Jpn. Kokai Tokkyo Koho JP 82 79,898.

29. Wu, C.Y., Chen, S.T., Chiou, S.H. and Wang, K.T., Specific peptide-bond cleavage by microwave irradiation in weak acid solution, *J. Protein Chem.*, 11, 45-50, 1992.

30. Cuenoud, B., Tarasow, T.M. and Schepartz, A., A new strategy for directed protein cleavage, *Tetrahedron Lett.*, 33, 895-898, 1992.

31. Hancock, W.S. (ed.), *High performance liquid chromatography in biotechnology*, John Wiley and Sons, New York, 1990, 564 pp.

32. Deutscher, M.P., ed., *Meth. Enzymol.*, vol. 182, Guide to protein purification, Academic Press, London, 1990.

33. Coligan, J.E., Dunn, B.M., Ploegh, H.L., Speicher, D.W. and Wingfield, P.T. (eds.), *Current protocols in protein science*, John Wiley and Sons, New York, 1995.

34. Frenz, J., Quan, C.P., Hancock, W. and Bourell, J., Characterization of a tryptic digest by high performance displacement chromatography and mass spectrometry, *J. Chromatogr.*, 557, 289-305, 1991.

35. Karger, B.L., Capillary electrophoresis, *Curr. Opin. Biotechnol.*, 3, 59-64, 1992.

36. Bongers, J., Offord, R.E., Felix, A.M., Lambros, T., Liu, W., Ahmad, M., Campbell, R.M. and Heimer, E.P., Comparison of enzymatic semisyntheses of peptide amides: human growth hormone releasing factor and analogs, *Biomed. Biochim. Acta*, 50, 10/11 S 157-162, 1991.

37. Loo, J.A., Bioanalytical mass spectrometry: many flavors to choose, *Bioconj. Chem.*, 6, 644-665, 1995.

38. Baca, M. and Kent, S.B.H., Direct observation of a ternary complex between the dimeric enzyme HIV-1 protease and a substrate-based inhibitor, *J. Amer. Chem. Soc.*, 114, 3992-3993, 1992.

39. Vestal, M.L., Juhasz, P. and Martin, S.A., Delayed extraction matrix-assisted laser desorption time-of-flight mass spectrometry, *Rapid Commun. Mass Spectrom.*, 9, 1044-1050, 1995.

40. Bongers, J., Felix, A.M., Campbell, R.M., Lee, Y., Merkler, D.J. and Heimer, E.P., Semisynthesis of human growth hormone-releasing factors by alpha-amidating enzyme catalyzed oxidation of glycine-extended precursors, *Peptide Res.*, 5, 183-189, 1992.

41. Markussen, J., Diers, I., Hougaard, P., Langkjaer, L., Norris, K., Snel, L., Sorensen, A.R., Sorensen, E. and Voigt, H.O., Soluble, prolonged-acting insulin derivatives. III. Degree of protraction, crystallizability and chemical stability of insulins substituted in positions A21, B13, B23, B27 and B30, *Protein Engineering*, 2, 157-166, 1988.

42. Breddam, K., Widmer, F. and Meldal, M., Amidation of growth hormone releasing factor (1-29) by serine carboxypeptidase catalysed transpeptidation, *Int. J. Peptide Protein Res.*, 37, 153-160, 1991.

43. Morihara, K., Thermolysin catalyzed semisynthesis of peptide hormones by introduction of Phe-NH_2 or Tyr-NH_2 at the carboxyl termini, *Biomed. Biochim. Acta*, 50, 10/11 S 15-18, 1991.

44. Henriksen, D.B., Breddam, K., Moeller, J. and Buchardt, O., Peptide amidation by chemical protein engineering. A combination of enzymic and photochemical synthesis, *J. Amer. Chem. Soc.,* 114, 1992, 1876-1877.

45. Means, G.E. and Feeney, R.E., Chemical modification of proteins: history and applications, *Bioconj. Chem.,* 1, 2-12, 1990.

46. Wong, S.S., *Chemistry of protein conjugation and cross-linking,* CRC, Boca Raton, USA, 1991, 328 pp.

47. Smith, R.A.G., Dewdney, J.M., Fears, R. and Poste, G., Chemical derivatization of therapeutic proteins, *Tibtech,* 11, 397-403, 1993.

48. Dewerchin, M. and Collen, D., Enhancement of the thrombolytic potency of plasminogen activators by conjugation with clot-specific monoclonal antibodies, *Bioconj. Chem.,* 2, 293-300, 1991.

49. Hayzer, D.J., Lubin, I.M. and Runge, M.S., Conjugation of plasminogen activators and fibrin-specific antibodies to improve thrombolytic therapeutic agents, *Bioconj. Chem.,* 2, 301-308, 1991.

50. Lyons, A., King, D.J., Owens, R.J., Yarranton, G.T., Millican, A., Whittle, N.R. and Adair, J.R., Site-specific attachment to recombinant antibodies via introduced surface cysteine residues, *Protein Engineering,* 3, 703-708, 1990.

51. Werlen, R.C., Lankinen, M., Offord, R.E., Schubiger, P.A., Smith, A. and Rose, K., Preparation of a trivalent antigen-binding construct using polyoxime chemistry: improved biodistribution and potential therapeutic application, *Cancer Res.,* 56, 809-815, 1996.

52. Packard, B., Edidin, M. and Komoriya, A., Site-directed labeling of a monoclonal antibody: targetting to a disulfide bond, *Biochemistry,* 25, 3548-3552, 1986.

53. Wetzel, R., Halualani, R., Stults, J.T. and Quan, C., A general method for highly selective cross-linking of unprotected polypeptides via pH-controlled modification of N-terminal alpha-amino groups, *Bioconj. Chem.,* 1, 114-122, 1990.

54. Rana, T.M. and Meares, C.F., N-terminal modification of immunoglobulin polypeptide chains tagged with isothiocyanato chelates, *Bioconj. Chem.,* 1, 357-362, 1990.

55. O'Shannessy, D.J. and Quarles, R.H., Labeling of the oligosaccharide moieties of immunoglobulins, *J. Immunol. Methods,* 99, 153-161, 1987.

56. Pochon, S., Buchegger, F., Pelegrin, A., Mach, J.-P., Offord, R.E., Ryser, J.E. and Rose, K., A novel derivative of the chelon desferrioxamine for site-specific conjugation to antibodies, *Int. J. Cancer,* 43, 1188-1194, 1989.

57. Geoghegan, K.F. and Stroh, J.G., Site-directed conjugation of nonpeptide groups to peptides and proteins via periodate oxidation of a 2-amino alcohol. Application to modification at N-terminal serine, *Bioconj. Chem.,* 3, 138-146, 1992.

58. Alouani, S., Gaertner, H.F., Mermod, J.-J., Power, C.A., Bacon, K.B., Wells, T.N.C. and Proudfoot, A.E.I., A fluorescent interleukin-8 receptor probe produced by targeted labelling at the amino terminus, *Eur. J. Biochem.,* 227, 328-334, 1995.

59. Tam, J.P. and Spetzler, J.C., Chemoselective approaches to the preparation of peptide dendrimers and branched artificial proteins using unprotected peptides as building blocks, *Biomed. Pept. Prots. Nucl. Acids,* 1, 123-132, 1995.

60. Zhang, L. and Tam, J.P., Thiazolidine formation as a general and site-specific conjugation method for synthetic peptides and proteins, *Anal. Biochem.,* 233, 87-93, 1996.

61. Zara, J.J., Wood, R.D., Boon, P., Kim, C.H., Pomato, N., Bredehorst, R. and Vogel, C.W., A carbohydrate-directed heterobifunctional cross-linking reagent for the synthesis of immunoconjugates, *Analyt. Biochem.,* 194, 156-162, 1991.

62. Gaertner, H.F., Offord, R.E., Cotton, R., Timms, D., Camble, R. and Rose, K., Chemo-enzymic backbone engineering of proteins, *J. Biol. Chem.,* 269, 7224-7230, 1994.

63. Mikolajczyk, S.D., Meyer, D.L., Starling, J.J., Law, K.L., Rose, K., Dufour, B. and Offord, R.E., High yield, site-specific coupling of N-terminally modified beta-lacta-mase to a proteolytically derived single-sulfhydryl murine Fab, *Bioconj. Chem.,* 5, 636-646, 1994.

64. Werlen, R.C., Lankinen, M., Rose, K., Blakey, D., Shuttleworth, H., Melton, R. and Offord, R.E., Site-specific conjugation of an enzyme and an antibody fragment, *Bioconj. Chem.,* 5, 411-417, 1994.

65. Wilcheck, M., Schwarz, A., Wandrey, C., and Bayer, E.A., Use of carboxypeptidase Y for the introduction of probes into proteins via their carboxy terminus, in *Peptides: chemistry, structure and biology,* J.E. Rivier and G.R. Marshall, eds., ESCOM, Leiden, The Netherlands, 1990, pp 1038-1040.

66. Schwarz, A., Wandrey, C., Bayer, E.A. and Wilchek, M., Enzymic C-terminal bioti-nylation of proteins, *Methods Enzymol.,* 184, 160-162, 1990.

67. Vilaseca, L.A., Rose, K., Werlen, R., Meunier, A., Offord, R.E., Nichols, C.L. and Scott, W.L., Protein conjugates of defined structure: synthesis and use of a new carrier molecule, *Bioconj. Chem.,* 4, 515-520, 1993.

68. Offord, R.E. and Rose, K., Press-stud protein conjugates, in *Protides of the biological fluids,* Peeters, H., ed., Pergamon, Oxford, 1986, pp 35-38.

69. Offord, R.E., Pochon, S. and Rose, K., Press-stud protein conjugates, in *Peptides 1986,* Theodoropoulos, D., ed., W. de Gruyter, Berlin, 1987, pp 279-281.

70. Maeda, Y., Ueda, T. and Imoto, T., Effective renaturation of denatured and reduced immunoglobulin G *in vitro* without assistance of chaperone, *Protein Engineering,* 9, 95-100, 1996.

3 Stereoselective and Chemoselective Religation

Carmichael J.A. Wallace

CONTENTS

3.1 INTRODUCTION

In the review of semisynthesis methodology presented in Chapter 1, the difficulties inherent in the use of conventional peptide chemistry for fragment condensation were alluded to; that impression will be reinforced by the case studies of individual proteins that follow. Indeed, in all cases, semisynthesis has evolved to avoid this type of chemistry altogether. In the previous, and in subsequent chapters it will be seen that structural analogs for functional investigations are obtained by stepwise (insulin, myoglobin, PLA_2), non-covalent (ribonuclease S, cytochrome c), or -S-S-bridged (1L2) semisynthesis, in all of which fragment condensation is redundant, or else by the selective ligation methods that are the subject of this chapter.

To recapitulate, these approaches remove the necessity for much of the functional group protection and deprotection reactions that complicate the process and lead to substantial losses of scarce materials, avoid exposure to the harsh conditions characteristic of both those reactions and that of peptide bond formation, and are usually far more efficient. The cumulative effect of all these factors is generally to increase the final recovery of purified analog from often <1% of starting protein mass to, typically, 30% or more. That quantitative advantage has led to the virtual extinction of the "classical" methods.

Selective ligations come in two types. In the conformationally-directed class, the chemistry of ligation is conventional: a nucleophilic free amino group attacks an "activated"—highly electrophilic—carbonyl carbon, usually esterified by an electron-withdrawing group. The selectivity for the groups that will participate in the peptide bond formation is provided by the protein itself, requiring that the fragments to be religated associate and fold in a near-normal manner so that the amino group of residue n + 1 is the only one in proximity with the activated carbonyl carbon of residue n. For this reason we will term these approaches "stereoselective" methods.

It comes as a surprise to some that large protein fragments will form non-covalent complexes with such facility. However, often, the only element of the protein folding code missing in such systems is a single peptide bond. There are thus many examples known of functionally competent non-covalent complexes among smaller proteins, and although at certain sites the lack of a physical link between small structural elements might prevent an early crucial nucleation reaction, this property of proteins in pieces is likely to be universal. It would be of interest to ascertain whether the property also extends to larger proteins, perhaps mediated by chaperonins.

This move to take advantage of the properties of macromolecules has changed semisynthesis from a chemically to a biochemically based technology, and the appropriateness of this shift, when the desired products are themselves biochemicals, is emphasized by the recent results. The development had its origins, as is so often the case in science, in serendipitous discoveries and inspired insight, followed by painstaking experimentation.

Inspiration, with the realization that, to make a functional contiguous protein analog from fragments, not all the linking bonds need be peptidic, was at the origin of the other approach to ligation we will consider. Chemoselective methods have evolved rapidly in the past few years and are finding a place in both semisynthesis and peptide synthesis. In the latter case the fragments may not reassociate and conformationally directed religation is not feasible. A number of successful techniques have emerged, all of which rely on the introduction into the peptide structures of uniquely reactive groupings at the termini to be linked. These groups have a high mutual affinity; so reaction is rapid even at concentrations that are low relative to those used in conventional peptide chemistry; and no reactivity with normal side-chain functional groups or water. Thus, ligation can be achieved in aqueous solution with minimal protection.

Although predicated on the notion that other types of covalent links can be a perfectly good substitute for a peptide bond in the final structure, the evolution of this methodology has culminated in schemes where the initial chemoselective linkage is an intermediate in the ultimate, selective, formation of a pseudopeptide or even a true peptide bond. The methods will be detailed below.

The stereoselective methods have a longer history and have developed on two parallel tracks, in one of which the activating agent is a proteolytic enzyme *(reverse proteolysis)* whereas in the other it is an internal esterification *(lactonization)*.

This lactone structure is, in fact, automatically generated at the C-terminus of all but the carboxyl-terminal peptide when a protein is specifically cleaved at methionine residues by cyanogen bromide.[1]

$$
\begin{array}{ccccccc}
CH_3 & & CH_3 & & & & R \\
| & & | & & & & | \\
S & +\ CNBr & {}^{+}S{-}CN & +\ Br^{-} & +\ CH_3{-}S{-}CN & & +\ H_2N{-}C{-} \\
| & & | & & & & \\
CH_2 & \longrightarrow & CH_2 & \longrightarrow & CH_2 & \xrightarrow{H_2O} & CH_2 \\
| & & | & & \diagup\ \diagdown & & \diagup\ \diagdown \\
CH_2\ \ O\ \ \ \ R & & CH_2\ \ O\ \ \ \ R & & CH_2\ \ O\ \ \ \ R & & CH_2\ \ O \\
|\ \ \ \|\ \ \ \ | & & |\ \ \ \|\ \ \ \ | & & |\ \ \ |\ \ \ \ | & & |\ \ \ | \\
{-}CH{-}C{-}NH{-}C{-} & & {-}CH{-}C{-}NH{-}C{-} & & {-}CH{-}C{=}NH{-}C{-} & & {-}CH{-}C{=}O \\
& & & & +
\end{array}
$$

(R3.1)

The discovery that peptide bonds could spontaneously reform between protein fragments generated by this reaction was sufficiently serendipitous that it was in fact not immediately recognized for what it was. Corradin and Harbury reported that CNBr fragments could give a functional non-covalent complex in 1971,[2] but religation of these fragments was so unexpected that a complete understanding only emerged three years later.[3] This occurred after the discovery of a second example of the phenomenon was reported in 1974,[4] and led almost immediately to a program of exploitation, reported below and in Chapter 8, that is still ongoing.

The use of proteolytic enzymes as activating agents originated in Laskowski's laboratory and was built on three principles, all reported in 1969,[5,6] developed with the protease inhibitors. One is manipulation by pH change of the hydrolysis/dehydration equilibrium,[5] another employment of mass action by including high concentrations of the nucleophile,[5,6] and the third is that of the 'molecular trap' in which equilibrium is pulled in the direction of synthesis by the selective binding of the resynthesized product.[6]

While limited to the rather special case of the inhibitors specifically evolved to interact with proteases, these discoveries stimulated the search for generalized methods of shifting the equilibria of protease-catalyzed hydrolyses in the synthetic direction, based on these principles and others.

The serendipitous observation in this trail was published in 1973:[7] it was noted that the presence of 60% glycerol caused an eight-fold shift in the equilibrium in the desired direction (Figure 3.1). Subsequent work[8] proved that the effect was general, and not limited to inhibitors, and revealed its mechanistic basis. The seminal application to an entirely non-covalent complex of protein fragments was also undertaken by Laskowski's group, with ribonuclease S.[9] The organic cosolvent shifted the equilibrium for the hydrolysis of the Ala20-Ser21 bond in favor of synthesis by encouraging the formation of the acyl-enzyme intermediate (with an esterified Ala20 carboxyl group) rather than its hydrolytic breakdown. The formation of the complex between the fragments then overcomes the entropic barrier to resynthesis by uniquely placing the appropriate amino group in position for nucleophilic attack on that ester.

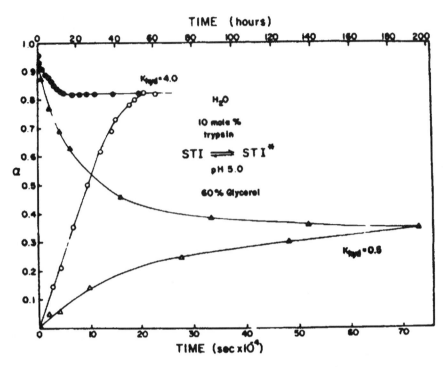

FIGURE 3.1 The influence of glycerol concentration on the equilibrium constant for hydrolysis of the Arg63-Ile64 peptide bond of soybean trypsin inhibitor. Either the unmodified or the fully hydrolyzed inhibitor (STI*) was treated with 10 mole % trypsin in fully aqueous buffer at pH 5 (o, forward; •, reverse) or in the same buffer containing 60% glycerol (Δ, forward; ▲ reverse), and the fraction of the modified inhibitor (α) determined by disc gel electrophoresis. Reprinted with permission from Homandberg, G.A., et al., *Biochemistry,* 176, 5220, 1978. Copyright 1978 American Chemical Society.

General exploitation of this methodology followed, in which the only limitation is the availability of proteolytically derived, non-covalent complexes of a small number of fragments. Recent further developments, to be detailed below, are the discovery that this reaction can apparently[10] operate at at least four peptide bonds simultaneously, and the modification by mutagenesis of a protease so that its intrinsic proteolysis equilibrium is altered in favor of dehydration.[11]

3.2   SPONTANEOUS RELIGATION OF CNBR FRAGMENTS

Reaction (R3.1) showed the mechanism of generation of C-terminal homoserine lactone from methionine during protein fragmentation by CNBr. The reaction and some side-reactions are thoroughly discussed by Fontana and Gross.[12] The cleavage reaction is performed in strong acid, and if fragment separation and purification are also performed under acidic conditions, no hydrolytic conversion of lactone to the open-ring form occurs. This is crucial, since the free acid will not participate in the religation reaction. Model studies on the aminolysis that fragment complexation promotes have been undertaken using protected amino acids.[13] Reaction is promoted by high pH, presumably by deprotonation of the amino group, and is detectable only in organic solvent systems. When water is present, aminolysis is not competitive with hydrolysis. Even at high concentrations of reagents, reaction is many orders of magnitude slower than with comparable systems employing active esters. Thus, the catalysis provided by fragment complexation is highly effective if not fully understood.

Religation occurs in an aqueous environment, so the effect is to promote aminolysis relative to hydrolysis. This does not occur by shielding reacting groups from solvent. Recent work has shown that surface exposure of the groups gives better yields than burying them.[14] It seems likely that the close proximity enforced by adoption of the native conformation increases aminolysis rate in the case of cytochrome c about 200-fold relative to the optimum model conditions.[3,13] In the other well studied examples, reaction rates appear a little slower, although conditions were not exactly comparable.[4,15] The net effect for cytochrome c is that standing of a dilute [10^{-4} M] equimolar mixture of the two fragments in phosphate buffer at pH 7 will give a maximal 60–70% yield of religated product after 24 h. Figure 3.2 shows the restoration of electron transfer activity associated with religation observed by Corradin and Harbury.[3] Presumably, the upper limit is determined by competing hydrolysis of the lactone.

Although this procedure sounds simple, there are two further complexities. The fragments employed were 1–65 and 66–104. The former retains the covalently attached heme, and the latter includes the other methionine residue of the horse cytochrome c sequence, residue 80. This residue's sulfur atom acts as a ligand of the heme iron, and must remain intact, so it is necessary to use conditions that ensure

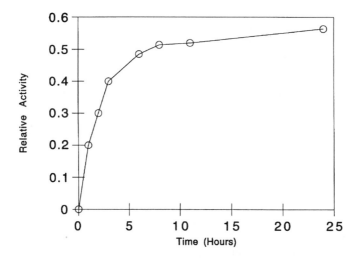

FIGURE 3.2 Spontaneous religation of CNBr fragments of cytochrome c. Equimolar amounts of fragments 1–65 and 66–104 were mixed to give a final concentration of 10^{-4} M. The complex was reduced with sodium dithionite and kept anaerobic at pH 5.7 and room temperature for 24 h. The progress of covalent religation of the 65–66 peptide bond was followed by withdrawing samples for assay in the succinate oxidase system.[14] Data from Ref. 3.

limited cleavage and overlap fragments. The resulting complex digest must be resolved, and the required fragments purified, by multiple chromatography steps. The other observation is that religation occurs under strongly reducing conditions, but not otherwise. These conditions are achieved by sodium dithionite and keeping the solution anaerobic for 24 h. Differences in the polypeptide chain fold between the two oxidation states of cytochrome c are minimal,[16] so some of the structural influences on catalytic efficiency must be extremely subtle.

Meeting these additional requirements is not difficult, so the phenomenon was rapidly seized on as a facile means to cytochrome c semisynthesis, in ways described fully in Chapter 8.

The phenomenon has also been used in other circumstances. First, it has been used in pancreatic trypsin inhibitor, where it was first recognized. In that case, as in cytochrome c, the cleaved methionine is a variable residue and its replacement in the structure by homoserine has no effect on function. However, experiments on semisynthesis were not directed at replacing the entire C-terminal hexapeptide but at introducing a shorter, labeled sequence.[17] Because the natural fragment is held in place by disulfide bonding and not just covalent forces, its ability to compete was eliminated by the use of denaturing solvents, and the catalytic effect of native structure formation replaced by a mass-action approach in the use of a 1 M concentration of the short peptide.

This study confirmed that religation between components at low concentration absolutely required the native conformation, and furthermore showed that the precise orientation of reacting groups was crucial. The removal of the N-terminal residue of the hexapeptide was enough to prevent religation under normally productive circumstances.[17] This observation was also made for cytochrome c, where the fragments are aligned by non-covalent forces only.[18]

The ability to "force" a coupling by mass action has been exploited in other ways. CNBr fragments can be attached to amino group-bearing insoluble resins for subsequent sequencing by solid-phase Edman degradation. Using high concentrations of peptide in organic solvent, a large excess of free amino groups and basic conditions, attachment of 80–100% of the peptide could be obtained.[19] We also have used this approach, in free solution, to condense two small non-complexing fragments (65–80 and 81–104) of cytochrome c, when up to 20% of 65–104 can be obtained by incubation of saturated solutions of the peptides in basic DMSO at 37° for 72 h.[20]

One consequence of the need for precise alignment of the fragments in the catalyzed process is that we know of a number of cases where complexes between CNBr fragments form, but no peptide bond formation follows. Examples are thioredoxin,[21] barnase,[22] superoxide dismutase and lactalbumin (C.J.A. Wallace, unpublished observations). Constructs have also been prepared where homoserine lactone has been introduced at the C-terminus of a component of a complex derived using other proteolytic agents. Again, neither the RNAase S analog (8%)[23,24] nor a tryptic fragment complex (1–38: 39–104) analog of cytochrome c (1%)[25] supported substantial religation. Most strikingly, the fragments derived from cytochrome c of yeast, which has a methionine residue at position 64, do not recombine, although the break point is just one residue away from that in the horse protein.[26]

The combination of the evident rarity of methionine residues, their frequently inconvenient location or functional conservation, and this unexplained variability in religation efficiency, constitutes a major limiting factor in the general exploitation of the phenomenon in protein engineering. But its manifest advantages in simplicity and efficiency have led to efforts both to define the structural factors that promote autocatalytic ligation and to increase accessibility to this property. This work is described below.

3.3 REVERSE PROTEOLYSIS

The early discoveries of Laskowski's group described above initiated first a rigorous examination of the phenomenon and then a program of exploitation that rapidly spread to other laboratories.

It was the protease inhibitors that spelled out the message that peptide bond hydrolysis is not an irreversible process; simply that entropic factors usually mean it is effectively so for protein-sized structures. So it was with these systems, reviewed in Chapter 7, that much of that work was first undertaken. Of course, work dating back to 1938 had shown that small peptides could be efficiently synthesized using enzymic catalysis, particularly when some suitable product trap (e.g. insolubility) was available. For an admirable discussion of these achievements, see Kullman's monograph.[27]

The principles described above were elucidated in experiments summarized in Figure 3.3. The use of a pH-mediated shift in equilibrium position, and overcoming the entropic problem by covalent and non-covalent complexation of the reactants, were later supplemented by employing the effects of organic cosolvents. As well as the more obvious property of reducing the concentration of water, a key component in the reaction, the data in Figure 3.1 imply an additional effect. This was shown to be due to the influence of cosolvent on the pK_as of reacting group ionizations, primarily that of the carboxyl group.[8] Indeed, upward pK shifts for this group in model compounds of more than 3 pH units could be measured, and such shifts will have profound effects on the thermodynamics of the reaction.

In the model experiments, the most pronounced shift in Ksyn was experienced with 1,4-butanediol, although other diols of similar size were almost equally effective. Glycerol, with which the effect was first observed was comparatively ineffective. Regrettably, though, butanediol can be a denaturant of both the non-covalent complexes and the proteases that might religate them; whereas glycerol is often used as a stabilizer of tertiary structure! Thus, the application of the set of principles now developed was always undertaken in glycerol.

The first non-protease inhibitor target was ribonuclease S, as described above.[9] Subsequent work extended the applicability to staphylococcal nuclease[28] and somatotropin,[29] but the supply of complexing products of limited proteolysis that could be induced to religate this way soon ran dry. An attempt with cytochrome c appeared to have succeeded,[30] but later work showed that the fragments were cross-linked by contaminating acrolein in the glycerol.[31] Despite the recent excitement of the discovery of a system for the simultaneous resynthesis of four peptide bonds in triosephosphate isomerase,[10] it is clear that the opportunities for semisynthetic

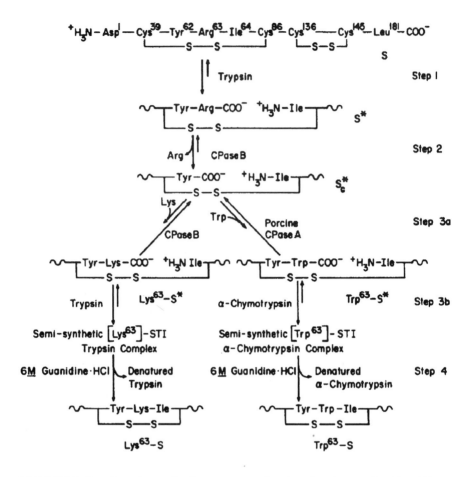

FIGURE 3.3 Reaction schemes for the protease-mediated semisynthesis of Lys63 and Trp63 soybean trypsin inhibitor. In this protocol, the internal C-terminal Arg63 residue of the modified inhibitor is removed by carboxypeptidase B. Then either Cpase A or Cpase B can be used to elongate the truncated form, using the mass action principle. However, this reaction is inherently so thermodynamically unfavorable that product would be undetectable without the use of a trap. This is provided by either chymotrypsin or trypsin, since the elongated forms will bind strongly to these enzymes, but not the truncated forms. These enzymes then also stimulate the resynthesis of the 63–64 peptide bond in the presence of glycerol, as above. Finally, the semisynthetic inhibitor can be released intact from the enzyme-inhibitor complex by using denaturing concentrations of guanidine hydrochloride. Figure taken from Ref. 51 with permission.

exploitation of these principles, too, are severely limited. The requirements of strictly limited proteolysis of native protein structures to give appropriately folded complexes, using enzymes that will obey these principles, and operating on bonds whose Ksyn is sufficiently high that the poorly effective glycerol will work, are simply too restrictive.

3.4 SEPARATING THE STEPS: REVERSE PROTEOLYSIS PRECEDES THE STEREOSELECTIVE RELIGATION

Although the reverse proteolysis approach has yielded disappointingly few semisyntheses from non-covalent complexes, the principles established and the methodology derived have had much wider application, in particular the mass action principle, in peptide and protein chemistry. Such application is generally not in the scope of this chapter but has been well reviewed elsewhere,[27,28] and some examples occur in the following chapters. But there is one important exception: its use as a preparatory step for a subsequent spontaneous stereoselective coupling.

The concept has its origins in the idea of preparing a synthetic peptide with C-terminal homoserine as a replacement for a natural component of a two-fragment non-covalent complex. This was first attempted by Sheppard's group,[23] and later elaborated upon,[24] using ribonuclease S as the vehicle. Although this is quite a strong complex, and normally as active as ribonuclease A, a structure determination of the former has shown that the termini either side of the break point are not in close contact in the crystal, and this is probably the reason that religation yields are not greater than 8%. Nevertheless, useful quantities of protein analogs can be easily made this way.

We developed the idea for an analogous approach in which homoserine lactone would be added to the C-terminus of an enzymatically-generated protein fragment by reversing the action of that enzyme. To ensure that the complexing fragments remained coterminous, it was also necessary to truncate the complementary fragment at its N-terminus.

Elongation with homoserine lactone was achieved in two ways.[25] Either by incubation of the fragment (in this case the tryptic fragment 1–38 of horse cytochrome c) with a near-molar concentration of methionine methyl ester in the presence of trypsin and in 94% butanediol, followed by CNBr treatment of the product, or directly, under similar conditions, with homoserine lactone. In the latter case, the D,L-homoserine lactone was used, since the enzymatic step ensures incorporation of the L-form only. Although cytochrome c has many lysine residues that are potential cleavage sites, proteolysis is restricted to arginine residues by the N^ε-acetimidyl protection employed, and in fact is limited to a single cleavage at residue 38. Although such protection also potentially prevents cleavage during the reverse proteolysis step at sites other than the C-terminus, it turned out that this is not strictly necessary, and that under the reverse proteolysis conditions second-site cleavage does not occur even with unprotected fragments.[25] With both homoserine lactone and the methionine ester, product yield was greater than 90%.

Unfortunately, this success was not matched at the stereoselective religation stage. Only 1–2% yield of the coupled product was detected. The low efficiency at this site could be due to a stable nonproductive conformation of the complex, like the ribonuclease S case, or to a "looser" structure in the region of the break point. This latter view was supported by experiments that compared redox potentials and proteolytic sensitivity of different, but closely related, non-covalent complexes.[29] If correct, it followed that the lesser catalytic power of the complex might be compensated for by a more strongly activated electrophile. "Active esters" are, of course, the basis for many coupling strategies in peptide synthesis and do not require catalysis, but, equally importantly, those used in peptide chemistry would be unstable under the conditions of stereoselective religation. Thus, there appeared to be a requirement for an ester grouping of intermediate activating strength.

Careful work by Keith Rose and colleagues uncovered the ideal compromise between stability and activation, in the amino acyl 2,6-dichlorophenyl esters.[30] Not only could they be added to the C-terminus of enzymatically generated fragments in high yield, like the well-studied methyl esters, but the products appeared to be quite stable to hydrolysis through this procedure and subsequent work-up. Finally, when complexed with the truncated complementary peptide, raising the pH to 8.5 was sufficient to induce peptide bond formation in up to 60% yield. Thus, a two-step protocol of high efficiency had been evolved that had the promise of very wide applicability (Fig. 3.4).

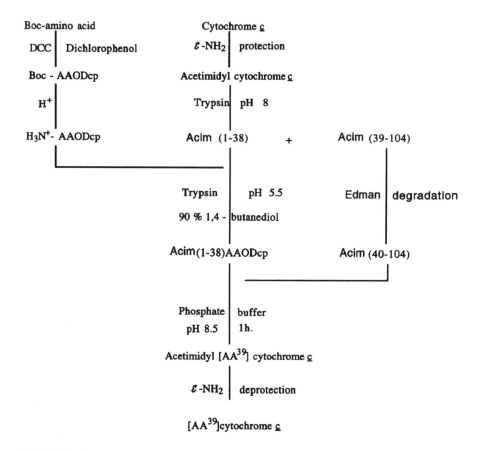

FIGURE 3.4 A typical scheme for the two-step approach to stereoselective religation. A protease-generated N-terminal fragment is elongated with an amino acid dichlorophenyl ester by reverse proteolysis in >90% organic cosolvent. The product is then complexed with a truncated complementary fragment in aqueous buffer. Raising the pH leads to rapid aminolysis of the dichlorophenyl ester and peptide bond formation.

This would be particularly true if it could be employed at many different enzymatic cleavage points. It was evolved with tryptic fragments, but we also used the lysine-specific *Achromobacter* protease to effect a stepwise semisynthesis in concert with trypsin.[25] Subsequently, we demonstrated that a range of other enzymes could be employed for fragment elongation by reverse proteolysis and that, in general, serine proteases performed best, though none so well as trypsin (Table 3.1[31]). It was also shown that the same enzyme could be effective at many different cleavage points in the same protein, and that many different amino acid esters could serve as nucleophile (Tables 3.2 and 3.3[25]). Some differences in efficiency were noted between them, as was also the case in the subsequent religation step. The final proof of the versatility of this approach came with the synthesis of the [des-40–55] analog of cytochrome *c*, through reaction of 1–39 ODcp with a non-contiguous fragment, 56–104 (Figure 3.5). The reactive termini are close in the 3D structure, but, clearly, the requirement for precise orientation that characterizes the CNBr fragment-based stereoselective method has been dispensed with.

3.5 SYNERGY: COMBINING SITE-DIRECTED MUTAGENESIS WITH STEREOSELECTIVE SEMISYNTHESIS

Even the increased versatility of the two-step combination of reverse proteolysis and autocatalytic ligation leaves the practitioner of semisynthesis at the mercy of the natural distribution of cleavage points, which may be too frequent for limited proteolysis, or unfavorably located if rare. We reasoned that we might be able to overcome this final limitation to the generality of stereoselective ligation, and better understand the structural principles governing the phenomenon, if we could exploit site-directed mutagenesis. A synergy between the techniques would develop if as a preliminary step mutagenesis were used to eliminate inconvenient residues and ideally locate cleavage sites. Then, a facile semisynthesis could be achieved between these designed fragments that would permit the introduction of residues that could not be done directly by the genetic approach.

The first test of this concept was initiated in collaboration with the University of British Columbia mutagenesis group, who were working with the *Saccharomyces cerevisae* iso-1 cytochrome *c*. The yeast protein has two methionine residues but differs from horse in having one at position 64 rather than 65. And whereas horse CNBr fragments 1–65 and 66–104 will religate in 60% yield, the yeast peptides 5–64 and 65–103 do not couple under identical circumstances. We therefore chose to move the methionine to position 65, in the double M64L, S65M mutant.[32] This proved to be simple to do by conventional methods, and the resulting mutant was functionally identical to the parent protein. Such a result is a desirable first criterion in the choice of location for introduced cleavage and religation sites.

The others are clean and complete cleavage, not hard to achieve with CNBr digestion; and that the homoserine residue in place after religation does not significantly compromise function, either. Finally, stereoselective coupling should be efficient. In the first trial, we reported yields of at least 50%, in subsequent experiments not yet published yields equal to those with the horse protein (60–70%) have been obtained. The extraordinary difference between the two adjacent sites is of

TABLE 3.1

The Efficiency of A Variety of Proteases in the Promotion of Reverse Proteolysis

Enzyme	Substrate (C-terminal residue)	Nucleophile	Diol Conc. %	PH	Yield %
Papain	CYC 1-38 (Arg)	PheO'Bu	95	5.8	2.5[b]
	CYC 1-39 (Phe)	PheO'Bu	95	5.8	1[b]
Chymopapain	CYC 1-38 (Arg)	PheO'Bu	95	5.8	7[b]
	CYC 1-39	PheO'Bu	95	5.8	12[b]
Subtilisin	CYC 1-36 (Phe)	GlyODcp	83	6.2	17
	CYC 1-38 (Arg)	AlaODcp	88	6.8	8
			88	6.5	8
	CYC 1-39 (Ala)	AlaO'Bu	86	5.2	18
				5.6	23
				6.0	32
				6.4	24
				6.8	15
	CYC 1-65 (Hse)	PheO'Bu	88	6.2	Proteolysis
	CYC 1-50 (Asp)	PheO'Bu	90	6.0	36[a]
Achromobacter protease	CYC 1-39 (Lys)	PheODcp	82	5.8	75
		ThrODcp	82	5.8	67
Thermolysin	CYC 1-36 (phe)	GlyODcp	70	6.2	7
		PheODcp	70	6.1	6
	CYC 1-38 (Arg)	PheODcp	70	5.9	6
			87	6.0	3
	CYC 1-39 (Ala)	PheO'Bu	70	5.7	5[a]
				6.1	10[a]
				6.4	16[a]
	CYC 1-65 (Hse)	PheO'Bu	70	6.1	5
Chymotrypsin	CYC 1-36 (Phe)	AlaODcp	87	6.8	21 (28°)
					12 (37°)
		PheODcp	87	6.8	45
		GlyODcp	87	6.8	21 (26°)
					12 (4°)
	CYC 1-38 (Arg)	AlaODcp	87	6.8	0
	CYC 1-39 (Ala)	PheO'Bu	87	6.8	0
	CYC 1-59 (Trp)	AlaODcp	87	6.8	33 (37°)

[a]Some proteolysis observed; [b]24 h incubations.

Conditions employed were similar to those in which trypsin can generally give 50-90% yields (Table 3.2). Table taken from Ref. 31 with permission.

TABLE 3.2

Maximal Coupling Yields for Enzymatic Addition of Amino Acid Derivatives to Fragments of Cytochrome c

Fragment	COOH- terminal residue	Enzyme	Nucleophile	Yield
1–38	Arginine	Trypsin	AlaODcp	86
			GlyODcp	60
			ε-TFA LysODcp[a]	90
			ε-Acim LysODcp	20
			PheODcp	95
			ThrODcp	63
			Homoserine lactone	92
			MetOMe	96
			ε-Boc LysOt-But	87
			Acim fragment (39–65)	24
1–39	Lysine	Trypsin	ThrODcp	40
			ε-TFA LysODcp	35
			PheODcp	45
			ValODcp	53
1–39	Lysine	Achromobacter	ThrODcp	67
		Protease	PheODcp	75
1–53	Lysine	Trypsin	AlaODcp	44
1–55	Lysine	Trypsin	AlaODcp	36

Note: In these reactions, the fragment, at 1 mM concentration, was incubated with 0.2–0.5 M amino acid dichlorophenyl ester in ≥90% 1,4-butanediol, adjusted to pH 5.5, and porcine trypsin (1:4 enzyme : substrate, w/w). Yields were quantitated by comparison of peak areas of esterified and unreacted fragments on HPLC or ion-exchange chromatography. Table taken from Ref. 25 with permission.

[a]TFA = trifluoroacetyl; Ot-But = *t*-butyl ester.

speculative interest. We noticed that, because they are located in an amphipathic helix, they differ in being buried (the 64–65 bond) or surface located (65–66) (Figure 3.6). Forming the former bond would require that the polar amino group penetrate the protein interior.

This hypothesis was one that was explored in the next stage of experimentation, which had multiple goals. In addition to the above, we wished to find out if the secondary structural context had any influence on efficiency, if all non-conservative sites could accommodate methionine or homoserine with impunity, and if such changes could be tolerated at conservative sites, too. To these ends, we constructed six new mutants, to give with the parent protein and the initial [Met65] form a total of eight distinct secondary structural contexts. The proteins' functional properties were carefully checked, and each member of the partial methionine scan put through cleavage and religation procedures.[33]

The results of one of the measures of functionality, and of the religation efficiency study, are given in Table 3.4. These results allow the following conclusions:

FIGURE 3.5 Computer-generated skeletal model of cytochrome *c* showing the relative positions of the reactive termini in the stereoselective religation of the loop-deleted form [des 40–55] cytochrome *c*. Although distant in the primary sequence, correct folding of the fragments into their positions in the native conformation will bring the esterified carboxyl of residue 39 into proximity with the amino group of residue 56. Figure prepared by Christian Blouin.

1. A wide range of efficiencies results from varying religation site.
2. Surface location of the peptide bond to be formed leads to moderate to high efficiency, but a buried site is inhibitory.
3. Efficiency at a surface located site seems not to depend on the nature of the secondary structure in which it is contained.
4. Generally, mutation of variable residues does not affect function, but there are some striking exceptions.
5. Mutations at conserved residues usually affects function, although frequently to a limited degree.
6. The consequences of introducing homoserine are not worse than for methionine, except at buried locations.

TABLE 3.3
Stereoselective Religations of Cytochrome *c* Fragments

α-Dichlorophenyl ester of Acetimidyl Fragment	Nucleophile	Coupling Yield %
[Gly³⁹] 1–39	40–104	5
[Ala³⁹] 1–39	40–104	48 ± 10
[ε-TFA Lys³⁹] 1–39[a]	40–104	53 ± 8
[ε-Acim Lys³⁹] 1–39	40–104	15
[Phe³⁹] 1–39	40–104	46 ± 16
[Phe³⁹] 1–39	41–104	55
[Thr³⁹] 1–39	41–104	42
[Val⁴⁰]1–40	41–104	30
[Thr⁴⁰] 1–40	41–104	31
[Phe⁴⁰] 1–40	41–104	35
[ε-TFA Lys⁴⁰] 1–40	41–104	20

[a]TFA = trifluoroacetyl

Note: Reverse proteolysis-extended fragment dichlorophenyl esters were incubated with contiguous C-terminal fragments in aqueous buffers, pH 8.5, for 1 h. Yields were estimated from relative peak areas on subsequent gel exclusion chromatography. Table taken from Ref. 25 with permission.

TABLE 3.4
Comparative Biological Activities and Religation Efficiencies of the Methionine–Scan Mutants

Cytochrome	Conservation of Residue	Cleavage Site Location	Succinate Oxidase Activity of Mutant	Religation Yield	Bioactivity of Product
P25M	V	At the surface-exposed extremity of an Ω-loop	100	40	91
V28M	V	Surface-exposed residue i + 1 of γ-turn	63	32	56
I35M	C	Buried residue i of ß-turn	100	10	34
K55M	V	Outer face of short a-helix within an Ω-loop	55	40	54
M64 (parent)	V	Hydrophobic inner face of amphipathic helix	100	0	n.a.
S65M	V	Hydrophobic inner face of amphipathic helix	100	60	100
L68M	I	Interface of amphipathic helix (partly buried)	73	14	65
I75M	C	Partly exposed residue i of ß-turn	48	35	82

V = variable, C = conserved, I = invariant

The changes were chosen to test the limits of the phenomenon, so it is clear that structure-based prediction of suitable sites with the clear aim of optimizing ligation while not compromising protein function should be relatively facile. The priority in choice of mutation site should be convenient access to the target residue for the ultimate semisynthesis, but with that in mind selection of a residue to which the C-terminal peptide bond is exposed, and which has structural similarity to methionine, should virtually guarantee success.

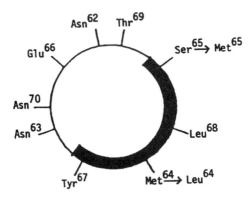

FIGURE 3.6 Helical wheel representation of residues 62–70 of yeast iso-1 cytochrome *c*. The bold area is the hydrophobic face of this ideal amphipathic helix, the remainder is surface exposed. The mutations shown shift the site of the peptide bond to be religated by homoserine lactone aminolysis from within the buried face to a surface location.

As a consequence of this exploratory study, we are currently undertaking the previously impossible semisynthesis of yeast cytochrome *c*, using synthetic homologs of fragments 76–103, 66–103, and 56–103.

3.6 FUTURE DIRECTIONS

The success of this strategy for cytochrome *c* and its probable applicability to other small proteins should not shut the door on attempts at further developments and other directions to the goal of optimizing fragment religation in semisynthesis.

One virtue of the stereoselective processes described above, compared to chemical coupling methods, is the mild conditions encountered throughout the process. With the exception of the CNBr cleavage, the lengthy exposure to strong acid might be harmful to some proteins. We are, therefore, currently applying the same synergistic strategy evolved with the methionine-scan mutants to enzymatic cleavages. Initial trials are focussed on using trypsin for cleavage and reverse proteolysis. Like the horse cytochrome, acetimidylated yeast iso-1-cytochrome *c* is cleaved by trypsin only at arginine 38: both Arg13 and Arg91 are resistant. We are therefore using mutagenesis to replace this former residue with lysine and to introduce arginine at a range of potentially useful sites well distributed throughout the cytochrome sequence. Specific cleavage will be followed by a two step activation-ligation protocol. Although inevitably more complex than the methionine scan protocol, the range of enzymes available and the benign conditions of every step should make this strategy widely appealing.

Perhaps the most exciting recent development in proteolytic enzyme application was the discovery that triose phosphate isomerase (and now lysozyme also) could be simultaneously cleaved at three peptide bonds by subtilisin, yielding a functional complex, and then reformed with the same enzyme in the presence of organic cosolvent.[10]

Cleavage was effected by a 1:100 ratio of enzyme to substrate during a 24 h incubation and appeared to be confined to three bonds, and to be incomplete, so that overlap fragments in addition to the basic four were seen on SDS gels. On non-denaturing size-exclusion chromatography, however, a single peak of molecular mass of the entire protein was seen, demonstrating strong and complete complexes, which proved to have full biological activity. The initial observation was that upon addition of acetonitrile to 60%, the same enzyme, unresolved from the complex, catalyzed a majority reconversion to apparently intact triose phosphate isomerase in, amazingly, just 10 min. Later experiments showed that in 90% glycerol religation was nearly quantitative.

If it can be shown that the complex(es) can be disassembled and reconstituted prior to religation this way, then it will be possible to replace a fragment with a structural homolog, and a strikingly effective and, hopefully, general new methodology is available for semisynthesis.

The other area of current exploration that has real potential to benefit semisynthesis is chemoselective ligation. Selectivity for unique reaction between the two targeted termini is achieved through the use of chemistry not normally encountered in the reactions of amino acids and proteins. A number of schemes have evolved.

An approach called *press-stud conjugation*[34,35] or *dovetailing*[36] of protein fragments was based on the premise that non-peptidic linkages within the polypeptide backbone could be acceptable, perhaps even desirable.[37] In the former, the unusual chemistry was hydrazone formation between hydrazide and aldehyde groups introduced at the termini to be linked using both chemical modification and reverse proteolysis. The *dovetailing* procedure employed the specificity of the reaction between thiol and bromoalkyl functions, much used in the chemical modification of proteins. In the initial trial, two totally synthetic fragments of the HIV protease were prepared and the uniquely reactive moieties built in to the synthetic protocol to give:

$$P^1 - I^{50} - NH.CH_2.CO.SH + Br.CH_2.CO-F^{53} - F^{99}$$
$$\downarrow$$
$$P^1 - I^{50} \ G^{51}.CO.S.G^{52} \ F^{53} - F^{99} + HBr$$

in which the product differs from native HIV protease in having a thioester instead of an amide bond between residues 51 and 52.[36] In both this and the *press-stud conjugation* method,[35] the mutual affinity of the reactive species is sufficient to ensure high coupling yields at fragment concentrations of less than 10 mM, even without complex formation.

The press-stud method has been used in the first application of chemoselective ligation methods in a semisynthesis protocol.[35] This innovative study also employed the tactic of prior sequence manipulation of the protein, granulocyte colony-stimulating factor (GCSF), by site-directed mutagenesis, that was developed for CNBr fragment religation. Here, unique Lys-Ser sequences were created in two separate mutants, to create lysyl endopeptidase-susceptible sites (62–63 and 75–76) either side of a loop structure occupying positions 64–74. Natural fragment 1–62, derived from one mutant and 76–174, from the other, were joined by a synthetic analog of the loop in the following scheme:

$$Ser^{76} \underline{\hspace{2cm}} Pro^{174}$$

$$\downarrow \; NaIO_4$$

$$Ser^{63} \underline{\hspace{1cm}} Leu^{75} \; NH\cdot NH_2 \;+\; HCO\cdot CO\cdot Asn^{77} - Pro^{174}$$

$$\downarrow$$

$$Ser^{63} \underline{\hspace{1cm}} Leu^{75} \cdot NH\cdot N = CH\cdot CO\cdot Asn^{77} - Pro^{174}$$

$$\downarrow \; NaBH_3CN$$

$$Ser^{63} \underline{\hspace{1cm}} Leu^{75} \cdot NH\cdot NH\cdot CH_2\cdot CO \; Asn^{77} - Pro^{174}$$

$$\downarrow \; NaIO_4$$

$$Thr^1 \underline{\hspace{1cm}} Lys^{62}\cdot NHNH_2 + HCO\cdot CO\cdot Cys^{64} - Leu^{75}\cdot NH\cdot NH\cdot CH_2\cdot CO\cdot Asn^{77} - Pro^{174}$$

$$\downarrow \; NaBH_3CN$$

$$Thr^1 \underline{\hspace{1cm}} Lys^{62}\cdot NH\cdot NH\cdot CH_2\cdot Cys^{64} - Leu^{75}\cdot NH\cdot NH\cdot CH_2\cdot CO\cdot Asn^{77} - Pro^{174}$$

Periodate oxidation of N-terminal Ser or Thr is specific, smooth, and quantitative, and the hydrazide group is either incorporated in the synthetic peptide or added by reverse proteolysis in a near-quantitative manner. Hydrazone formation between the two chemoselective groups was found to be generally 60–80% at equilibrium, and subsequent reduction was 95% complete, so that overall efficiency of the scheme is quite high. Although the starting mutant had substantially lower activity than the wild-type GCFS, the semisynthetic products were better, suggesting that the reduced hydrazone, having greater flexibility then the peptide bond it replaces, can compensate for the steric effects of the Leu to Lys alteration.[35]

The chemoselective tactic can be used as a preliminary to coupling via a peptide (or peptide-like) bond in a strategy called *entropy activation*. One route is via thiol capture:

$$Polypeptide - CO.ORSH + NH_2.CH[CH_2S.SR].Polypeptide$$

$$\downarrow$$

$$Polypeptide - CO.ORS\text{-}SCH_2.CH[NH_2].Polypeptide + R\;SH$$

The linkage of the two peptides via the disulfide brings the esterified C-terminus into proximity with the free amino terminus, providing the *entropy activation* for aminolysis of the ester. The HORSH group can then be easily removed from the cysteine residue by reduction. The potential of the method was first demonstrated through the reaction of a 13-residue N-terminal fragment with either a 12- or 26-residue C terminal portion prepared by solid phase methods.[38]

Entropy activation has also been pursued in the *segment ligation* strategy,[39] which works on the following principles:

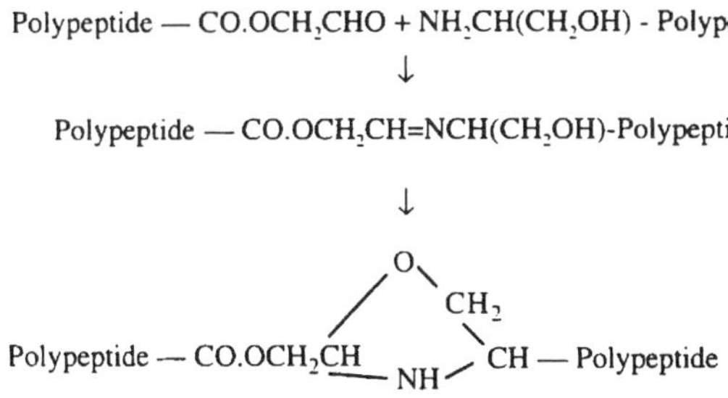

Polypeptide — CO.OCH$_2$CHO + NH$_2$CH(CH$_2$OH) - Polypeptide

↓

Polypeptide — CO.OCH$_2$CH=NCH(CH$_2$OH)-Polypeptide

↓

The chemoselective reaction first generates a schiff's base, but then condensation with the β-hydroxyl group of the serine (threonine or cysteine can also function in this role) leads to the cyclic structure shown.

As in the thiol capture protocol described above, this linkage brings into proximity the nucleophilic imino nitrogen of the oxazolidine ring and the carbonyl carbon of the linking ester, leading to O-to-N acyl transfer and the formation of an amide bond, and giving this structure.

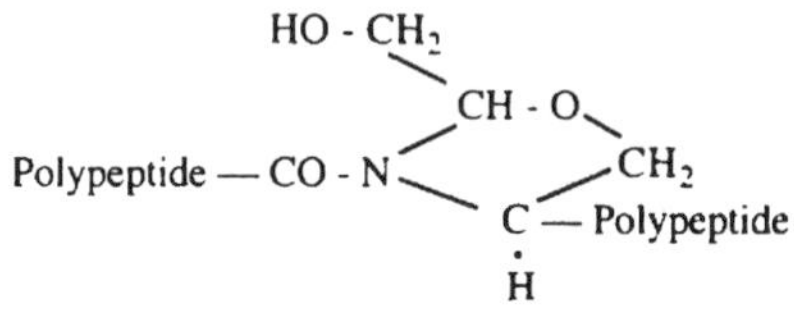

The *imino acid* residue (hydroxymethylthiazolidine) thus generated should be acceptable in many locations in protein structures, although the initial substantial product, a 50-residue epidermal growth factor-like peptide, was not tested for biological activity.

The simplest and most direct of these approaches to date was first described in 1994.[40] It requires a minimalist departure from unmodified peptide structures to achieve quantitative reaction in less than five minutes.

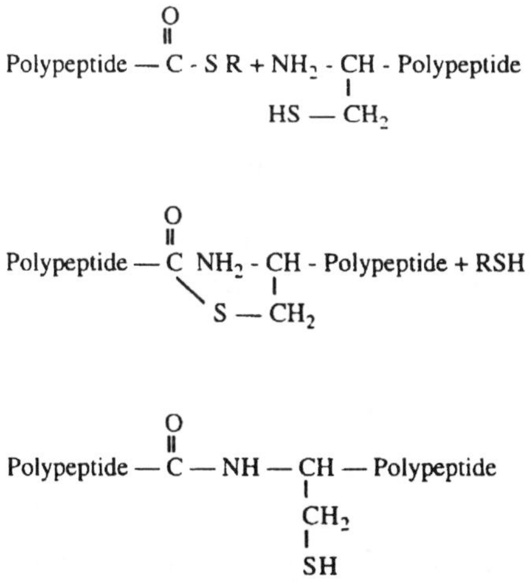

The requirements are simply that the N-terminal fragment be thiolesterified and that the N-terminus of the C-terminal fragment be provided by a cysteine residue. Since the protocol results in a completely authentic polypeptide structure there are no concerns about the effects of alternatives to peptide bonds or natural residues using this scheme.

The Kent group's introduction of native chemical ligation was followed by an almost identical scheme from Tam and co-workers,[41] along with a variant of it, in which the initiating thiol capture step occurs through reaction of a C-terminal thiocarboxylic acid with a bromoalkane side chain of the N-terminal residue of the other fragment. This results in the same intermediate as in the original scheme, with a thiolester linking the two polypeptides, and the subsequent S-to-N acyl shift yields a peptide bond. The use of β-bromoalanine, however, may make this version marginally less convenient than the original scheme. In this instance,[41] peptides of up to 54 residues were created, while, in the original report, the ultimate product was a fully active analog of a 72-residue cytokine [Ala33] IL-8.[40] Since then, Stephen Kent's group has used the technique to produce a range of fully active small proteins, including HIV-1 protease, barnase, the third domain of turkey ovomucoid, eglin C, and, most recently, human secretory phospholipase A_2.[42–44]

They have also extended the applicability of native chemical ligation to sites other than the original -X-Cys- type. With the new approach,[43] -Gly-X- and -X-Gly- sites are accessible, but if both α-carbon atoms are substituted, the crucial rearrangement step does not take place. The principle of the method differs in having the thiol capture group now part of a substituent on the α-NH$_2$ group, thus:

$$\text{Polypeptide} - \text{CO} . \text{SR}^1 + \text{HS.CH}_2.\text{CH}_2.\text{O.NH.CHR} - \text{Polypeptide}$$

$$
\begin{array}{c}
\text{Polypeptide} - \text{CO NH.CHR.Polypeptide} \\
\quad\quad\quad\quad\quad | \quad | \\
\quad\quad\quad\quad\quad \text{S} \quad \text{O} \\
\quad\quad\quad\quad\quad | \quad | \\
\quad\quad\quad\quad \text{H}_2\text{C- CH}_2
\end{array}
$$

$$
\begin{array}{c}
\text{Polypeptide} - \text{CO.N.CHR} - \text{Polypeptide} \\
\quad\quad\quad\quad\quad\quad | \\
\quad\quad\quad\quad\quad\quad \text{O} \\
\quad\quad\quad\quad\quad\quad | \\
\quad\quad\quad\quad\quad (\text{CH}_2)_2 \\
\quad\quad\quad\quad\quad\quad | \\
\quad\quad\quad\quad\quad \text{SH}
\end{array}
$$

The oxy substituent on the peptide bond can be readily removed by treatment with zinc in a mildly acidic medium. Thus far, no protein products have been made by this methodology, although 20-residue peptides have been prepared in 60–70% yield. The requirement that one of the terminal residues should be a glycine is rationalized by the observation that the -S-to-N- acyl shift above requires formation of a six-member ring, with consequent steric hindrance to approach of the reactive atoms due to the presence of the side-chains.

These developments have important implications for solid-phase peptide synthesis, in particular in rendering facile the creation of full-size proteins, a task that is still challenging for a totally stepwise approach. But will they have an impact on semisynthesis?

One of the advantages advanced for chemoselective methods is that unprotected fragments can be ligated. However, fragments for ligation prepared by solid-phase methods are likely to be highly protected already. Thus, this advantage is more relevant to the ligation of proteolytically derived peptides. In this context, the advantages of these methods over the ones we have specifically evolved for semi-synthesis, which are described above, are that they can efficiently condense non-complexing fragments. The incorporation, then, of such tactics into the semisynthetic repertoire could considerably enhance it. How can this be done? The majority of semisyntheses include a synthetic fragment, so the design of this could include appropriate N- or C-termini, or both, for chemoselective methods. Tailoring pro-teolytic cleavages to ensure suitable N-terminal residues is less simple but could be achieved by preliminary site-directed mutagenesis if nature is not obliging. More simple, perhaps, will be manipulation of the C-terminal residues derived from proteolysis. If that is enzymatic, it may not be problematic to create the necessary thioester by reverse proteolysis. After a CNBr cleavage, the C-terminal homoserine lactone could be treated with high concentrations of thiol in basic media to cause thiolysis of the ring to a thioester.

In fact, the latest developments in *native chemical ligation* address these issues.[45] In these procedures, the bulk of the peptide fragments are constructed on a solid-phase support in which the peptide tether is a thioester linkage. Both synthesis and HF deprotection proceed normally on this matrix leaving the fully unprotected peptide attached by the same type of linkage required for the mechanism of chemose-lective ligation discussed above. The peptide is thus simultaneously freed from the resin and ligated to its C-terminal partner under the same sorts of conditions, and the process can be repeated until an entire protein is constructed (Figure 3.7). Muir's group have used the SH3 domain from a protein tyrosine kinase (c-Abl) as a model, constructing this 56-residue segment from a 21-residue and a 35-residue fragment, and then further ligating this domain to a 22-residue proline-rich ligand and linker region at its N-terminus.[45] The products were very clean, and the avoidance of intermediate purification steps significantly reduces handling losses. However, the intermolecular ligations are quite slow, and it was found necessary to include aro-matic thiols to catalyse them, so that in these cases the ligation is no longer strictly solid-phase, but does remain one-pot, with all the attendant benefits.

Another innovation in the peptide synthesis field is that of a *designed peptide ligase* derived from the well known protease subtilisin. This was subjected to a double mutagenesis to convert Ser221 to Cys and Pro225 to Ala. The former change means that the acyl-enzyme intermediate (formed via the intermediacy of a peptide ester) is a thioester and, hence aminolysis is favored over hydrolysis. The other mutation reduces steric crowding in the active site. Although a preference is retained for bulky hydrophobic side chains at the C-terminus, only proline or negatively charged residues are disfavored in the N-terminal positions. With appropriate choice of six fragments prepared by solid-phase methods, the Wells group was able to achieve a complete synthesis of ribonuclease A with full biological activity.[11] Although the overall yield was modest, for each step it was quite high, so that

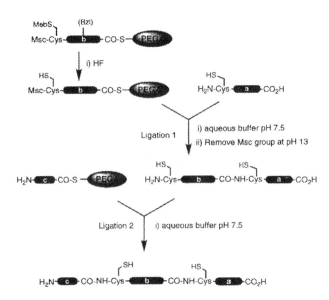

FIGURE 3.7 A scheme for sequential solid-phase chemical ligation based on the native chemical ligation strategy. Reprinted from Ref. 45 with permission, © 1998 Munksgaard International Publishers Ltd., Copenhagen, Denmark.

subtiligase has great potential as a catalyst in the ligation of pairs of non-complexing fragments in typical semisynthetic schemes. It seems likely that fragments derived from subtilisin and chymotypsin digests would meet the terminal residue criteria, and the necessary pretreatment (some α-NH$_2$ protection and esterification of the ligating C-terminus) should be facile.

Yet another very recently developed methodology relies on the intermediacy of thiolester derivatives of peptides. The basis of the technique is the phenomenon of protein splicing, most recently reviewed by Paulus.[46] Much of the work on deciphering this biochemical curiosity, in which a precursor polypeptide is post-translationally transformed into two protein products by the autocatalytic splicing of the N and C-terminal fragments (the exteins), with the simultaneous excision of the central portion (the intein), has been undertaken by the New England Biolabs group. The accepted mechanism is shown in panel A of Figure 3.8.

In 1995, I suggested[47] that the phenomenon might be exploited as a tool in semisynthesis, when reviewing semisynthesis and chemoselective methods. With an eye to commercial application of the phenomenon, NEB first developed a method for a very clean and specific affinity purification method, now known as the IMPACT system (New England Biolabs). In this, the target recombinant protein is expressed as a fusion protein with an intein and a chitin-binding domain at its C-terminus, in which the downstream splice site has been disabled by an Asn→Ala mutation.[48] This change renders impossible step 3 of the splicing process (Figure 3.8) but does not prevent the N→S acyl shift. The resulting intermediate, in which the target protein is linked to the carrier (itself bound to an affinity column chitin ligand) by a thiol ester bond, may be cleaved by thiolysis using small-molecule sulfhydryls

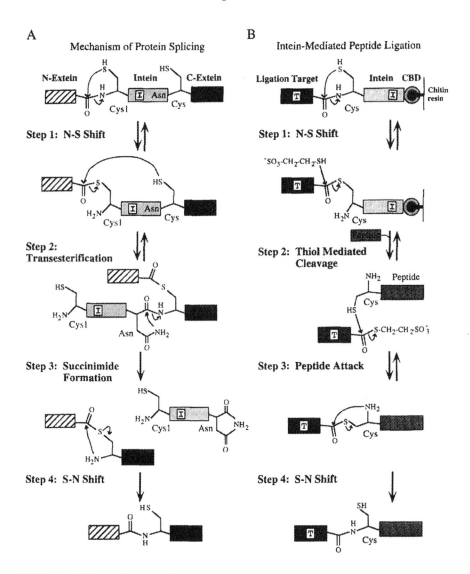

FIGURE 3.8 The mechanism of protein splicing (A), and its adaptation to a chemoselective peptide ligation strategy for semisynthesis (B). In the latter, the downstream splice site is disabled, so that only the initial N→S acyl shift step in splicing can occur. However, the presence of an excess of added low molecular weight thiol causes thiol exchange and the release of the target biosynthetic fragment from the affinity matrix. This is then the major component in a chemoselective ligation to a synthetic peptide. Figure taken from Ref. 50 with the permission of Cambridge University Press.

such as mercaptoethanol, whereupon the thiolester product may be eluted in pure form from the affinity column.

Such products can be, of course, intermediates in the native chemical ligation strategy of fragment condensation in total synthesis of proteins, so it was not long before two groups simultaneously proposed employing the methodology in two variant semisynthetic schemes.[49,50] The study of Kinsland et al. was aimed at using the NEB IMPACT system to generate substantial quantities of the thiocarboxylate form of ThiS, a 7.3 Kd protein intermediate in the biosynthesis of thiamin. They reasoned that by using ammonium sulfide as the cleavage agent in the IMPACT methodology, they would directly generate the thiocarboxylate rather than the normal thiolester. Although they wanted this product for studies of thiamin metabolism, they suggested that such thiocarboxylates could be used as intermediates in protein semisynthesis via chemoselective ligation with synthetic peptides with appropriate N-termini.[49]

The NEB group[50] have actually used their system to generate two cytotoxic proteins, choosing RNase A and the restriction endonuclease HpaI as models. The cytotoxicity prevents direct expression of such recombinant proteins, so semisynthesis is a useful tactic even in the preparation of protein of native sequence, and is, of course, essential to preparing certain derivatives. The bulk of the sequence in both cases is devoid of activity, so it can be overexpressed in vivo in the IMPACT construct and then, as the thiolester cleavage product, ligated to a relatively small and easily-made synthetic peptide (15 residues for RNase A, 31 residues for Hpa I) with N-terminal cysteine, immediately after elution from the chitin affinity column (Figure 3.8B).

In both cases, products with high activity were obtained in good yield, and there is no doubt that peptide analogs could be exploited in this system of semisynthesis to yield useful protein variants. The semisyntheses described in subsequent chapters have laid a solid technical foundation for this type of protein engineering. It is my belief that the innovations reported in this chapter will permit any interested protein engineer to adapt this practice to a specific target and achieve unusual engineering goals with simplicity.

REFERENCES

1. Gross, E. and Witkop, B., The cyanogen bromide reaction, *J. Am. Chem. Soc.*, 83, 1510, 1961.

2. Corradin, G. and Harbury, H.A., Reconstitution of horse heart cytochrome *c*: interaction of the components obtained upon cleavage of the peptide bond following methionine residue 65, *Proc. Natl. Acad. Sci.,* USA 68, 3036, 1971.

3. Corradin, G. and Harbury, H.A., Reconstitution of horse heat cytochrome *c*: reformation of the peptide bond linking residues 65 and 66, *Biochem. Biophys. Res. Commun.,* 61, 1400, 1974.

4. Dyckes, D.F., Creighton, T. and Sheppard, R.C., Spontaneous reformation of a broken peptide chain, *Nature* (London), 247, 202, 1974.

5. Niekamp, C.W., Hixson, H.F., Jr., and Laskowski, M., Jr., Peptide-bond hydrolysis equilibria in native proteins, *Biochemistry*, 8, 16, 1969.

6. Sealock, R.W. and Laskowski, M., Jr., Enzymatic replacement of the arginyl by a lysyl residue in the reactive site of soybean trypsin inhibitor, *Biochemistry* 8, 3703, 1969.

7. Mattis, J.A. and Laskowski, M., Jr., pH dependence of the equilibrium constant for the hydrolysis of the Arg63-Ile reactive-site peptide bond in soybean trypsin inhibitor, *Biochemistry,* 12, 2239, 1973.

8. Homandberg, G.A., Mattis, J.A. and Laskowski, M., Jr., Synthesis of peptide bonds by proteinases. Addition of organic cosolvents shifts peptide bond equilibria toward synthesis, *Biochemistry,* 176, 5220, 1978.

9. Homandberg, G.A. and Laskowski, M., Jr., Enzymatic resynthesis of the hydrolyzed peptide bond(s) in Ribonuclease S, *Biochemistry,* 18, 586, 1979.

10. Vogel, K. and Chmielewski, J., Rapid and efficient resynthesis of proteolyzed triose phosphate isomerase, *J. Am. Chem. Soc.,* 116, 11163, 1994.

11. Jackson, D.Y., Burnier, J., Quan, C., Stanley, M., Tom, J. and Wells, J.A., A designed peptide ligase for total synthesis of ribonuclease A with unnatural catalytic residues, *Science*, 266, 243, 1994.

12. Fontana, A., and Gross, E., Fragmentation of polypeptides by chemical means in Practical Protein Chemistry - A handbook (A. Darbre, Ed.) John Wiley and Sons Ltd., pp 67-120, 1986.

13. Wallace, C.J.A., Chemical studies on cytochrome *c*. Ph.D. thesis, University of Oxford, 1976.

14. Woods, A.C., Guillemette, J.G., Parrish, J.C., Smith, M. and Wallace, C.J.A., Synergy in Protein engineering, *J. Biol. Chem.,* 271, 32008, 1996.

15. Galpin, I.J. and Hoyland, D.A., Semisynthesis III - the homoserine 12,105 analogue of hen egg-white lysozyme, *Tetrahedron,* 41, 907, 1985.

16. Berghuis, A.M. and Brayer, G.D., Oxidation state-dependent conformational changes in cytochrome *c*, *J. Mol. Biol.,* 223, 959, 1992.

17. Dyckes, D.F., Kini, H. and Sheppard, R.C., Studies on the partial synthesis of protein analogues by direct coupling to terminal homoserine lactone derivatives, *Int. J. Peptide Protein Res.,* 9, 340, 1997.

18. Wilgus, H., Ranweiler, J.S., Wilson, G.S. and Stellwagen, E., Spectral and electro-chemical studies of cytochrome *c* peptide complexes, *J. Biol. Chem.,* 253, 3265, 1978.

19. Horn, M.J. and Laursen, R.A., Solid-phase Edman degradation: attachment of car-boxyl-terminal homoserine peptides to an insoluble resin, *FEBS Lett.,* 36, 285, 1973.

20. Wallace, C.J.A., The semisynthesis of some structural analogs of cytochrome *c*, *Proc. Amer. Pept. Symp.,* 6, 609, 1979.

21. Holmgren, A. and Reichard, P., Thioredoxin 2: cleavage with cyanogen bromide, Eur. *J. Biochem.*, 2, 187, 1967.

22. Sancho, J. and Fersht, A.R., Dissection of an enzyme by protein engineering, *J. Mol. Biol.,* 224, 741, 1992.

23. Sheppard, R.C., Selective chain cleavage and combination in protein partial synthesis, *Proc. Amer. Pept. Symp.,* 6, 577, 1979.

24. Hoogerhout, P. and Kerling, K.E.T., Synthesis of [Ile-13, Hse-20] - S-peptide lactone and semisynthesis of [Ile-13, Hse-20] - RNase A, *Recl. Trav. Chim. Pays-Bas,* 101, 246, 1982.

25. Proudfoot, A.E.I., Rose, K. and Wallace, C.J.A., Conformation-directed recombina-tion of enzyme-activated peptide fragments: a simple and efficient means to protein engineering, *J. Biol. Chem.,* 264, 8764, 1989.

26. Wallace, C.J.A., Corradin, G., Marchioni, F. and Borin, G., Cytochrome *c* chimerae from natural and synthetic fragments: significance of the biological properties, *Biopolymers,* 25, 2121, 1986.

27. Kullman, W., Enzymatic peptide synthesis, CRC Press, Boca Raton, 1987.

28. Bongers, J. and Heimer, E.P., Recent applications of enzymatic peptide synthesis, *Peptides* 15, 183-193, 1994.

29. Proudfoot, A.E.I., Wallace, C.J.A., Harris, D.E. and Offord, R.E., *Biochem. J.*, 239, 333, 1986.

30. Rose, K., Herrero, C., Proudfoot, A.E.I., Offord, R.E. and Wallace, C.J.A., Enzyme-assisted semisynthesis of polypeptide active esters and their use, *Biochem. J.,* 249, 83, 1988.

31. Wallace, C.J.A., Developing a general method for protease-promoted fragment condensation semisynthesis, *Proc. Eur. Pept. Symp.*, 21, 260, 1991.

32. Wallace, C.J.A., Guillemette, J.G., Hibiya, Y. and Smith, M., Enhancing protein engineering capabilities by combining mutagenesis and semisynthesis, *J. Biol. Chem.* 266, 21355, 1991.

33. Woods, A.C., Guillemette, J.G., Parrish, J.C., Smith, M. and Wallace, C.J.A., Synergy in protein engineering, *J. Biol. Chem.* 271, 32008, 1996.

34. Offord, R.E., Pochon, S. and Rose, K., Press-stud protein conjugates, *In Peptides, 1986* (Theodoropoulos, D., Ed.) de Gruyter, Berlin, 279-281.

35. Gaertner, H.F., Offord, R.E., Cotton, R., Timms, D., Camble, R. and Rose, K., Chemo-enzymic backbone engineering of proteins, *J. Biol. Chem.*, 269, 7224, 1994.

36. Schnolzer, M. and Kent, S.B.H., Constructing proteins by dovetailing unprotected synthetic peptides: backbone-engineered HIV protease, *Science,* 256, 221, 1992.

37. Offord, R.E., Chemical approaches to protein engineering. In *Protein design and the development of new therapeutics and vaccines* (Hook, J.B. and Poste, G, Eds.), Plenum, New York, 253-282.

38. Kemp, D.S. and Carey, R.I, Synthesis of a 39-peptide and a 25-peptide by thiol capture ligations, *J. Org. Chem.,* 58, 2216, 1993.

39. Liu, C-F. and Tam, J.P., Peptide segment ligation strategy without use of protecting groups, *Proc. Natl. Acad. Sci.,* USA 91, 6584, 1994.

40. Dawson, P.E., Muir, T.W., Clark-Lewis, I. and Kent, S.B.H., Synthesis of proteins by native chemical ligation, *Science*, 266, 776, 1994.

41. Tam, J.P., Lu, Y-A., Liu, C-F and Shao, J., Peptide synthesis using unprotected peptides through orthogonal coupling methods, *Proc. Natl. Acad. Sci. USA,* 92, 12485, 1995.

42. Lu, W., Qasim, M.A. and Kent, S.B.H., Comparative total syntheses of turkey ovo-mucoid third domain by both stepwise solid phase peptide synthesis and native chemical ligation, *J. Am. Chem. Soc.*, 118, 8518, 1996.

43. Canne, L.E., Bark, S.J. and Kent, S.B.H., Extending the applicability of native chemical ligation, *J. Am. Chem. Soc.,* 118, 5891, 1996.

44. Hackeng, T.M., Mounier, C.M., Bon, C., Dawson, P.E., Griffin, J.H. and Kent, S.B.H., Total chemical synthesis of enzymatically active human type II secretory phospholipase A_2, *Proc. Natl. Acad. Sci. USA,* 94, 7845, 1997.

45. Camarero, J.A., Cotton, G.J., Adeva, A. and Muir, T.W., Chemical ligation of unprotected peptides directly from a solid support, *J. Peptide Res.,* 51, 303, 1998.

46. Paulus, H., The chemical basis of protein splicing, *Chem. Soc. Rev.,* 27, 375, 1998.

47. Wallace, C.J.A., Peptide ligation and semisynthesis, *Curr. Opin. Biotechnol.,* 6, 403, 1995.

48. Chong, S., Mersha, F.B., Comb, D.G., Scott, M.E., Landry, D., Vence, L.M., Perler, F.B., Benner, J., Kucera, R.B., Hirvonen, C.A., Pelletier, J.J., Paulus, H. and Xu, M-Q., Single-column purification of free recombinant proteins using a self-cleavable affinity tag derived from a protein splicing element, *Gene,* 192, 271, 1997.

49. Kinsland, C., Taylor, S.V., Kelleher, N.L., McLafferty, F.W. and Begley, T.P., Over-expression of recombinant proteins with a C-terminal thiocarboxylate: implications for protein semisynthesis and thiamin biosynthesis, *Protein Sci.,* 7, 1839, 1998.
50. Evans, T.C., Benner, J. and Xu, M-Q., Semisynthesis of cytotoxic proteins using a modified protein splicing element, *Protein Sci.,* 7, 2256, 1998.
51. Laskowski, M., Jr., The use of proteolytic enzymes for the synthesis of specific peptide bonds in globular proteins. *In Semisynthetic peptides and proteins* (Offord, R.E. and DiBello, C., Eds.) Academic Press, London, 255-262, 1978.

4 Insulin Semisynthesis

J. Gwynfor Davies and Carmichael J.A. Wallace

CONTENTS

4.1 INTRODUCTION

Insulin is a special case. In almost every major advance in protein chemistry, this small polypeptide, only on the borders of being classed as a protein due to its size (the human version has a relative molecular mass of around 5800), has had an important part to play. Despite its modest size, insulin's structure is relatively complex as it has two peptide chains linked by two disulphide bonds as well as an intra-chain disulphide in the A chain (see below). It is also a special case in that i it has been used to treat diabetics for over 70 years, ever since the courageous and pioneering work of Banting, Best, Macleod, and Collip in 1921.[1] One of the major achievements of semisynthesis was the production of insulin with the human sequence from porcine insulin, on an industrial scale, by Novo-Nordisk. This semisynthetic insulin has played an important role in the modern management of insulin-dependent diabetes.

In normal humans and animals, insulin is produced in the beta cells of the islets of Langerhans of the endocrine pancreas. Stored in electron-dense secretory granules, it is released in response to fluctuating levels of blood glucose and is itself an essential element in blood glucose homeostasis. The primary gene product is a pre-prohormone with an amino-terminal extension of 24 amino-acid residues on the proinsulin sequence, which is cleaved in the Golgi system to give the single chain proinsulin. Proinsulin has dibasic processing sites that are cleaved by specialized enzymes giving insulin and C-peptide in the storage granules. After release into the blood, insulin travels to its target organs, principally the liver, muscle, and adipose tissue, where its major function is to bind to specific cell-surface receptors. The ligand is internalized in association with the receptor, and ultimately the two molecules dissociate either to release the intact insulin back into the fluid surrounding the tissue or to lead to proteolytic degradation of the hormone.

A major challenge in the study of insulin since its discovery has been to describe how insulin brings about the metabolic changes that are observed in its target cells. Receptor binding produces intracellular signals that interact with a diverse range of enzymatic activities, producing both stimulations and inhibitions. As second messengers in other systems were identified, all were investigated for insulin, but indications that a tyrosine kinase system was involved also showed puzzling inconsistencies.[2] These were resolved by both structural studies and the discovery that the insulin receptor operates by a near-unique second messenger signalling route. Autophosphorylation is following by kinase action on insulin receptor substrate, which in turn interacts with a number of the more familiar intracellular signalling systems.[3,4]

Insulin's effects on metabolism are most dramatically illustrated by what happens in its absence. Almost everybody has an acquaintance who has diabetes in one form or another, and most of us know someone who depends on insulin injections to stay alive. As a brief summary, in the absence of insulin, the transport of glucose into peripheral tissues is restricted, and blood glucose rises. At the same time, lipogenesis is restricted, and breakdown of triglyceride to fatty acids and glycerol occurs, leading to increased plasma fat levels. The liver catabolism of fatty acids forms ketone bodies to an increasing extent and, as the capacity for their oxidation in peripheral tissue is exceeded, ketosis develops. Protein breakdown is increased, protein synthesis is diminished, and liver gluconeogenesis increases at the expense of protein. The results is wasting of the diabetic's body and, ultimately, death.

Maintenance of metabolic control requires insulin. Its primary site of action is the insulin receptor, and thus structure-function studies (the relationship between structure and mechanism of action) of insulin are chiefly directed toward this interaction. A second, important "function" of insulin is its clearance from the circulation and degradation, as this, along with release from the pancreas, controls levels of the hormonal signal. We shall describe below the role of semisynthesis in the investigation of these functions, and how ideas of insulin's interactions have ultimately led to design and production of insulins, partly by semisynthesis, with novel sequences and special properties that may well be the therapeutic forms of insulin of the future.

4.2 INSULIN STRUCTURE

Insulin was the first protein to have its primary structure determined by the pioneering work of Sanger and colleagues in Cambridge.[5] It has a two-chain structure with an A chain of 21 amino-acid residues joined to a B chain of 30 amino-acid residues by two interchain disulphide bridges (Figure 4.1). In addition, there is an intrachain disulphide between Cys^{A6} and Cys^{A11}. For semisynthesis, this structure poses the problem of distinguishing between the two peptide chains when either cleavage or synthesis of peptides bonds is desired. In terms of side-chain protection, for the porcine or human sequences, the only amino group is the lysine at position B29.

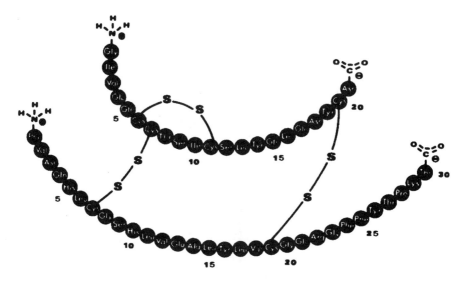

FIGURE 4.1 The amino acid sequence of human insulin. The uppermost of the two disulphide-bridge polypeptides is the A-chain. Taken from Ref. 117 with permission.

There are four glutamic acid residues, no aspartic acid, and two histidine imidazoles. However, tryptophan and methionine residues, which can present problems in certain deprotection conditions, are absent.

Comparison of sequences from different species (there are currently 56 listed in the Swiss-Prot data base) shows that the half-cystines are conserved, as would be expected, to maintain the basic structure of the molecules. The residues around the half-cystines are also frequently highly conserved, suggesting the need to keep the half-cystines in the correct environment to form the normal disulphide-bonding pairs. The amino-terminus of the A chain is conserved, whereas that of the B chain shows much more variation. The central region of the B chain, residues 11 to 16, shows little interspecies difference, as does the sequence at B23 to B25.

Insulin was also the first protein for which the three-dimensional structure in the crystal was determined by Hodgkin's X-ray crystallography group in Oxford (Figure 4.2).[6] Insulin has a hexameric structure in the crystal, consisting of three dimers. The monomer has a hydrophobic core consisting of cystine A7–B7, leucines A16, B11, and B15, and the isoleucine A2.[7] The other conserved residues (with the exception of B6, B8, and B18, which are next to interchain disulphides) are clustered on or close to a surface of the insulin molecule involved in dimer-dimer interaction in the crystal structure. Pullen et al.[8] proposed that this represents the receptor binding surface of the molecule. This proposal has since been amply confirmed, in many cases by the semisynthetic studies cited below. Implicated in this region are the hydrophobic residues phenylalanine B24, phenylalanine B25, tyrosine B26, valine B12, and tyrosine B16. In addition, residues not involved in dimerization, glycine A1, tyrosine A19, and asparagine 21, form part of the receptor-binding region. The effect of chemical modification of the amino-terminus of the A chain and circular dichroism studies supported these conclusions.

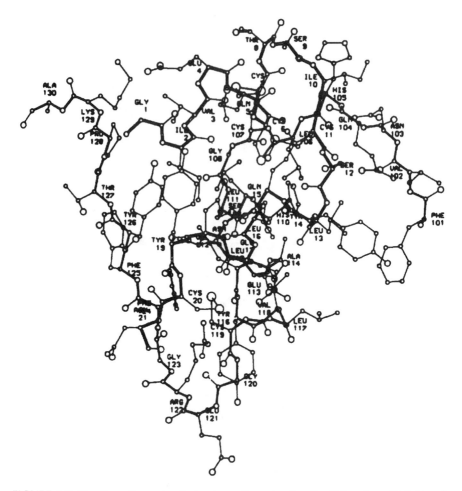

FIGURE 4.2 The three-dimensional structure of porcine insulin. Taken from Ref. 7 with permission.

In physiological conditions, the concentrations of insulin are very low, and the monomeric form of the molecule predominates.[7] Recent studies of the molecule in solution as the monomer by NMR have generally confirmed the structure obtained by X-ray crystallography but have shown some conformational differences. Weiss et al.[9] find more flexibility in the amino-terminal and carboxy-terminal regions of the B chain, and Roy et al.[10] find a change in the conformation of tyrosine B26 on dimerization. Overall, however, the receptor binding region proposed from crystallographic studies appears to be also present in the monomer in solution.

The increasing use of reversed-phase H.P.L.C. has been a major step forward in many areas of protein research and particularly for insulin where high recoveries and retention of biological activity are possible. This technique can produce separations where only minor changes in insulin structure have occurred. Examples are the separation of human and porcine insulin, which differ only in one of the 51 amino-acid residues (threonine in the human sequence vs. alanine in the porcine sequence[1]) or of intact insulin and the des-Gly[A1] derivative[12].

The residues at the amino terminus of the A chain and the carboxy terminus of the B chain can be modified by semisynthesis, and their role in receptor binding should be studied in more detail. The lack of direct involvement of the amino terminus of the B chain in the proposed receptor-binding region, and the apparent tolerance of this region to changes in side-chain structure, suggest that this position, easily accessible by semisynthetic techniques, would be an appropriate site to introduce labeled groups or other bulky substituents such as photoactivatable groups.

4.3 SEMISYNTHESIS AND TOTAL SYNTHESIS OF INSULIN

Any discussion of insulin semisynthesis has the inherent problem of where semisynthesis stops and total synthesis starts. The individual chains of insulin are amenable to total synthesis and, as techniques and equipment have improved over the years, one needs to ask again whether the semisynthetic approach has real advantages. Even yields of resynthesis of the inter- and intra-chain disulphide bonds have improved and examples of such syntheses abound in the literature. However, it is clear that, where a limited number of operations are concerned, semisynthesis by chemical means has advantages in terms of minimizing side products and ease of manipulation. In addition, where enzymatic techniques are involved, the specificity of the semisynthetic reactions and the avoidance of chemical protection and deprotection that can often be achieved are major benefits. The conversion of porcine to human insulin is a good example of the latter. Thus, in some cases, simple limited-step processes are possible with the semisynthetic approach, and total synthesis cannot rival the technique. That said, there are regions of the molecule, the interior of either the A or B chain in between the interchain disulphide bonds, that are not easily accessible to semisynthetic techniques, and total synthesis has clear advantages. Both techniques have been extensively exploited to produce insulin analogs.

In the following discussion, we shall limit consideration to semisyntheses starting from the intact hormone and without cleavage of the interchain disulphide bonds. Although total synthesis of one chain and cross linking the synthetic product to the natural second chain through disulphide bonding can be considered as semisynthesis, it seems to us to fall more under the heading of synthetic rather than semisynthetic. It is clear that, for insulin, the lines between total and semisynthesis cannot easily be drawn, and both processes have played a major role in the current understanding of insulin structure function relationships. The choice is an individual one, and we apologize if any readers feel that their work has been incorrectly omitted.

For previous reviews on insulin semisynthesis, see Refs. 13 through 16.

4.4 AMINO-TERMINAL SEMISYNTHESIS

Protein semisynthesis developed much more rapidly for modifications at the amino terminus, largely owing to the availability of the Edman reaction for the sequential removal of amino acids from this end of the molecule. Chemical modification studies had also laid the basis for the reversible protection necessary for semisynthesis. Among the problems for insulin is its two-chain structure and, therefore, two amino termini that need to be distinguished chemically for site-specific modification. The single side-chain free amino group on lysine-B29 has also to be taken into consideration. Thus, the challenge for this kind of semisynthesis with insulin is first to find reagents and conditions capable of distinguishing between the amino termini and then between the alpha and epsilon amino groups.

Early work led to two schemes for these distinctions. Brandenburg (1969)[17] and Borras and Offord (1970)[18] showed that phenyl isothiocyanate had a greater reactivity toward Phe^{B1} than Gly^{A1} or Lys^{B29}. The Edman reaction could thus be directly carried out on the B chain. This work has been confirmed and extended more recently, giving conditions for relatively specific reaction at A1 or B1, depending mainly on the presence or absence of zinc ions.[19] However, the very high selectivity of the reagents used to introduce the Boc group for A1 and B29 over B1 demonstrated by Geiger et al. in 1971[20] has been extensively used for selective semisynthesis. The di-Boc-insulin has to be separated from small amounts of the mono and tri derivatives by ion-exchange chromatography. Phenylalanine-B1 has then the only free amino group where the Edman reaction can be carried out (Figure 4.3). The only problem with this scheme is that the Boc groups are labile under the conditions used for the cyclization step of the Edman reaction (anhydrous trifluoroacetic acid), and thus reprotection of the des-Phe^{B1} insulin and repurification of the A1, B29-di-Boc derivative is necessary before amino acids or other groups can be specifically coupled to the B2 amino group. This led to a search for alternative protecting groups. Many have been used, including the benzyloxycarbonyl and ethyloxycarbonyl groups [21–22], the methylsulphonylethyloxy-carbonyl (Msc) group[23] (Figure 4.3), the phthaloyl group[24], the citraconyl group[25], and the Fmoc group.[26] The relative reactivities of the insulin amino groups seems to be generally maintained with these different groups. However, Offord[14] has observed that, while Boc azide has the required specificity for A1 and B29 groups of insulin and des Phe^{B1}-insulin, Boc dicarbonate appears to lose some of the specificity for the latter. Acid-stable groups, of course, allow repeated Edman cleavage of the B-chain. Glutamine-B4 has a tendency to cyclize when it becomes amino terminal but can be kept in the open form, and the chain has been cleaved as far as des (B1–B5).[21,27]

Specific cleavage at the A1 position requires the use of protecting groups with different labilities. Thus, the relative specificity of the Boc group can be used to produce the A1, B29 Di-Boc derivative, and B1 protected with, for example, the Msc group. The Boc groups can then be removed. The relative specificity of the Edman reagent for alpha over epsilon amino groups can then be exploited and the amino terminus of the A chain converted to the Ptc form. Protection is then completed with the Msc group on Lys^{B29}. Trifluoroacetic acid treatment gives the des-A1 derivative and specific coupling at this site or further cleavage carried out (Figure 4.4).[28] Alternatively, Saunders has prepared the mono-Boc A1-derivative directly by reacting insulin with Boc-azide at pH 9.3, (slow reaction of B29) in aqueous methanol (slow reaction of B1) and purifying by ion-exchange.[29] The B1 and B29 groups can then be modified with, for example, the Z-Met group and the Boc removed, leaving A1 ready for addition or cleavage by the Edman reaction (Figure 4.5).

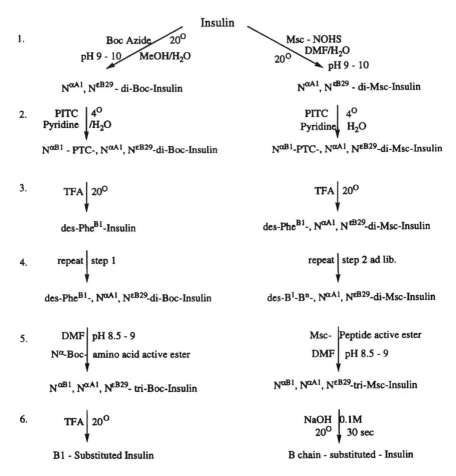

FIGURE 4.3 Two schemes for stepwise amino-terminal semisynthesis of the B-chain of insulin. To the left is the original route pioneered by Geiger et al.[20] using N-t-Boc protection, which suffered from the disadvantage of requiring fresh NH_2-protection at each degradative or synthetic step. To the right is an improved scheme employing Msc protection of amino groups[23] that permitted more facile access to internal residues.

These types of schemes have been extensively exploited for insulin to produce derivatives for almost all conceivable purposes, structure-function, specific labeling, attachment of groups with special reactivities, and so on. B1 was identified early on as being relatively unimportant for biological activity, and the site was tolerant of alterations and additions.[17,18,30] DesB1-insulin has full activity in *in vivo* blood sugar lowering in rabbits and 89% in the isolated fat cell.[30] Lys[B1]-insulin activity falls to 41% of the native hormone in the isolated fat cell but, in general, other substitutions retain high activity.[31] Addition of hydrophilic or hydrophobic amino acids at B0 decreases the activity of the resulting derivative to 40 to 80% of unmodified hormone in the isolated fat cell, which are relatively modest decreases in activity.[32] In contrast, the A1 glycine is important for activity and the site is generally more sensitive to

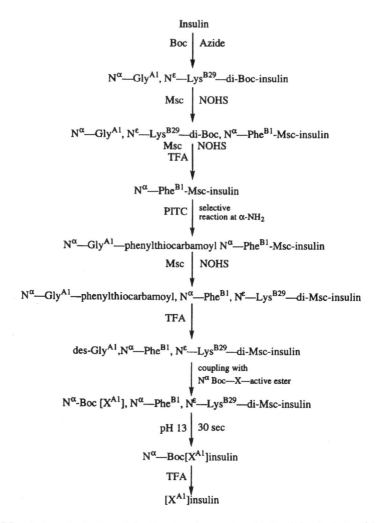

FIGURE 4.4 A route developed for the stepwise semisynthesis of the A-chain of insulin[28] that employs two different amino protection groups to maintain discrimination between the two chains.

modifications and additions.[33,34,28] Des-Gly[A1] insulin, for example, retains only around 1% of the activity and receptor binding affinity of the native hormone, confirming its role in the receptor binding region.[35-36] However, while substitution with L-amino acids leads to loss of activity, D-amino acids are tolerated, and the analogs retain relatively high activity.[27-28,37] Thus, for example, [L-Leu[A1]] insulin has around 12% activity in the isolated rat fat cell, whereas [D-Leu[A1]] insulin, and even the D-Trp[A1] derivative, has essentially full activity (97%). These results are consistent with the crystal structure of insulin (Fig. 4.2) where, in the L confor- mation, the amino-acid side chain would interfere with the receptor binding surface, whereas the D side chain would extend away from this surface. Addition of the basic

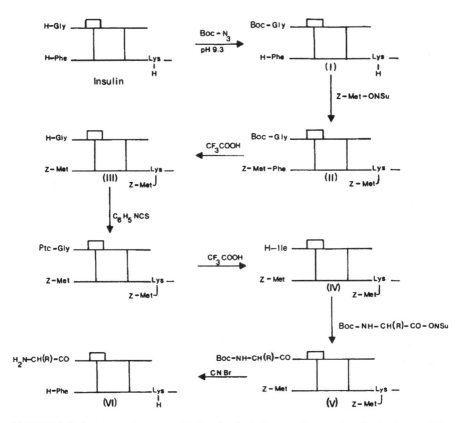

FIGURE 4.5 An alternative route for insulin A-chain stepwise semisynthesis that exploits the fact that the primary sequence of insulin contains no methionine residues. Taken from Ref. 22 with permission.

residues lysine or arginine at A0, mimicking partially processed proinsulins, reduces the activity to around 40% that of insulin in isolated rat fat cells.[38]

Using semisynthetic techniques, each chain has been shortened specifically by repeated Edman reactions and the effect on activity measured. Thus, the truncated B chain retains relatively high activity, 65% of normal *in vivo* blood sugar lowering in rabbits, even after the removal of four residues. When histidine B5 is removed the activity falls to 15%.[21,27,30] Removal of glycine A1 causes a major decrease in activity and repeated degradation was more in the interest of replacing amino-acid residues such as glutamic acid A4.[27] [Ala^{A4}] insulin retains 60% activity, indicating that the negative charge normally present at this position is not essential for activity.

Halban and Offord [39] used semisynthesis to label insulin specifically with tritium at position B1 by replacing the phenylalanine with its tritiated analog. A better technique, taking advantage of improvements in H.P.L.C. technology and avoiding losses in the work up of the coupling reaction, was reported more recently.[40] This labeled insulin has been extensively used in studies of insulin pharmacokinetics. It has been complemented by an equivalent insulin labeled at position A1 by replacement of the glycine with its tritiated analog using one of the semisynthetic schemes outlined above.[41] The use of these tracers will be discussed later. Labeling with stable isotopes gives derivatives that can be followed by mass spectrometry and such derivatives, [[^{2}H]Gly^{A1}] insulin and [[^{18}O]Lys^{B29}] insulin, have been used in studies of insulin degradation.[42]

Assoian and Tager[43] used Geiger's di-Boc insulin technique to prepare radio-iodinated insulin labeled specifically at position B1. This was employed along with iodo-insulin prepared by chemical modification to follow the degradation of insulin in isolated rat hepatocytes.[44]

Semisynthesis has played a major role in the study of the insulin receptor. Photoactivatable derivatives of insulin have been prepared by several groups and used to form permanent cross-links between labeled insulin and the receptor.[45,46,47] The now labeled receptor can be isolated for further characterization or for studies of the processing of insulin-receptor complexes in isolated cells using a variety of biochemical or histological techniques.[48–50] Knutsen[51] has prepared an insulin analog in which the bond linking the photoactivatable, and in this case radio-iodinated, cross linker to the insulin can subsequently be cleaved with glutathione. The insulin is thus released from the receptor leaving a labeled unoccupied receptor behind.

Koboyashi et al.[26] have combined techniques of amino-terminal and carboxy-terminal semisynthesis to produce Leu^{A3}-insulin, a naturally occurring mutant form of human insulin associated with a form of diabetes (see below), and the discovery has since stimulated the replacement of both A3 and A2 with a variety of amino acids, including several non-coded residues.[52] In this case, after initial formation of the A1 Ptc-insulin, the B1 and B29 amino groups were irreversibly blocked by acetylation, prior to the sequential degradation and resynthesis used to replace the A2 and A3 residues. Bis-acetylation only reduces receptor binding potentially by 24%, so the presence of these groups is considered compatible with meaningful structure-function results.

Geiger and co-workers[53] have extended their studies on the importance of the amino-terminal region of the B chain by preparing an IGF-1 (insulin-like growth factor-1)-insulin hybrid. To achieve this, they replaced the amino-terminal insulin B chain pentapeptide, Phe-Val-Asn-Gln-His with a tetrapeptide taken from the IGF sequence, Gly-Pro-Glu-Thr. The activity of this hybrid *in vivo* in the rabbit blood glucose test (12%) was even lower than that of the des-(B1-5) insulin. It was known from a totally synthetic analog that [Ala^{B5}] insulin was fully active in the rat, and that des-(B1-4) insulin retained 65% activity. They went on to prepare the [Ala^{B5}] insulin derivative by semisynthesis as well as replacing the amino terminal tetrapeptide with the tripeptide Gly-Pro-Glu from the IGF-1 sequence. Interestingly the [Ala^{B5}] insulin had to be prepared in two coupling steps, adding first Asn-Gln-Ala as the Boc protected HOOBt ester and, after deprotection of the Boc group, Boc-Phe-Val-OOBt was added. There was no reaction between the protected ester of the pentapeptide and the des-(B1-5)insulin. Despite the chemical coupling no racemization of Ala^{B5} could be detected. The biological assays showed that, while the [Ala^{B5}] insulin was fully active in the rat *in vivo* (blood sugar), it retained only about 50% in the mouse and 25% in the rabbit. The derivative with the IGF-tripeptide has no greater activity than the des-(B1-4)insulin. The activity of the earlier IGF-1-

insulin hybrid was thus due to loss of the tetrapeptide (B1-4) and the sequence change at position 5. The studies also underline that biological testing *in vivo* can be species dependent and care must be taken in comparing data. A similar species effect in vitro, however, was not observed.

4.5   CARBOXY-TERMINAL SEMISYNTHESIS

Modification of insulin at the carboxy terminus of the B chain carries with it one of the major goals of insulin semisynthesis: conversion of porcine insulin into human insulin. The porcine and human sequences differ by only one residue at position B30, alanine in the porcine sequence and threonine in the human. Nature, with apparent foresight, as if inviting the modification, has placed a lysine as the penultimate residue, one of the sites of specific cleavage by the endopeptidase trypsin. Not wishing to make things too easy, however, nature has also arranged a second trypsin specificity site at B22 arginine.

This goal should logically have stimulated C-terminal semisynthesis to be significantly ahead of amino-terminal modifications. However, the techniques were lacking and, up to the late 1970s, amino-terminal work was much more advanced. Despite this, a theoretical scheme for the conversion of porcine to human insulin was proposed much earlier. In this scheme, the amino groups are reversibly protected and the molecule cleaved with trypsin between residues Arg^{B22} and Gly^{B23}. The resulting amino-blocked, des-octapeptide insulin can be esterified to protect the free side chain and main chain carboxyl groups. Trypsin has esterase activity as well as peptidase activity and thus, under appropriate conditions of pH, the alpha-carboxyl group of Arg^{B22} can be regenerated from the ester. This single carboxyl group can now be activated by classical chemical techniques and coupled to a synthetic octapeptide of the human sequence or with modifications to the natural sequence. There are, however, serious problems with the deprotection conditions necessary to remove the ester groups that potentially would give unwanted side products.

The first report of the use of this scheme was by Ruttenberg in 1972.[54] Several groups attempted to repeat this work without success, and considerable modification to the described conditions was necessary before workable yields of desired product (as judged by the analytical techniques available at the time) were obtained. Weitzel and colleagues[55] were able to produce a considerable number of derivatives with modified octapeptide sequences. The degree of chemical and biological characterization of the products was, however, limited. The Chinese group[56] also used this basic scheme to investigate, for example, the role of arginine B22 in insulin activity and to conclude that this residue is not essential.

Obermeier and Geiger[57] tried a coupling scheme that avoided the carboxyl protection. The four side-chain carboxyl groups present in insulin could therefore potentially all be activated and undergo coupling with the synthetic octapeptide. Possibly assisted by non-covalent interactions between the octapeptide and the des-octapeptide insulin, they achieved reasonable yields of the desired product.

The demonstration that proteolytic enzymes can be manipulated by appropriate conditions to synthesize rather than cleave peptide bonds opened the way to serious study of modifications of the octapeptide region of insulin. The theoretical background to the technique is dealt with elsewhere in this book (see Chapter 3). In summary, the use of an organic co-solvent leads to reduction of the effective concentration of water and a shift in the pK_a of the carboxyl group, and one of the components (usually the amino) can be added at high concentration. This combination of factors leads to a shift in the equilibrium position of the hydrolysis reaction of a peptide bond and permits significant formation of the bond. The enzyme, acting as a catalyst, does not influence the position of the equilibrium, only the rate at which it is obtained. Enzymes generally retain their cleavage specificity under these conditions.

The significance of the sites of cleavage of insulin by trypsin now becomes apparent. As described above, trypsin acts only on the B-chain of insulin cutting it between Arg^{B22} and Gly^{B23} and between Lys^{B29} and Thr^{B30} (Ala in the porcine sequence). Hence, under reversed-proteolytic conditions, peptide bond synthesis at these sites becomes possible. The first successful application of this idea to insulin was by Inouye et al. in 1979.[58] The resynthesis was at Arg^{B22}, desoctapeptide insulin being prepared by the action of trypsin on porcine insulin under conditions favoring digestion. Possibly influenced by chemical semisynthetic techniques, they used a scheme where the amino groups of des-octapeptide insulin were protected with the Boc-group. This eliminates the risk of coupling between two molecules of the des-octapeptide insulin if the amino groups are left unprotected. The octapeptide or analog can then be added as the only amino-component present and, in the presence of trypsin as catalyst, gives a resynthesized human insulin or analog. Deprotection of the amino groups gives the desired product (Figure 4.6). An excess (10:1) of the amino component ([Boc-Lys^{B29}]octapeptide) over di-Boc-desoctapeptide insulin was used to drive the reaction toward synthesis of the peptide bond. In model studies, a pH of 6.5 was found to be optimum, and decreasing the water concentration by the addition of dimethylsulphoxide or dimethylformamide, although the increase in yield was greater than would be expected from the change in concentration. For insulin coupling, a Tris buffer 1:1 (v/v) with DMF was used. Trypsin:insulin derivative was 1:10 (w/w). After 20 h at 37° C, the mixture was applied to a Sephadex LH20 column and the isolated tri-boc insulin deprotected using TFA/anisole, and finally applied to Sephadex G50. An overall yield of 49% with respect to insulin was obtained and the product checked by reversed-phase H.P.L.C. and gel electrophoresis. This scheme represents a major improvement over the chemical techniques in terms of simplicity of the operation, yield of product, and freedom from the risk of racemization. It has been extensively used to synthesise derivatives of insulin in the octapeptide region.

Shortly thereafter, it was shown that trypsin can catalyze a resynthesis between des-Ala^{B30}-insulin (DAI), obtained by the action of carboxypeptidase A on porcine insulin, and threonine t-butyl ester, in good yield and without need to protect the amino groups of insulin.[59] Cleavage at Arg^{B22} was initially prevented by modification of the arginine side chain, but this was found to be unnecessary as cleavage did not occur in the conditions of resynthesis.[60] The reaction was carried out in a mixture of Tris buffer pH 6.5 DMF and ethanol with a 50:1 molar excess of Thr-OBut over DAI, and a ratio of 1:100 trypsin to DAI. 20h at 37°C gave 73% conversion to human insulin-OBut. The desired product was purified on Sephadex G-50 and DEAE-Sephadex A-25 and deprotected in TFA/anisole. The yield was 41%.

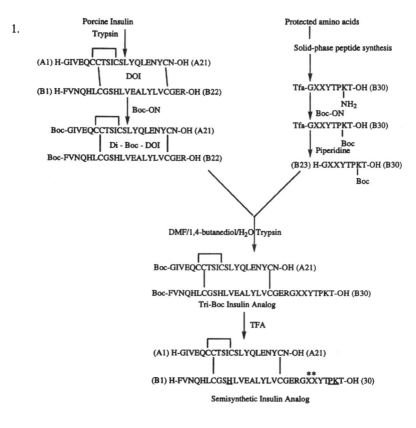

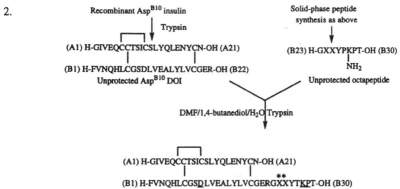

FIGURE 4.6 Strategy for the preparation of semisynthetic analogs of insulin modified at the C-terminus of the B-chain. Trypsin is used to excise the residues C-terminal to the sole arginine residue at B22 and again, under conditions of high organic cosolvent concentration, to effect the synthesis of a peptide bond between B22 of the des-octapeptide insulin and a synthetic octapeptide. In the original form of the procedure (1) all amino groups are protected by standard methods to ensure that potential side-products are avoided.[58] A more recent variant of the method is shown in (2), in which the added octapeptides have the trypsin-resistant sequence Lys-Pro at residues B28-29, and no amino protection is necessary.[98]

Kubiak and Cowburn[61] have shown that, under conditions favorable to trypsin-catalyzed resynthesis, incubation with trypsin of des-octapeptide insulin without amino protection can lead to oligomeric forms of the molecule and to the formation of a bond between Arg^{B22} and Gly^{A1}. Despite these side reactions, they found conditions where yields of desired products were comparable to those obtained with amino protection. They synthesized des-pentapeptide insulin, the best coupling yield being 56% at pH 7.0 in DMSO/1,4-butanediol 1:2 with 26.4% H_2O. A five-hour incubation time was used to minimize polymerization.

Achromobacter proteinase, which also has specificity for lysine residues, can replace trypsin in DAI couplings[62] (Figure 4.7). Jonczyk and Gattner[63] have shown that even the preparation of des-Ala^{B30}-insulin is unnecessary, and trypsin can catalyze a transpeptidation between porcine insulin and Thr-OBut. This reaction, under appropriate conditions, can proceed in remarkably high yield. Rose and co-workers obtained 99% recovery of coupled product (92% after deprotection) in only 2 h at 37° C in a mixture of butane-1,4-diol, DMSO and water.[64] The process became of

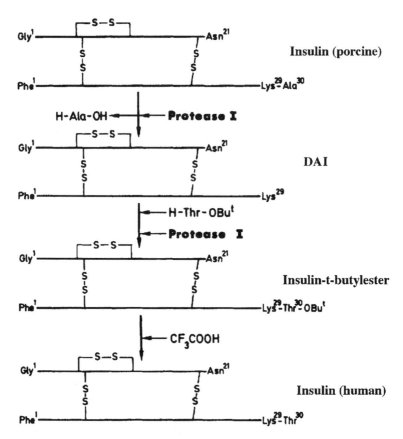

FIGURE 4.7 One of the simplest schemes developed for the conversion of porcine to human insulin. Protease I (Achromobacter Lys-C protease) is used for both degradative and synthetic steps[62]. Taken from Ref. 117 with permission.

great commercial interest and has been exploited particularly by Novo-Nordisk in Denmark (see below).

They also investigated the reasons for the preferential reaction with trypsin catalyst at B29 by preparing B29-arginine insulin and comparing its transamidation rates with those of native insulin where B29 is lysine. They conclude that the rate is not controlled by the pK or the structure of the basic amino-acid residue at B29 and that conformational effects are important.

After human insulin, among the first analogs of insulin to be prepared by enzyme-assisted semisynthesis were [Leu^{B24}] insulin and [Leu^{B25}] insulin.[65,66] A mutant form of insulin associated with diabetes had been identified in which the phenylalanine at either B24 or B25 was replaced by a leucine residue, but it was not possible to distinguish between the two. Synthesis of the analogs and comparison of their properties allowed the identification of the [Leu^{B24}] insulin as the mutant form associated with diabetes, as it acted as an antagonist of native insulin. A second mutant insulin associated with diabetes was also detected where phenylalanine-B24 was replaced by a serine residue. Semisynthetic preparation of the analog confirmed its properties.[67] Kobayashi et al. have prepared both the [Ser^{B24}] and [Ser^{B25}] insulin derivatives, which have 0.7–3% and 3–8% activity, respectively, in isolated rat hepatocytes.[68] The identification of another naturally occurring insulin associated with diabetes and with a leucine residue at position A3 instead of valine persuaded Kobayashi et al. to prepare this analog semisynthetically.[69] To do this, they combined the techniques of amino and carboxy-terminal semisynthesis. The first step was Geiger's di-Boc protection of the A1 and B29 amino groups. Trypsin was used to cleave the carboxy-terminal octapeptide of the B chain and B1 protected with the acid-stable Fmoc group. Trifluoroacetic acid treatment deprotects the glycine A1 amino group. The A chain can then be degraded by sequential Edman cycles. Protected peptides were then coupled onto the truncated A chain and the octapeptide replaced by reversed-enzyme techniques using trypsin. Deprotection gives the required insulin derivative. In this way, [Leu^{A3}] insulin and [Ala^{A3}] insulin were prepared. Both analogs had low binding affinities (0.3 to 0.5%), confirming the role of the amino-terminal region of the A chain in receptor binding. The fact that cleavage and replacement of the B chain octapeptide was preferred to the problems of chemical protection and deprotection of Lys^{B29} is a tribute to how routine and simple the technique has become, at least in experienced hands!

Kobayashi et al[70] have also produced a series of analogs of insulin at B30 using Achromobacter proteinase hoping to find analogs that were less immunoreactive in diabetic patients who develop circulating antibodies against insulin. The [Leu^{B30}] insulin had the lowest immunoreactivity of the analogs screened and retained full activity. Alterations at this position were shown to affect neither the receptor binding affinity nor the biological activity. Even an Ala-Ala-Ala extension on B29 did not significantly decrease the activity of the insulin derivative.

Tager and colleagues have used the approach of coupling synthetic octapeptides to di-Boc-des-octapeptide-insulin with trypsin as catalyst to produce a very large number of analogs in the octapeptide region.[65,71–77] These studies have revealed a great deal about the role of the various amino-acid residues in binding to receptors on isolated canine hepatocytes. Analogs include changes to other naturally occurring

L-amino acids as well as D isomers and non-natural amino acids, and tyrosine residues at various positions have been tested for solvent accessibility by radio-iodination. Stimulated by discoveries of naturally occurring mutant forms of insulin associated with diabetic conditions, they have particularly concentrated on the aromatic triplet Phe^{B24}-Phe^{B25}-Tyr^{B26},[71–73] which is thought, along with Tyr^{B16} and Val^{B12}, to interact directly with the receptor. Their studies show that none of the side chains is absolutely necessary for high affinity receptor interaction. The pentapeptide B26–B30 can be absent if the resulting C-terminal Phe is in the amide form,[71] which is superactive, (see also Ref. 78) and, although deletion of B24 and B25 decreases receptor affinity, L-alanine can replace both Phe^{B24} and Phe^{B25}, the latter as the amide. The Aachen group[78] has also synthesized and studied the truncated form Des-(B27–B30)-insulin for studies of the role of tyrosine B26.[79,80] In addition to the now conventional trypsin-catalyzed resynthesis, this group developed a chymotrypsin mediated route, using Des-(B26–B30) insulin as the substrate, to add a range of variant residues at the B26 position.[80] This approach has also been adopted by the Eli Lilly Insulin group.[81] It was earlier shown[82] that pepsin could be used to achieve a specific cleavage at Phe^{B25}. The resulting Despentapeptide (B26–B30) insulin is, however, a substrate for reverse proteolysis by chymotrypsin through which either amino acid derivatives[80] or tetrapeptides[81] were added. In the latter case, a Leu^{B26} analog resulted that manifested only 4% of normal insulin receptor binding potency, a result considered surprising in view of the reported full potency of a Glu^{B26} analog.[83]

However, in the full-length molecule, each position has a different influence on the receptor interaction.[72] Phe^{B24} appears to be tightly packed against the receptor, as L-Tyr is not tolerated, and is probably involved in steering the change in conformation in this region of the B-chain that is thought to occur on receptor binding. D-amino acids are tolerated well, some even having higher affinity than native insulin, and their presence may lead to a conformation closer to that in the receptor-bound state. At position B25, the shape and nature of the residue can lead to very low affinity ([D-Tyr^{B25},Phe^{B26}] insulin = 0.1%) or high affinity ([L-Tyr] insulin = 80%) and steric considerations seem to be the more important than the electronic nature of the side chain. B26 analogs can also influence affinity both positively and negatively. In conclusion, an important part of the role of these residues is thus directing the structure of the main peptide chain during receptor binding rather than necessarily direct interaction of the side chain with the receptor. As well as their use in studying the role of residues in this region in receptor association, analogs made by the desoctapeptide insulin route are still being used to investigate insulin self-association, an important issue in the pharmacodynamics of the hormone. The Eli Lilly group has made a series of analogs in the B24–29 region in which the conventional amide group of the C-terminal peptide bond is substituted by an aminomethylene, $\psi[CH_2NH]$, moiety. The substitution by this cationic moiety led to substantially less self-association, even among analogs with fully biological potency.[84] In a separate study, the same group performed an alanine scan of the B24–B30 peptapeptide stretch to investigate each residue's contribution to the self-association properties.[85]

Davies et al.[86] have prepared a radioanalog of porcine insulin by replacing alanine B30 with its tritiated analog. Achromobacter proteinase or *porcine* trypsin were capable of catalyzing the reaction between des-Ala^{B30}-insulin and alanine methyl ester. *Bovine* trypsin gave very poor yields, possibly due to increased sensitivity to free radicals generated by the radioactivity in the coupling solvents, and carboxypeptidase Y, with alanine amide, gave side products that were difficult to separate. 50 µL Ci of tritiated insulin at 1.14 Ci/mmol were obtained.

An alternative approach to producing analogs at the C-terminus of the B-chain is to specifically couple a chemically active group to a truncated insulin, allowing formation of new peptide (or other) bonds at this position. Carpenter's group has used trypsin to couple phenylhydrazide to Boc-Gly^{A1}, Boc-Phe^{B1}-desoctapeptide insulin.[87] Oxidation of this derivative gives the phenyldiimide, which can be coupled to Boc-protected octapeptide or octapeptide analog. Redigestion with trypsin was essentially complete, suggesting that racemization, a potential problem with activated carboxyl groups, had not occurred to any significant extent. They produced a series of derivatives representing stepwise addition of residues B23, B24, B25, and B26 to the des-octapeptide insulin. Activity in the isolated rat fat cell increased as the aromatic residues were added (Phe^{B24}, Phe^{B25}, and Tyr^{B26}), with all four residues present (des-(B27-30)-insulin) 50 to 70% activity compared to the native hormone was obtained.

Rose and colleagues have shown that it is possible to enzymatically couple amino acid active esters directly to proteins under conditions where the amino group of the amino acid is free but protected by protonation. The active ester can then be used to couple other groups, usually peptides. Dichlorophenyl esters were found to have the most useful combination of stability and reactivity. A specifically activated form of porcine insulin was obtained by coupling alanine dichlorophenyl ester to lysine-B29.[88]

4.6 HUMAN INSULIN

Human insulin can be produced by total synthesis, by genetic engineering, or by semisynthesis, and more recently a combination of the two latter has been used to successfully produce a number of insulin analogs with interesting receptor binding affinities or properties of association.

The genetic engineering approach offers unlimited supplies of insulin but carries with it the fear of injection of trace amounts of other foreign protein. These fears seem to have been alleviated by extensive chemical characterization and clinical trials. The purely semisynthetic approach requires large quantities of porcine insulin precursor. However, fears that supplies of animal insulin were running short appear to have been premature. Novo-Nordisk in Denmark has developed an industrial process for the semisynthetic production of human insulin. As discussed above, the technique involves direct conversion of insulin, without protection or prior digestion with enzyme, and couples Thr-ester in its place. Deprotection leads to the desired human insulin.

In enzyme-assisted couplings, one component is normally in excess to drive the reaction toward completion. However, if the bond to be resynthesized is intramolecular, high yields of resynthesized product may be obtained. Chu et al. showed that, in a an insulin cross-linked between B29 and A1 and split between B22 and

B23, trypsin could resynthesize the split bond between B22 and B23.[89] Markussen and colleagues have shown that trypsin can catalyze an intramolecular transpeptidation where GlyA1 replaces alanine at position B30, leading to a single-chain insulin. This single-chain insulin precursor with a peptide link between LysB29 and GlyA1 will fold correctly and in higher yield than even natural proinsulin.[90] This led to the possibility of simple and rapid conversion of such an intermediate to human insulin.[91] The single-chain precursor is acylated at the amino terminus before reaction with threonine ester or amide using trypsin as catalyst. By combining this approach with that of site directed mutagenesis of the single-chain precursor, a very large number of analogs can be produced. This is a very powerful example of combined genetic engineering and semisynthesis that is liable to be used more and more in other systems as the necessary techniques are extended. Thus, derivatives have been made with increased positive charge at the carboxy terminus of the B chain and have an isoelectric point near neutral pH. These are soluble at slightly acid pH but crystallize instantly at pH 7.0 and, as a result, have prolonged action *in vivo*. Increased pI can also be obtained by replacing glutamic acid residues with glutamines. Also, insulin analogs have been obtained which remain monomeric even at pharmacological concentrations and are absorbed from the subcutaneous depot two to three times faster than current fast-acting insulins. This was achieved by introducing charge repulsion between the surfaces that come together to form the dimer or by introducing steric hindrance to dimer formation. These insulins may form the basis for improved diabetic treatment.

Human insulin is now widely used to treat diabetics. While it seems mostly to live up to expectations, it has recently been suggested that diabetics lose the awareness of approaching hypoglycemia which is characteristic with animal insulins.[92] This clearly would be a dangerous situation, and reports of unexpected deaths among young diabetics using human insulin have appeared. However, it is unclear whether the phenomenon is a real one, and it remains extremely rare.[93] A very large number of diabetics receive human insulin without complications. A major recent advance has been the development of a fast-acting but equipotent insulin, now in commercial use, in which the naturally occurring sequence at Positions B28 and B29 is reversed. The resulting (Lys^{B28}, Pro^{B29})-insulin, known as "Lys-Pro" was modeled on the inversion of the insulin sequence noted in the homologous insulin-like growth factor I (IGF-I).[94] The original analog was prepared by solid-phase peptide synthesis; subsequently, it and a large series of derivatives with alternative residues at position B28, or additional substitutions, were prepared by a variety of methods including trypsin-catalyzed semisynthesis.[95] Extensive clinical trials have shown that this structural variant provides much improved pharmacodynamics;[96] in addition to more rapid onset of action, the duration has proved to also more closely match the normal physiological response to a meal.[94]

The physicochemical basis of these altered properties relative to native human insulin, when administered by subcutaneous injection, appears to be a substantial drop in the propensity to self-associate. These data have been obtained by NMR and CD spectroscopy in the Weiss and Shoelson Laboratory, using Eli Lilly's Lys-Pro, recombinant Asp^{B10} human insulin from Novo, and a semisynthetically derived combination of the two, called DKP-insulin. This was generated by trypsin-catalyzed fragment condensation using Des-octapeptide insulin from the Asp^{B10} mutant (Figure 4.6).[97] DKP-insulin was then used as a background in which further semisynthetic substitution at positions B24 and B25 was used to probe both self-association and receptor association, with the conclusion that the three structural elements modified could be regarded as independent targets for protein design.[98]

4.7   LABELED INSULINS

Semisynthesis has found a major role in insulin research as a technique for specific labeling of the molecule. Most tracer work with insulin has relied on radio-iodination of the tyrosine residues and this technique has a long history in the literature. It is now relatively easy to isolate each of the four possible mono-iodotyrosyl insulins at high specific activity. However, apart from the A14 derivative, the modified insulins all show quite different behavior from the native molecule in for example their receptor binding affinity. Thus, alternative labeling techniques have been sought. There are also advantages in interpreting the fragmentation of insulin by enzymatic degradation systems in having the label at the ends of the chains, for which step-wise semisynthesis is eminently suited.

Early on, heavy metal derivatives were prepared for X-ray crystallography and a [^{13}C-GlyB1]-insulin prepared for NMR studies.[99] Krail et al. produced iodo-PheB1-insulin[31,100] and Ellis et al. B1-3,5-diiodotyrosine insulin,[109] in view of the relative insensitivity of position B1 to substitution. The latter behaved like native insulin in tests in vitro and *in vivo*. Halban and Offord prepared a tritiated insulin specifically labeled at the B1 phenylalanine.[39] This material found use in numerous degradation and pharmacokinetic studies. Thus, it has been used to investigate degradation rates by the perfused mouse liver, uptake from the subcutaneous depot in rats, including the effects of exercise.[101] Some work has been carried out in human volunteers, but the desirability of injecting any radioactive product is more and more in question, and it seems likely that stable-isotope techniques will be necessary for this kind of work in future (see below). Degradation at the subcutaneous injection site and the circulating species generated from the tracer after subcutaneous injection in rats have also been studied.[102] [[^{3}H]Phe^{B1}] insulin has also been used to investigate insulin degradation by IM-9 lymphocytes and the role of the insulin-receptor compartment in the clearance of insulin in streptozotocin diabetic rats.[101] Where direct comparison can be made, this tracer has always behaved as native insulin.

To study the sequential degradation of insulin in isolated hepatocytes, Assoian and Tager prepared a B1 iodo-insulin by semisynthesis and compared its degradation to that of the A14-iodotyrosyl insulin derivative.[43,44] The use of two labeled insulins led to much more information than could have been obtained with either one alone. Thus, the fragments generated, and their order of appearance, were described, although it was not possible always to tell exactly at which peptide bonds cleavage had occurred nor the precise nature of the fragments (Figure 4.8).

To have tritiated insulins with labeling sites on either chain, Davies and Offord prepared [[^{3}H]Gly^{A1}]insulin.[41] C-terminal labeling of the B-chain via enzymatic semisynthesis has also been achieved.[86] This material has been used in a series of studies on the action of the enzyme insulin proteinase on insulin.[102,103,104] Fragments

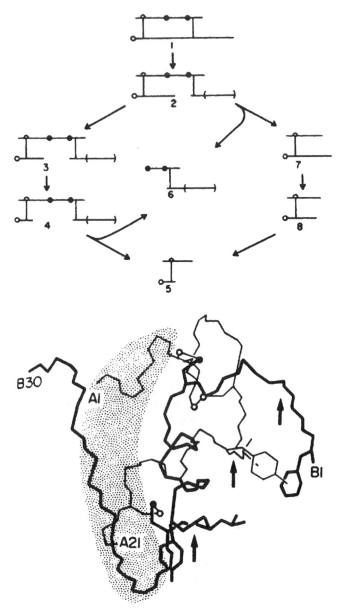

FIGURE 4.8 Proposed pathway for insulin metabolism suggested by studies of labeled intermediates isolated from hepatocytes in a time-dependent manner. The closed circles show the
sites of radiolabel in the A-chain of insulin iodinated with ^{125}I. The open circle is the site of
label (^{125}I-Tyr) introduced by semisynthesis at the N-terminus of the B-chain. By using both
forms independently in identical cell cultures, the branched sequence of proteolysis events
shown in the upper part of the figure could be established. The lower part of the figure shows
the polypeptide backbone of insulin with the three primary sites of proteolysis. It is of
significance that these sites are on the other side of the molecule from the receptor binding
surface (shaded). Taken from Ref. 44 with permission.

can be characterized by a combination of paper electrophoresis and reversed-phase
H.P.L.C. using authentic unlabeled fragments of the A and B chains as markers. In
this manner, a considerable degree of precision as to the chemical structure of such
fragments can be obtained (Figure 4.9). The work has been extended to partially-
degraded, circulating forms of insulin in the rat showing a similarity, but not full
identity, with fragments obtained by the action of insulin proteinase *in vitro*.[104]

Fragments from insulin proteinase have also been characterized by mass-spectrometric analysis of stable-isotope labeled insulins.[42] Thus, ^{2}H-GlyA1 and ^{18}O-
Lys^{B29} insulins were prepared by chemical and by enzymatic semisynthesis, respectively. The results confirm and extend those obtained by use of the tritiated tracers
and are in agreement with analyses carried out by techniques not involving semi-

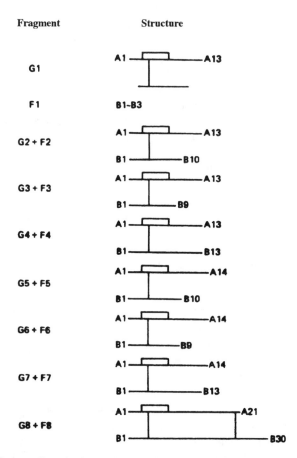

FIGURE 4.9 Intermediates in the *in vitro* metabolism of insulin by insulin proteinase established by the use of insulins labeled on either the N-terminus of the A-chain or the N-terminus
of the B-chain with ^{3}H-Gly (G series) or ^{3}H-Phe (F series), respectively. This study permitted
a most precise knowledge of the peptide bonds cleaved and of additional intermediate steps.[102]
Reproduced with permission from J.G. Davies et al., 1988, *Biochemical Journal* 249, 209–214,
.© the Biochemical Society.

synthetic products. Markussen, Rose, and co-workers have made a number of stable-isotope labeled insulins[105,106,107] that have been used in the study of mechanism of trypsin-catalyzed semisynthesis.[105,108]

Semisynthetic tritiated insulins have also been extensively used by Radziuk's group for pharmacokinetic studies *in vivo* in dogs.[110] The steady-state turnover and metabolic clearance have been measured. Non-steady-state kinetics have also been measured with the tracer used to assess the linearity of appearance of insulin in the peripheral circulation. The results suggest that caution is necessary in assuming linearity. The use of two labeled insulins, A1 and B1, gives the possibility of measuring splanchnic extraction by infusing one tracer into the splanchnic circulation with the other systemically and the peripheral appearance calculated. The tracers can be distinguished by H.P.L.C. purification followed by hydrolysis and quantification of the radioactive amino acid. Again, insulin kinetics were shown to be nonlinear through physiological insulin concentrations. Splanchnic extraction was confirmed as being a major determinant in the metabolic clearance rate of insulin, but peripheral removal mechanisms are also important. Rates of absorption of insulin through different routes after intraperitoneal injection have also been measured using the double-label technique. I.p. insulin was rapidly absorbed, and the absorption was almost evenly split between splanchnic and peripheral routes of entry.

Jones et al.[111] have prepared proinsulin labeled at the amino-terminal phenylalanine that has potential use as a tracer for the prohormone that, under some circumstances, can be present in relatively high concentrations in the circulation. Earlier attempts to make semisynthetic analogs of prolinsulin were made by the Aachen group. Initially, native bovine proinsulin was used as the starting material for these derivatives[112,113] but, eventually, a more ambitious scheme for the semisynthesis of human proinsulin derivatives from porcine insulin was elaborated.[114] While containing many innovative features, it is believed that the project was ultimately abandoned due to the technical complexity.

Major contributions to the study of insulin-receptor structure and physiology have been made with photoactivatable insulin derivatives prepared by semisynthesis.[45] The photoactivatable group has been inserted at the amino group of glycine-A1, in place of phenylalanine-B1 or valine-B2[46] or on the epsilon amino group of lysine-B29.[47] As had already been established from earlier semisynthetic work, the amino terminus of the B chain is less sensitive to modification than that of the A chain but is not part of the receptor binding region of insulin. However, photolabels on the B chain do bind to the receptor. The most commonly used has been the [2-nitro-4-azido-phenylacetyl-B2]insulin. The B29-modified insulins retain good binding affinity and have also proved useful. Cross-linked receptors gave permanently activated lipogenesis in rat fat cells.[48,49] Internalization of insulin-receptor complexes has been followed in isolated cells by permanently cross-linking radio-iodinated insulin to its receptor, the types of membrane vesicles involved in the process have been identified, and recycling of the receptor to the cell surface has also been demonstrated.[50] Knutsen has prepared a photoactivatable insulin where the radio-iodine is present on the crosslinker (as well as the insulin), which contains a disulphide bond.[51] This disulphide can be cleaved under relatively mild conditions, which were shown not to affect the receptor structure, leaving a covalently-labeled receptor free of its ligand. Wedekind et al. have extended the use of photoaffinity labels by preparing, using semisynthetic techniques, an analog with a biocytin (N^ϵ-biotinyl-L-lysine) group attached to the photoactivatable group coupled to lysine B29.[115] The insulin derivative was iodinated and the B26 monoiodo-insulin isolated. In this way, subsequent degradation by trypsin leaves the radio-label attached to the photoactivatable group. Cross-linking of the derivative to receptor and subsequent digestion with trypsin gives a radiolabeled peptide from the insulin receptor that can be purified by exploiting the affinity of the biocytin group for streptavidin. The authors thus identified a part of the receptor binding site for insulin as a sequence near the amino terminus of its alpha chain from residues 20 to approximately 120.

Most recently, the photoactivatable amino acid benzoyl-phenylalanine (Bpa), was incorporated in place of phenylalanine at position B25.[116] This substitution gave a product with approximately 40% of native affinity for the receptor, a value which was unchanged by the incorporation of a biotin handle on the side chain of Lys^{B29}. The derivative could be further modified by iodination at the Tyr^{A14} position. Cross-linking to the receptor occurred with very high efficiency, resulting in a full and sustained set of physiological consequences that could be readily analyzed with the aid of the incorporated handle and label.

4.8   SEMISYNTHESIS AND INSULIN ACTION

From the above discussion, it should be clear that semisynthesis has played a major role in the production of insulin analogs for the study of the mechanism of insulin action. If relatively little has been learned about post-receptor signaling from such studies, much information has been obtained on the interaction of insulin with its specific cell-surface receptor, the first step in insulin action. Thus, semisynthesis has contributed to understanding the receptor-binding region of the insulin molecule and how residues such as the aromatic triplet in the carboxy-terminal octapeptide of the B-chain direct the main-chain structure into a tight-binding conformation. The binding region of the receptor for insulin has also been probed with the help of semisynthetic analogs.

A second important feature of insulin action, internalization of the hormone-receptor complex and hormone degradation, has also benefited from the availability of semisynthetic labeled insulin analogs. While the pathways of degradation remain incompletely understood, the important role of the enzyme insulin proteinase in physiological insulin degradation is becoming established.

Thus, among the hundreds of insulin analogs produced by chemical modification techniques, those produced by semisynthesis, having the advantages of the semisynthetic approach, have been widely used and have made major contributions to the current understanding of insulin physiology.

ACKNOWLEDGMENTS

We wish to thank our past collaborators in our contribution to the work described above, particularly Prof. R.E. Offord, Dr. K. Rose, Ms. A.V. Muir, Mme. B. Dufour, and Mr. C.G. Bradshaw, and the Swiss National Research Foundation for financial support.

REFERENCES

1. Banting, F.G., Best, C.H., Collip, J.B., Campbell, W.R. and Fletcher, A.A., Pancreatic extracts in the treatment of diabetes mellitus, *Canad. Med. Ass. J.*, 2, 141, 1922.
2. Czech, M.P. Molecular basis of insulin action, *Plenum Press*, New York, 1985.
3. Lee, J. and Pilch, P.F., The insulin receptor: Structure, function and signaling, Am. J. Physiol., 266, 319, 1994.
4. White, M.F. and Kahn, C.R., The insulin signalling system, *J. Biol. Chem.*, 269, 1, 1994.
5. Ryle, A.P., Sanger, F., Smith, L.F. and Katai, R., The disulphide bonds of insulin, *Biochem. J.*, 60, 541, 1955.
6. Blundell, T.L., Cutfield, J.F., Cutfield, S.M., Dodson, E.J., Dodson, G.G., Hodgkin, D.C., Mercola, D.A. and Vijayan, M., Atomic positions in rhombohedral 2-zinc insulin crystals, *Nature (London)*, 231, 506, 1971.
7. Blundell, T., Dodson, G., Hodgkin, D. and Mercola, D., Insulin: the structure in the crystal and its reflection in chemistry and biology, Advances in protein chemistry, 26, Anfinson, C.B., Edsall, J.T. and Richards, F.M., Eds. *Academic Press*, New York, 1972, 279.
8. Pullen, R.A., Lindsay, D.G., Wood, S.P., Tickle, I.J., Blundell, T.L., Wollmer A., Krail, G., Brandenburg, D., Zahn, H., Gliemann, J. and Gammeltoft, S., Receptor-binding region of insulin, *Nature (London)*, 259, 369, 1976.
9. Weiss, M.A., Nguyen, D.T., Khait, I., Inouye, K., Frank, B.H., Beckage, M., O'Shea, E., Shoelson, S.E., Karplus, M. and Neuringer, L.J. Two-dimensional NMR and photo-CIDNP studies of the insulin monomer: assignment of aromatic resonances with application to protein folding, structure and dynamics, *Biochemistry*, 28, 9855, 1989.
10. Roy, M., Lee, R.W.K., Brange, J. and Dunn, M.F., ^{1}H NMR Spectrum of the native human insulin monomer, *J. Biol. Chem.*, 265, 5448, 1990.
11. Terabe, S., Konaka, R., and Inouye, K., Separation of some polypeptide hormones by high-performance liquid chromatography, *J. Chromatog.*, 172, 163, 1979.
12. Davies, J.G. and Offord, R.E., The preparation of tritiated insulin specifically labelled by semisynthesis at glycine-A1, *Biochem. J.*, 231, 389, 1985.
13. Offord, R.E. and DiBello, C. Semisynthetic peptides and proteins, *Academic Press*, London and New York, 1978, 141.
14. Offord, R.E., Semisynthetic proteins, *John Wiley*, Chichester, 1980.
15. Offord, R.E., Protein engineering by chemical means?, *Protein Engineering*, 1, 151, 1987.
16. Brandenburg, D. Insulin chemistry, in Handbook of experimental Pharmacology, Vol. 92, Cuatrecasas, P. and Jacobs, S., Eds., *Springer-Verlag*, Berlin 1990, 1.
17. Brandenburg, D. Des-PheB1-insulin, ein kristallines Analogon des Rinderinsulins, *Hoppe-Seyler's Z. Physiol. Chem.* 350, 741, 1969.
18. Borras, F. and Offord, R.E., Protected intermediate for the preparation of semisynthetic insulins, *Nature (London)*, 227, 716, 1970.
19. Drewes, S.E., Magojo, H.E.M., and Gliemann, J., (N-methylpyridinium) insulins-modification at the A19-Tyrosine and B16-Tyrosine, *Hoppe-Seyler's Z. Physiol. Chem.*, 364, 461, 1983.

20. Geiger, R., Schone, H.H., and Pfaff, W., Bis(tertbutyloxycarbonyl) insulin, *Hoppe-Seyler's Z. Physiol. Chem.*, 352, 1487, 1971.

21. Saunders, D.J. and Offord, R.E., Semisynthetic analogues of insulin, *Biochem. J.*, 165, 479, 1977.

22. Saunders, D.J. and Offord, R.E., Novel semisynthetic route to insulin analogs modified at the N-terminus of the A-chain, *Hoppe-Seyler's Z. Physiol. Chem.*, 358, 1469, 1977.

23. Tesser, G.I., and Balvert-Geers, I.C., The methylsulfonylethyloxycarbonyl group: a new and versalite amino protective function, *Int. J. Pept. Protein Res.*, 7, 295, 1975.

24. Geiger, R. and Langner, D., Insulin-analoga mit N-terminal verkurzter B-kette. Selektine Edmann-Abbau an der B-kette des Insulins, *Hoppe-Seyler's Z. Physiol. Chem.*, 354, 1285, 1973.

25. Naithani, V.K. and Gattner, H.G., Preparation and properties of citroconyl insulins, *Hoppe-Seyler's Z. Physiol. Chem.*, 363, 1443, 1982.

26. Kobayashi, M., Takata, Y., Ishibashi, O., Sasaoka, T., Iwasaki, M., Shigeta, Y. and Inouye, K., Receptor binding and negative cooperativity of a mutant insulin, [Leu^{A3}]insulin, *Biochem. Biophys. Res. Commun.*, 137, 250, 1986.

27. Geiger, R., Teetz,V., Konig,W. and Obermeier, R. Semisynthetic amino acid and peptide exchange in insulin, in Semisynthetic peptides and proteins, Offord, R.E. and DiBello, C., Eds., *Academic Press*, London, 1978, 141.

28. Geiger, R., Geisen, K., Summ, H.D. and Langner, D. [A1-D-Alanin]Insulin, *Hoppe-Seyler's Z. Physiol. Chem.*, 356, 1635, 1975.

29. Saunders, D.J., Semisynthetic analogues of insulin modified at the N-terminus of the A-chain, in Semisynthetic peptides and proteins, Offord, R.E. and DiBello, C., Eds., *Academic Press*, London, 1978, 213.

30. Kerp, L., Steinhilber, S., Kasemir, H., Han, J., Henrichs, H.R. and Geiger, R., Changes in immunospecificity and biologic activity of bovine insulin due to subsequent removal of the amino acids B1, B2 and B3, *Diabetes*, 23, 651, 1974.

31. Krail, G., Brandenburg, D. and Zahn, H., Insulinanaloga mit N-terminal verlangerter B-Kette, *Die Makromolekulaire Chemie, Suppl.* 1, 7, 1975.

32. Yeung, C.W.T., Moule, M.L. and Yip, C.C., Functional role of the N-terminal region of the B-chain of insulin, in *Insulin. Chemistry, structure and function of insulin and related hormones*, Brandenburg, D. and Wollmer, A. Eds., *Walter de Gruyer*, Berlin, 1980, 417.

33. Brandenburg, D., Biela, M., Herbertz, L. and Zahn, H., Partial synthesis and properties of des-A1-glycine-insulin, *Hoppe-Seyler's Z. Physiol. Chem.*, 356, 961, 1975.

34. Krail, G., Brandenburg, D. and Zahn, H. Insulin analogs with permuted A-chain N-terminus, *Hoppe-Seyler's Z. Physiol. Chem.*, 356, 981, 1975.

35. Gliemann, J. and Gammeltoft, S., The biological activity and the binding affinity of modified insulins determined on isolated rat fat cells, *Diabetologia*, 10, 105, 1974.

36. Berndt, H., Gattner, H-G. and Zahn, H. Semisynthetic Sheep Des-A1-glycine insulin, *Hoppe-Seyler's Z. Physiol. Chem.*, 356, 1472, 1975.

37. Geiger, R., Geisen, K. and Summ, H.D. Austausch von A1-Glycin in Rinderinsulin gegen L- und D- Tryptophan, *Hoppe-Seyler's Z. Physiol. Chem.*, 363, 1231, 1982.

38. Rosen, P., Simon, M., Reinauer, H. Brandenburg, D., Friesen, H.J. and Diaconescu, C., A1-modified insulins: receptor binding and biological activity, in *Insulin. Chemistry, structure and function of insulin and related hormones*, Brandenburg, D. and Wollmer, A. Eds., *Walter de Gruyer*, Berlin, 1980, 403.

39. Halban, P.A. and Offord, R.E., The preparation of a semisynthetic tritiated insulin with a specific radioactivity of up to 20 Curies per millimole, *Biochem. J.* 151, 219, 1975.

40. Grant, K.I. and von Holt, C., Improved preparation of semisynthetic PheB1-tritiated insulin, *Biol. Chem. Hoppe-Seyler*, 368, 239, 1987.

41. Davies, J.G. and Offord, R.E., The preparation of tritiated insulin specifically labelled by semisynthesis at glycine-A1, *Biochem., J.*, 321, 389, 1985.

42. Savoy, L.A., Jones, R.M.L., Pochon, S., Davies, J.G., Muir, A.V., Offord, R.E. and Rose, K., Identification of fast atom bombardment mass spectrometry of insulin fragments produced by insulin proteinase, *Biochem. J.*, 249, 215, 1988.

43. Assoian, R.K. and Tager, H.S., [(^{125}I)-iodotyrosyl B1]insulin:semisynthesis, receptor binding and cell-mediated degradation of a B chain-labelled insulin, *J. Biol. Chem.*, 256, 4042, 1981.

44. Assoian, R.K. and Tager, H.S., Peptide intermediates in the cellular metabolism of insulin *J. Biol. Chem.*, 257, 9078, 1982.

45. Thamm, P. Saunders, D. and Brandenburg, D., Photoreactive insulin derivatives: preparation and characterization, in *Insulin. Chemistry, structure and function of insulin and related hormones*, Brandenburg, D. and Wollmer, A. Eds., *Walter de Gruyer*, Berlin, 1980, 309.

46. Rees, A.R. and Whittle, M., The preparation and application of an A1-substituted photo-reactive insulin analogue, in *Insulin. Chemistry, structure and function of insulin and related hormones*, Brandenburg, D. and Wollmer, A. Eds., *Walter de Gruyer*, Berlin, 1980, 327.

47. Yip, C.C., Yeung, C.W.T. and Moule, M.L., Photoaffinity labeling of insulin receptor proteins of liver plasma membranes, in *Insulin. Chemistry, structure and function of insulin and related hormones*, Brandenburg, D. and Wollmer, A. Eds., *Walter de Gruyer*, Berlin, 1980, 337.

48. Saunders, D. and Brandenburg, D. Photoactivatable insulins and receptor photoaffinity labelling, in *Methods in Diabetes Research*, 1, Larner, J and Pohl, S. Eds., Wiley, New, York, 1984, 3.

49. Brandenburg, D., Diaconescu, C., Klotz, G., Mucke, P. Neffe, J., Saunders, D. and Schuttler, A., Biological action and fate of photoaffinity-labelled insulin-receptor complexes, *Biochimie*, 67, 1111, 1985.

50. Carpentier, J.L., Gazzano, H., Van Obberghen, E., Fehlmann, M., Freychet, P. and Orci, L., Intracellular pathway followed by the insulin receptor covalently coupled to ^{125}I-photoreactive insulin during internalization and recycling, *J. Cell. Biol.*, 102, 989, 1986.

51. Knutsen, V.P., The covalent tagging of the cell surface insulin receptor in intact cells with the generation of an insulin-free, functional receptor. A new approach to the study of receptor dynamics, *J. Biol. Chem.*, 262, 2374, 1987.

52. Nakagawa, S.H. and Tager, H.S., Importance of aliphatic side-chain structure at positions 2 and 3 of the Insulin A chain in insulin-receptor interactions, *Biochemistry*, 31, 3204, 1992.

53. Qui-Ping, C., Geiger, R., Langner, D. and Geisen, K., Biological activity in vivo of insulin analogues modified in the N-terminal region of the B-chain, *Biol. Chem. Hoppe-Seyler*, 367, 135, 1986.

54. Ruttenberg, M.A., Human Insulin: facile synthesis by modification of porcine insulin, *Science*, 177, 623, 1972.

55. Weitzel, G., Bauer, F.U. and Eisele, K., Structure and activity of insulin. XIV. Further studies on the three-step increase in activity due to the aromatic amino acids B24-26 (-Phe-Phe-Tyr-), *Hoppe-Seyler's Z. Physiol. Chem.*, 357, 187, 1976.

56. The Shanghai Insulin Research Group, C-terminal semisynthesis of porcine insulin: the role of arginine residue B22, *Sci. Sin.*, 16, 61, 1973.

57. Obermeier, R. and Geiger, R., A new semisynthesis of human insulin, *Hoppe-Seyler's Z. Physiol. Chem.*, 357, 759, 1976.

58. Inouye, K., Watanabe, K., Morihara, K., Tochino, T. Kanaya, T., Emwa, J. and Sakakibara, S., Enzyme assisted semisynthesis of human insulin, *J. Am. Chem. Soc.*, 101, 751, 1979.

59. Morihara, K., Oka, T. and Tsuzuki, H., Semisynthesis of human insulin by trypsin-catalyzed replacement of Ala B-30 by Thr in porcine insulin, *Nature (London)*, 280, 412, 1979.

60. Morihara, K., Oka, T., Tsuzuki, H., Inouye, K., Tochino, Y., Kanaya, H., Masaki, T., Soejima, M., and Sakakibara, S., *Peptide Chemistry* 1979, Yonehara, H., Ed., Protein Research Foundation, Osaka, Japan, 1980, 113.

61. Kubiak, T. and Cowburn, D., Enzymatic semisynthesis of porcine despentapeptide (B26-30) insulin using unprotected desoctapeptide (B23-30) *Int. J. Pept. Protein Res.*, 27, 514, 1986.

62. Morihara, K., Oka, T., Tsuzuki, H., Tochino, Y. and Kanaya, T., Achromobacter protease I-catalyzed conversion of porcine insulin into human insulin, *Biochem. Biophys. Res. Commun.*, 92, 396, 1980.

63. Jonczyk, K.A. and Gattner, H.G., Eine neue semisynthese des Humaninsulins. Tryptisch-katalysierte Transpeptidierung von Schweineinsulin mit L-Threonin-tert-butyl-ester, *Hoppe-Seyler's Z. Physiol. Chem.*, 362, 1591, 1981.

64. Rose, K., de Pury, H. and Offord, R.E., Rapid preparation of human insulin and insulin analogues in high yield by enzyme-assisted semisynthesis, *Biochem. J.*, 211, 671, 1983.

65. Tager, H., Thomas, M., Assoian, R., Rubenstein, A., Saekow, M., Olefski, J., and Kaiser, E.T., Semisynthesis and biological activity of porcine (Leu[B24]) and (Leu[B25]) insulin, *Proc. Natl. Acad. Sci. USA*, 77, 3181, 1980.

66. Kobayaski, M., Ohgaku, S., Iwasaki, M., Maegawa, H., Strigeta, Y. and Inouye, K., Supernormal insulin: [D-Phe[B24]]-insulin with increased affinity for insulin receptors, *Biochem. Biophys. Res. Commun.* 107, 329, 1982.

67. Shoelson, S.E., Polonsky, K.S., Zeidler, A., Rubinstein, A.H. and Tager, H.S., Identification of a mutant human insulin predicted to contain a serine-for-phenylalanine substitution, *Proc. Natl. Acad. Sci. USA*, 80, 7390, 1983.

68. Kobayashi, M., Haneda, M., Maegawa, H., Watanabe, N., Takada, Y., Shigeta, Y. and Inouye, K., Receptor binding and biological activity of [Ser[B24]]-insulin, an abnormal mutant insulin, *Biochem. Biophys. Res. Commun.* 119, 49, 1984.

69. Kobayashi, M., Takata, Y., Ishibashi, O., Sasaoka, T., Iwasaki, M., Shigeta, Y., and Inouye, K., Receptor binding and negative cooperativity of a mutant insulin [Leu[A3]]-insulin, *Biochem. Biophys. Res. Commun.* 137, 250, 1986.

70. Kobayashi, M., Ohgaku, S., Iwasaki, M., Maegawa, H., Shigeta, Y., and Inouye, K., Characterization of [Leu B-24] and [Leu B-25]-insulin analogues: receptor binding and biological activity, *Biochem. J.*, 206, 597, 1982.

71. Nakagawa, S.H. and Tager, H.S., Role of the phenylalanine B25 side chain in directing insulin interaction with its receptor, *J. Biol. Chem.*, 261, 7332, 1986.

72. Mirmira, R.G., Nakagawa, S.H., and Tager, H.S., Importance of the character and configuration of residue B24, B25, and B26 in insulin-receptor interactions, *J. Biol. Chem.*, 266, 1428, 1991.

73. Mirmira, R.G., Nakagawa, S.H. and Tager, H.S., Disposition of the phenylalanine B25 side chain during insulin-receptor and insulin-insulin interactions, *Biochemistry* 30, 8222, 1991.

74. Brems, D.N., Brown, P.L., Nakagawa, S.H., and Tager, H.S., The conformational stability and flexibility of insulin with an additional intramolecular cross-link, *J. Biol. Chem.*, 266, 1611, 1991.

75. Nakagawa, S.H., and Tager, H.S., Role of the COOH-terminal β-chain domains in insulin-receptor interactions, *J. Biol. Chem.*, 262, 12054, 1987.

76. Nakagawa, S.H. and Tager, H.S., Perturbation of insulin-receptor interactions by intramolecular hormone cross-linking, *J. Biol. Chem.* 264, 272, 1989.

77. Mirmira, R.G. and Tager, H.S., Role of the phenylalanine B24 side-chain in directing insulin interaction with its receptor, *J. Biol. Chem.* 264, 6349, 1989.

78. Fischer, W.H., Saunders, D., Brandenburg, D., Wollmer, A. and Zahn, H., A shortened insulin with full in vitro potency, *Hoppe-Seyler's Z. Physiol. Chem.*, 366, 521, 1985.

79. Lenz, V., Gattner, H.G., Sievert, D., Wollmer, A., Engels, M. and Hocker, M. Semisynthetic Des-(B27-B30)-insulins with modified B26-tyrosine, *Biol. Chem. Hoppe-Seyler* 372, 495, 1991.

80. Sievert, D., Gatther, H.G., Lenz, V. and Hocker, H., Enzyme assisted semisynthesis of shortened B26- modified insulins, *Biomed. Biochim. Acta*, 50, 5197, 1991.

81. Fan, L., Alter, L.A., Ellis, R.M., Korbas, A.M., Brooke, G.S. and Chance, R.E., Chymotrypsin-catalyzed semisynthesis—an alternative approach for synthesis of insulin analogs, *Proc. Amer. Pept. Symp.*, 12, 549, 1992.

82. Gattner, H.G., B-chain shortening of matrix-bound insulin by pepsin. 1. Preparation and properties of bovine despentapeptide (B26-30) insulin, *Hoppe-Seyler's Z. Physiol. Chem.* 356, 1397, 1975.

83. Brange, J., Owens, D., King, S. and Volund, A., Monomeric insulins and their experimental and clinical implications, *Diabetes Care*, 13, 923, 1990.

84. Zhao, M. and DiMarchi, R.D., Amino-methylene, ψ (CH_2NH) substitution of amide bonds in the C-terminal portion of the insulin B chain, *Proc. Amer. Pept. Symp.*, 14, 641, 1996.

85. Zhao, M., Fan, L., Long, H.B., Pekar, A.H., Slicker, L.J., Chance, R.E. and DiMarchi, R.D., Sequential replacement of the C-terminal residues of the human insulin B chain with alanine, *Proc. Amer. Pept. Symp.*, 14, 643, 1996.

86. Davies, J.G., Rose, K., Bradshaw, C.G. and Offord, R.E., Enzymatic semisynthesis of insulin specifically labelled with tritium at position B30, *Protein Eng.*, 1, 407, 1987.

87. Canova-Davies, E. and Carpenter, F.H.C., Specific activation of the arginine carboxyl group of the β-chain of bovine desoctapeptide-(B23-30) insulin, in Insulin, Brandenburg, D.W. and Wollmer, A., Eds., de Gruyter, Berlin and New York, 1980, 107.

88. Rose, K., Herrero, C., Proudfoot, A.E.I., Offord, R.E. and Wallace, C.J.A., Enzyme-assisted semisynthesis of polypeptide active esters and their use, *Biochem. J.* 249, 83, 1988.

89. Chu, S.C., Wang, C.C., and Brandenburg, D., Intramolecular enzymatic peptide synthesis: trypsin-mediated coupling of the peptide bond between B22-arginine and B23-glycine in a split cross-linked insulin, *Hoppe-Seyler's Z. Physiol. Chem.*, 363, 647, 1981.

90. Markussen, J., Jorgensen, K.M., Sorensen, A.R. and Thim, L., Single-chain des-(B30) insulin, *Int. J. Peptide Protein Res.*, 26, 70, 1985.

91. Markussen, J., Diers, I., Hongaard, P., Langkjar, L., Norris, K., Snel, L., Sorensen, A.R., Sorensen, E., and Voigte, H.O., Soluble prolonged-acting insulin derivatives. 3. Degree of protraction, crystallizability and chemical stability of insulins substituted at positions A21, B13, B23, B27 and B30, *Protein Eng.* 2, 157, 1988.

92. Teuscher, A. and Berger, W.G., Hypoglycemia un-awareness in diabetics transferred from beef or porcine to human insulin, *Lancet*, 2 (8555), 382, 1987.

93. Woff, S.P., Trying times for human insulin, *Nature (London)*, 356, 375, 1992.

94. DiMarchi, R.D., Mayer, J.P., Fan, L., Brems, D.N., Frank, B.H., Green, L.K., Hoffman, J.A., Howey, D.C., Larg, H.B., Shawn, W.N., Shields, J.E., Slicker, L.J., Su, K.S.E., Sundall, K.L. and Chance, R.E., Synthesis of a fast-acting insulin based on structural homology with insulin-like growth factor I, *Proc. Amer. Pept. Symp.* 12, 26, 1992.

95. Long, H.B., Baker, J.C., Belagaje, R.M., DiMarchi, R.D., Frank, B.H., Green, L.K., Hoffman, J.A., Muth, W.L., Pekar, A.H., Reams, S.G., Shaw, W.N., Shields, J.E., Slieker, L.J., Su, K.S.E., Sundell, K.L. and Chance, R.E., Human insulin analogs with rapid onset and short duration of action, *Proc. Amer. Pept. Symp.*, 12, 88, 1992.

96. Andersen, J.H., Jr., Chance, R.E., Slieker, L.J., Vignati, L. and DiMarchi, R.D., Design of an optimal mealtime insulin (Lys-Pro) and its assessment in phase III clinical studies, *Proc. Amer. Pept. Symp.*, 14, 617, 1996.

97. Weiss, M.A., Hua, Q.X., Lynch, C.S., Frank, B.H. and Shoelson, S.E., Heteronuclear 2D NMR studies of an engineered insulin monomer: assignment and characterization of the receptor-binding surface by selective ^{2}H and ^{13}C labeling with application to protein design, *Biochemistry*, 30, 7373, 1991.

98. Shoelson, S.E., Lu, Z.X., Parlautan, L., Lynch, C.S. and Weiss, M.A., Mutations at the dimer, hexamer and receptor-binding surfaces of insulin independently affect insulin-insulin and insulin-receptor interactions, *Biochemistry*, 31, 1757, 1992.

99. Saunders, D.J. and Offord, R.E., The use of semisynthetically-introduced ^{13}C probes for nuclear magnetic resonance studies on insulin, *FEBS Lett.*, 26, 286, 1972.

100. Krail, G., Brandenburg, D., Zahn, H. and Geiger, R., [B1-paraiodophenylalanine] insulin, a homogeneous monoiodoinsulin, Hoppe-Seyler's Z. *Physiol. Chem.*, 352, 1595, 1971.

101. Duckworth, W.C. and Kitabchi, A.E., Insulin metabolism and degradation, *Endocrinol. Rev.*, 2, 210, 1981.

102. Davies, J.G., Muir, A.V. and Offord, R.E., Identification of some cleavage sites of insulin by insulin proteinase, *Biochem. J.*, 240, 609, 1986.

103. Muir, A.V., Offord, R.E. and Davies, J.G., The identification of a degradation of insulin by insulin proteinase, *Biochem. J.*, 237, 631, 1986.

104. Davies, J.G., Muir, A.V., Rose, K. and Offord, R.E., Identification of radioactive insulin fragments liberated by insulin proteinase during the degradation of semisynthetic [[^{3}H] Gly^{A1}] insulin and [[^{3}H] Phe^{B1}]insulin, *Biochem. J.*, 249, 209, 1988.

105. Markussen, J. and Schaumburg, K., Reaction mechanism in trypsin catalyzed synthesis of human insulin studied by ^{17}O-NMR spectroscopy, in *Peptides* 1982, Blaha, K. and Malon, P., (Eds.), *W. de Gruyter*, Berlin, 1983, 387.

106. Rose, K., Gladstone, J. and Offord, R.E., A mass-spectrometric investigation of the mechanism of the semisynthetic transformation of pig insulins into an ester of insulin of human sequence, *Biochem. J.*, 220, 189, 1984.

107. Stocklin, R., Rose, K., Green, B.N. and Offord, R.E., The semisynthesis of [octadeutero-Phe^{B1}–octadeutero–Val^{B2}]–porcine insulin and its characterization by mass spectrometry, *Protein Eng.*, 7, 285, 1994.

108. Rose, K., Stocklin, R., Savoy, L.A., Regamey, P-O., Offord, R.E., Vuagnat, P. and Markussen, J., Reaction mechanism of trypsin-catalysed semisynthesis of human insulin studied by fast atom bombardment mass spectrometry, *Protein Eng.*, 4, 409, 1991.

109. Ellis, M.J., Jones, R.H., Thomas, J.H., Geiger, R., Teetz, V. and Sonksen, P.H., B1-3,5-diiodotyrosine insulin: a valid tracer for insulin, *Diabetologia*, 13, 257, 1977.

110. Morishima, T., Bradshaw, C. and Radziuk, J., Measurement using tracers of steady-state turnover and metabolic clearance of insulin in dogs, *Am. J. Physiol.*, 248, E203, 1985.

111. Jones, R.M.L., Rose, K. and Offord, R.E., Semisynthetic [[^{3}H]$_2$Phe^1] proinsulin, *Biochem. J.*, 247, 785, 1987.

112. Naithani, V.K., Bullesbach, E.E. and Zahn, H., Semisynthesis of a des-(1-21)-prepro-insulin derivative, *Hoppe-Seyler's Z. Physiol. Chem.*, 360, 1363, 1979.

113. Bullesbach, E.E. and Naithani, V.K., Darstellung und anwendung von N$^{\epsilon 29}$, N$^{\epsilon 59}$–bis-(methylsulfonyl-ethoxycarbonyl)–proinsulin, *Hoppe-Seyler's Z. Physiol. Chem.*, 361, 723, 1980.

114. Naithani, V.K. and Gattner, H.G., Semisynthesis of human proinsulin, I: preparation of arginyl-A-chain cyclic bis-disulfide, *Hoppe-Seyler's Z. Physiol. Chem.*, 362, 685, 1981.

115. Wedekind, F.F., Baer-Pontzen, K., Bala-Mohan, S., Choli, D., Zahn, H. and Brandenburg, D., Hormone binding site of the insulin receptor: analysis using photoaffinity-mediated avidin complexing, *Hoppe-Seyler's Z. Physiol. Chem.*, 370, 251, 1989.

116. Shoelson, S.E., Lee, J., Lynch, C.S., Backer, J.M. and Pilch, P.F., Bpa^{B25} insulins: photoactivatable analogues that quantitatively cross-link, radiolabel and activate the insulin receptor, *J. Biol. Chem.*, 268, 4085, 1993.

117. Kullman, W., *Enzymatic Peptide Synthesis,* CRC Press, Boca Raton, Fl., 1987.

5 Ribonuclease Semisynthesis

Marilynn Scott Doscher

CONTENTS

5.1 INTRODUCTION

Owing to its presence in the relatively simple mixture of digestive proteins present in bovine pancreas and pancreatic juice, ribonuclease A (RNase A[1]) was purified and characterized early in the history of protein chemistry.[1] It is a small enzyme (13,683 Da; 124 amino acid residues, see Figure 5.1) that functions without the benefit of coenzyme or metal cofactor to digest RNA in the intestine, an important recycling event in ruminants.[2,3] It catalyzes two distinct reactions:

1. A transphosphorylation that ruptures a P-O$^{5'}$ bond in the RNA and forms a 2',3'-cyclic phosphodiester followed by hydrolysis of this ester to form a 3'-phosphomonoester (Figure 5.2).
2. Transphosphorylation is catalyzed regardless of which of the four bases is on the 5'-side of the phosphodiester bond, but the base must be uracil or cytosine on the 3'-side of the bond.

During the many years since its discovery, RNase A has been the subject of unremitting study as well as serving as a paradigmatic enzyme for testing new

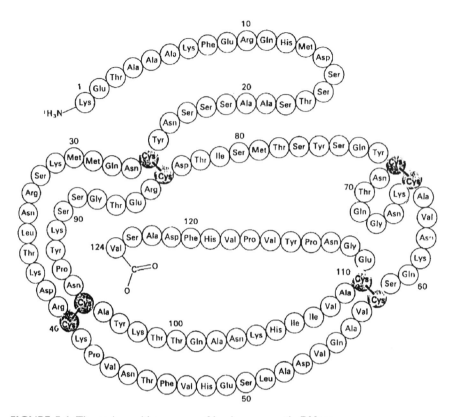

FIGURE 5.1 The amino acid sequence of bovine pancreatic RNase.

techniques (see Refs. 4–7 for reviews). As one result of this activity, its mode of action is now considered to be a suitable subject for textbook presentations about enzyme mechanisms.[8,9] Among the amino acid residues that are proposed to play major roles in its mechanism of action are His-12, Lys-41, and His-119.[10] Nonetheless, studies exploring the mechanistic subtleties of the enzyme continue, with both experimental[11–17] and theoretical[18,19] approaches being applied.

In keeping with the extraordinary range and number of the studies that have been made with RNase A, two distinct semisynthetic systems have been extensively utilized to study this enzyme. The first was provided by the reversibly dissociating, fully active RNase S, a product of the partial proteolysis of RNase A by the bacterial endoprotease subtilisin.[20,21] The two chains formed by the proteolysis remain tightly associated at neutral pH values, but dissociate under acidic conditions; neither chain alone exhibits any enzymatic activity, but when allowed to reassociate at neutral pH, full activity is restored (Figure 5.3). Hofmann and co-workers reported that the addition to an S-protein preparation of synthetic RNase 1–13, prepared by classic fragment condensation methods, produced a noncovalent complex that exhibited 68–72% of the activity of RNase S against yeast RNA and 62–65% activity against uridine 2',3'-cyclic phosphate.[22] These authors noted that this was the first partial laboratory synthesis of an enzyme.

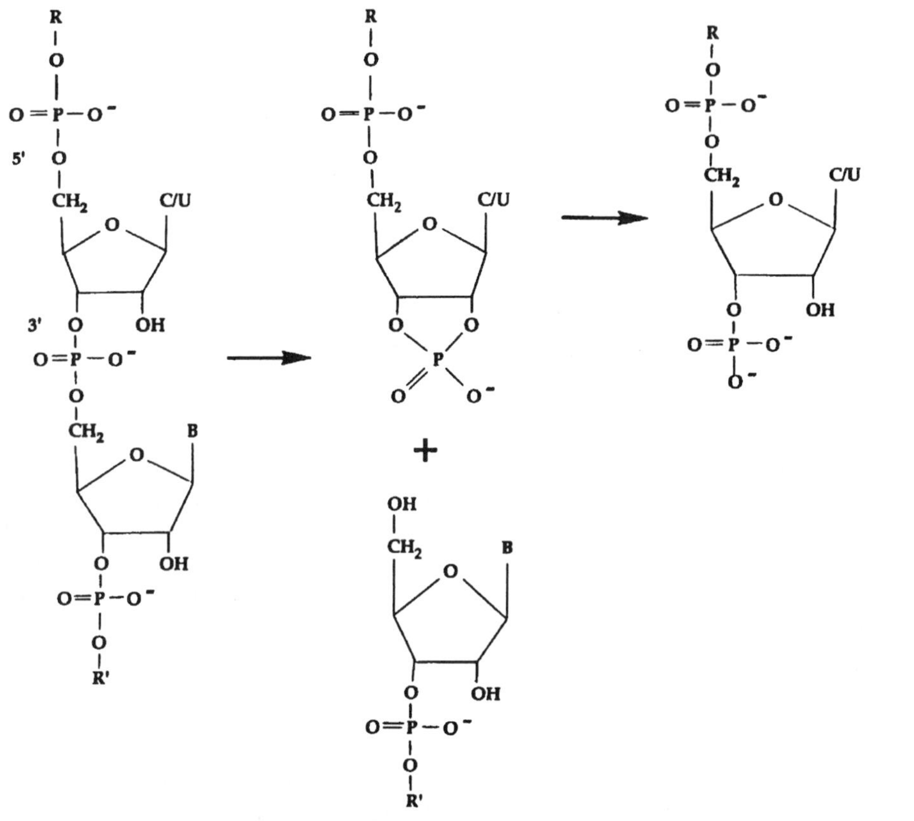

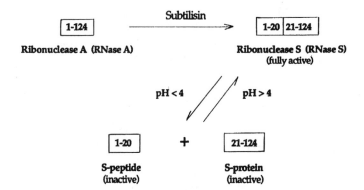

FIGURE 5.2 Reactions catalyzed by RNase.

FIGURE 5.3 Preparation and properties of RNase S. Limited proteolysis of RNase A by subtilisin cleaves primarily the bond between Ala-20 and Ser-21. Although no disulfide bonds connect the two resulting fragments, noncovalent interactions provide a stable fully active complex at neutral pH values. Under acidic conditions, the complex dissociates. Neither of the individual chains exhibits detectable enzymatic activity.

The second system emerged from studies of RNase 1–120, a product of the partial pepsin digestion of RNase A.[23] It was initially thought to be devoid of activity, but Lin later reported that the scrupulously purified product exhibited 0.5% activity against cytidine 2',3'-cyclic phosphate.[24] In view of the implications of this finding for the proposed role of His-119 as an essential active site residue (25, 26), Lin prepared and characterized RNase 1–118 by subjecting RNase 1–120 to exhaustive carboxypeptidase A digestion.[24] The product was devoid of activity, but could be fully reactivated by the addition of a synthetic tetradecapeptide comprised of residues 111–124 of RNase A.[27,28] (See Figure 5.4.)

Refined high-resolution structures have been determined for crystals of both RNase S[29] and RNase 1–118:111–124.[30,31] Based on these structures, the respective locations of the N-terminal (S-peptide) and C-terminal (RNase 111–124) peptides of the two systems are shown schematically in Figure 5.5.

Since they were first reported, the above semisynthetic systems have been utilized in many laboratories to study the structural basis for the catalytic efficiency of RNase against a number of substrates and, in addition, to investigate the structural basis for the strong and specific noncovalent interactions between the large and small peptide moieties that constitute each of the two semisynthetic enzymes. During the course of these studies, a number of observations have been made that are also applicable to a more generalized understanding of structure-function relationships in proteins.

5.2 RIBONUCLEASE S

In view of the possible modifications that might occur in the S-peptide and S-protein components when separated from one another, such as air oxidation of exposed

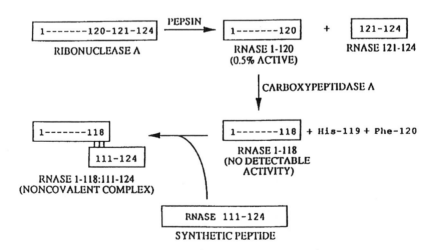

FIGURE 5.4 Preparation and assembly of RNase 1-118:111-124. Successive limited proteolysis of RNase A by pepsin and by carboxypeptidase A provides RNase 1-118, which is devoid of activity. Combination of this chain with a synthetic peptide containing residues 111-124 of RNase results in the formation of a noncovalent complex exhibiting essentially full activity.

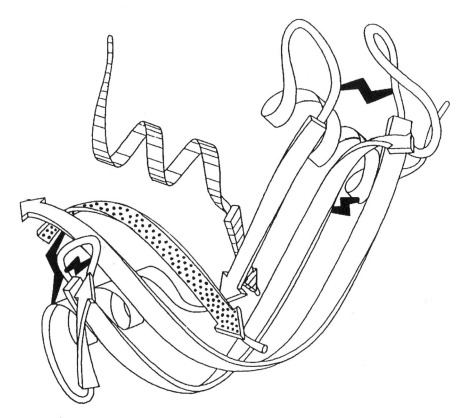

FIGURE 5.5 Schematic depiction of the relative locations of the S-peptide (crosshatched) and RNase 111–124 (stippled) in the three-dimensional structure of RNase. Filled zigzag shapes depict disulfide bonds.

methionine residues or a localized irreversible conformation change in S-protein, the complex formed by recombining these components at neutral pH has often been designated RNase S' to distinguish it from the material isolated directly from the subtilisin digest. Aside from the partial oxidation of Met-13 in S-peptide preparations to a mixture of the d-and l-sulfoxides,[32] no such modifications have actually been observed. Nevertheless, the occurrence of significant hydrolysis in some subtilisin digests at the Ser[21]-Ser[22], rather than at the Ala[20]-Ser[21] bond, does provide an additional basis for differences between the original and recombined complexes.[32,33] Such heterogeneity proved to be a complicating factor in experiments investigating the enzymatic reformation of RNase A from RNase S'.[34]

Chemical syntheses of the 20-residue S-peptide by both classical[35] and solid-phase techniques[36] have been described. Comparison of the X-ray diffraction patterns from Y-form crystals of semisynthetic RNase S and those grown from the enzyme as isolated from a subtilisin digest show no observable differences.[36,37]

Residues 16–20 of S-peptide can be removed by digestion with carboxypeptidase A to provide a 15-residue peptide, RNase 1–15; the RNase S analog formed by the combination of this truncated peptide with S-protein was found to exhibit no detectable differences in activity or stability from that of the parent RNase S complex, that is, RNase 1–20:21–124.[38] The nonessentiality of residues 16–20 was consistent with their disordered state in the crystal.[37] Moreover, it was shown that at a resolution of 2.6 Å an electron density map for RNase 1–15:21–124 did not differ noticeably in "overall backbone structure and active site configuration" from that of RNase S.[39] Consequently, many semisynthetic analogs of RNase S have been prepared using synthetic peptides containing only the first 15-, 14-, or 13-residues of the S-peptide. Nevertheless, Niu et al. subsequently found the dissociation constant for the complex of RNase 1–15 with S-protein to be five times larger than that for S-peptide and S-protein.[40] They tentatively ascribed this difference to the possible presence of hydrogen bonding between Ser-16 and His-48. Also, quite recently, it was reported that electron density maps for RNase S, that is, RNase 1–20:21–124, are of a higher quality than those for RNase 1–15:21–124.[29]

5.2.1   STRUCTURAL SOURCES OF THE BINDING ENERGY BETWEEN S-PEPTIDE AND S-PROTEIN

Initial measurements of the strength of binding between S-peptide and S-protein were calculated from the enzymatic activity against RNA generated by the addition of successively higher levels of S-peptide to a constant amount of S-protein, and they indicated a dissociation constant approaching the nanomolar range.[21] Subsequent measurements made in the absence of substrate by ^{13}C NMR spectroscopy,[40] titration calorimetry,[41,42] UV difference spectroscopy,[43,46] or molar ellipticity change at 222 nm[47] have provided dissociation constants with values between 10^{-8} M and 7×10^{-5} M. The basis for the tremendous range of these values is not evident. Nonetheless, all parties are agreed that thermodynamically useful constants are obtained only in the absence of substrate.

5.2.1.1   The Role of Methionine-13

The conversion of Met-13 in S-peptide to its sulfone, its S-carboxymethyl, or its S-carboxamidomethyl derivative, greatly reduced the strength of binding between S-peptide and S-protein but did not affect the enzymatic activity of the resulting complexes.[48] This finding not only pinpointed one of the two residues in S-peptide responsible for the majority of the binding energy between it and the S-protein, it also provided the first experimental verification of the ideas of Kauzmann and others concerning the possible contribution of hydrophobic interactions to the stabilization of protein structures.[49]

Early studies of semisynthetic analogs in which the sulfur atom of Met-13 had been oxidized to the sulfoxide or sulfone level, or in which the residue was absent or had been replaced by Aba or Nle, confirmed its central role in the binding energetics of S-peptide and S-protein[46,50,51] (see Table 5.1).

A recent calorimetric study by Richards and Sturtevant and co-workers of analogs in which Met-13 was replaced by Gly, Ala, Aba, Val, Ile, Leu, Nle, and Phe has now provided a detailed thermodynamic analysis of the structural stabilization afforded by this residue[42,52] (Table 5.1). This thorough study was undertaken in

TABLE 5.1
Synthetic S-Peptide Analogs that Examine the Role of Met-13 in the Binding Between S-Peptide and S-Protein

Analog	K_D with Substrate	Ref.	K_D Without Substrate	Ref.
RNase 1–20 [Met-13 d,1-sulfoxide][a]	b	1	—	—
RNase 1–20[Met-13 d-sulfoxide]	b	1	2.1×10^{-4} M; c	2
RNase 1–14[Nle-13]	—	—	1.4×10^{-5} M; d	2
RNase 1–13[Met-13 sulfone]	e	1	—	—
RNase 1–13[Aba-13]	f	1	—	—
RNase 1–12-NH$_2$	g	1	—	—
RNase 1–20[Orn-10, Nle-13]	h	3	—	—
RNase 1–15-NH$_2$	—	—	1.1×10^{-7} M; i; 2.9×10^{-7} M; j	4
RNase 1–15-NH$_2$[Gly-13]	—	—	5.0×10^{-4} M; i	4
RNase 1–15-NH$_2$[Ala-13]	—	—	1.1×10^{-4} M; i	4
RNase 1–15-NH$_2$[Aba-13]	—	—	1.8×10^{-6} M; i; 4.2×10^{-6} M; j	4
RNase 1–15-NH$_2$[Val-13]	—	—	1.9×10^{-7} M; i; 2.7×10^{-7} M; j	4
RNase 1–15-NH$_2$[Ile-13]	—	—	1.5×10^{-7} M; i	4
RNase 1–15-NH$_2$[Leu-13]	—	—	3.1×10^{-7} M; i; 4.5×10^{-7} M; j	4
RNase 1–15-NH$_2$[Nle-13]	—	—	3.6×10^{-7} M; i	5
RNase 1–15-NH$_2$[Phe-13]	—	—	1.1×10^{-5} M; i; 2.1×10^{-5} M; j	4

[a]From naturally occurring S-peptide; all other peptides are synthetic.

[b]A mole ratio of analog to S-protein of 0.8 was required for 50% activation against RNA; S-protein concentration, 3.5–5.0 μM; pH 5.0, 25°.

[c]From UV difference spectra at pH 5.0, 25°; the corresponding value for unoxidized S-peptide was found to be 1.0×10^{-5} M.

[d]From UV difference spectra at pH 5.0, 25°; the corresponding value for unoxidized RNase 1–14 was found to be 1.3×10^{-5} M.

[e]A mole ratio of analog to S-protein of 32 was required for 50% activation against RNA; S-protein concentration, 3.5–5.0 μM; pH 5.0, 25°.

[f]A mole ratio of analog to S-protein of 9 was required for 50% activation against RNA; S-protein concentration, 3.5–5.0 μM; pH 5.0, 25°.

[g]A mole ratio of analog to S-protein of 88 was required for 50% activation against RNA; S-protein concentration, 3.5–5.0 μM; pH 5.0, 25°.

[h]A mole ratio of analog to S-protein of 10 provided 55% activation against RNA.

[i]By titration calorimetry at pH 6.0, 25°.

[j]By titration calorimetry at pH 5.0, 25°.

[1]Finn, F.M., and Hofmann, K. (1965). *J. Am. Chem. Soc.* 87, 645–651.

[2]Finn, F.M. (1972) *Biochemistry* 11, 1474–1478.

[3]Rocchi, R., Scatturin, A., Moroder, L., Marchiori, F., Tamburro, H.M., and Scoffone, E. (1969). *J. Am. Chem. Soc.* 91, 492–496

[4]Connelly, P.R., Varadarajan, R., Sturtevant, J.M., and Richards, F.M. (1990). *Biochemistry* 29, 6108–6114.

[5]Thompson, J., Ratnaparkhi, G.S., Varadarajan, R., Sturtevant, J.M., and Richards, F.M. (1994). *Biochemistry* 33, 8587–8593.

appreciation of the relative ideality of such a dissociating system for the quantitative determination of the factors involved in protein stabilization compared to the more commonly used systems that require denaturing conditions.[42] It included a determination of the refined high-resolution crystal structure of each of the analogs[52,53] as well as a determination of the heat capacity changes accompanying the formation of each of the complexes.[54] With the three smallest replacements, namely, Gly, Ala, and Aba, a water molecule is found in the cavity near residue 13, and a surface hydrophobic residue, Leu-51, has moved in each case in a direction that would tend to decrease the cavity volume. The only other major change is the movement of the 66–69 loop, which is about 25 Å from position 13. Interestingly, this loop moves 1.5 Å away from the rest of the molecule whether the replacement is smaller than methionine, namely, Gly, or larger, namely, Phe.

When compared to the enthalpy changes observed upon binding of the Met-containing "wild-type" peptide to S-protein, large decreases in ΔH values were observed for the analogs. The Nle analog, which has a side chain isosteric with that of Met, provided the basis to address the question whether these changes were due to differences in the shapes of the analog side chains or to the chemical difference between a sulfur atom and a methylene group.[52] The Nle side chain in the complex was found to exist in a single conformation that was coincident with that of the Met side chain in the "wild-type" complex. Moreover, the positions of the atoms in the cavity surrounding the Nle side chain were undisturbed. Nevertheless, the temperature factors (B-factors) for the Nle side chain atoms were significantly higher, while those for the surrounding atoms in the cavity were lowered. The authors concluded that the large $\Delta\Delta H$ (7.9 kcal mol^{-1} at 25°, pH 6.0) observed with this analog does not arise from differences in geometry with respect to the "wild-type" complex, but rather reflects differences in the chemical properties of the -S- and -CH$_2$- groups, differences that may be manifested in the overall dynamic behavior of the two RNase S complexes.

The heat capacity changes observed for the various mutants[52,54] did not correlate in any straightforward way with the nonpolar surface area that was buried upon binding, leading to the suggestion that there may be differences in the vibrational dynamics of the "wild-type" and mutant complexes, a phenomenon previously invoked in the interpretation of heat capacity changes in other processes involving proteins.[55] The thermodynamic and structural complexities revealed by the studies of Richards and Sturtevant and their co-workers will serve as a cautionary tale for protein engineers for the foreseeable future. Nevertheless, these studies have also confirmed earlier work that indicated that the interactions of the side chain of Met-13 with various residues in S-protein provide slightly over half of the energy of stabilization between S-peptide and S-protein.

The availability of high-resolution X-ray structures as well as experimental thermodynamic data for RNase S analogs in which Met-13 had been replaced by Leu, Ile or val[42, 52–54] allowed Simonson and Brünger to make a rigorous analysis of some current theoretical protocols for calculating the change in the free energy of binding to be expected from such structural modifications.[56] From this analysis, it appears that uncertainty about the degree to which all pertinent local energy minima are sampled by the simulations remains a formidable problem.

5.2.1.2   The Role of Phenylalanine-8

Although there is a relative paucity of experimental data about the role of Phe-8 in the generation of the binding energy between S-peptide and S-protein, it is clear that the interactions of this residue with S-protein constitute a major factor in the thermodynamics. The replacement of Phe-8 with Gly, Ala, Ile, or cyclopentylglycine (cpg) virtually destroys binding[47,57] (Table 5.2). It is sometimes stated that Phe-8

TABLE 5.2
Synthetic S-Peptide Analogs that Examine the Role of Phe-8 in the Binding between S-Peptide and S-Protein

Analog	K_D with Substrate	Ref.	K_D without Substrate	Ref.
RNase 1–20[Phe-8, Orn-10]	a	1	0.6×10^{-6} M; b	2, 3
RNase 1–20[Gly-8, Orn-10]	c	1	d	1
RNase 1–20[Ala-8, Orn-10]	—	—	e	2
RNase 1–20[Ile-8, Orn-10]	—	—	f	2
RNase 1–20[Tyr-8, Orn-10]	—	—	8×10^{-6} M; g	2
Guanidinated RNase 1–20 [Tyr-8, Orn-10]	—	—	h	3
RNase 1–20[Cpg-8, Orn-10]	b	3	i, j	1, 2
RNase 1–20[Cha-8, Orn-10]	—	—	4.3×10^{-6} M	3
RNase 1–20[Fphe-8, Orn-10]	—	—	0.19×10^{-6} M	3

[a]A maximal activity against RNA of 45% at a mole ratio of analog:S-protein of 9, pH 6.8, 25°.

[b]Complete complex formation at a mole ratio of analog: S-protein of 4 as detected by molar ellipticity change at 222 nm. S-protein concentration, 2.81 μM, pH 6.8, 25°.

[c]No activity against RNA, even at a mole ratio of analog:S-protein of 50:1.

[d]No detectable UV difference spectrum at a mole ratio of analog:S-protein of 20:1. S-protein concentration, 60–75 μM, pH 6.7, 20°.

[e]Nil binding at a mole ratio of analog:S-protein of 68 as detected by molar ellipticity change at 222 nm. S-protein concentration, 2.81 μM, pH 6.8, 25°.

[f]Barely detectable binding at a mole ratio of analog:S-protein of 60 as detected by molar ellipticity change at 222 nm. S-protein concentration, 2.81 μM, pH 6.8, 25°.

[g]Only 50% complex formation at a mole ratio of analog:S-protein of 20 as detected by molar ellipticity change at 222 nm. S-protein concentration, 2.81 μM, pH 6.8, 25°.

[h]At a mole ratio of analog:S-protein of 2, 50% complex formation as detected by molar ellipticity change at 222 nm. S-protein concentration, 2.81 μM, pH 6.8, 25°.

[i]At a mole ratio of analog:S-protein of 10, only 10% of the $\Delta\varepsilon_{287}$ seen with S-peptide was observed. S-protein concentration, 60–75 μM, pH 6.7, 20°.

[j]Less than 10% complex formation at a mole ratio of analog:S-protein of 35 as detected by molar ellipticity change at 222 nm. S-protein concentration, 2.81 μM, pH 6.8, 25°.

[1]Rocchi, G., Borin, G., Marchiori, F., Moroder, L., Peggion, E., Scoffone, E., Crescenzi, V., and Quadrifoglio, F. (1972). *Biochemistry* 11, 50–57.

[2]Filippi, B., Borin, G., and Marchiori, F. (1976). *J. Mol. Biol.* 106, 315–324.

[3]Borin, G., Filippi, B., Moroder, L., Santoni, C., and Marchiori, F. (1977). *J. Peptide Prot. Res.* 10, 27–38.

can only be replaced by another aromatic residue, but its replacement by cyclohexylalanine (cha) does not drastically reduce binding (Table 5.2), and the complex exhibits 21% of the activity of RNase A against RNA.[47]

The best retention of binding is found with replacements by aromatic residues, however. The 4-fluoro-phenylalanine analog actually binds better than Phe (0.19 µM versus 0.6 µM) and generates a complex that is fully active[47] (Table 5.2). This analog provides one of the few examples of a substitution that increases the strength of binding of S-peptide to S-protein (for others, see Section 5.2.1.3). The guanidinated RNase 1–20[Tyr-8, Orn-10] analog, where Orn-10 has been converted to Arg (and Lys-1 has been converted to homoarg), binds almost as well as the RNase 1–20[Phe-8, Orn-10] analog and provides a complex with 73% activity against RNA[47] (Table 5.2). (The pertinence of the Orn-10 to Arg-10 conversion is discussed in Section 5.2.1.3)

The binding of $1^\varepsilon,7^\varepsilon$-diguanidino-RNase 1–20[Tyr-8] to S-protein generates a UV difference spectrum characteristic of a buried tyrosine,[58] a finding consistent with the extremely hydrophobic environment found for Phe-8 in the crystal structure of RNase S.[29,37] Filippi et al. note that the difference spectrum is distinct from that seen with a buried tyrosine such as Tyr-25, where the phenolic hydroxyl can form hydrogen bonding interactions.[58]

There appears to be a synergistic effect in the contributions made by Phe-8 and Met-13 to the strength of binding between S-peptide and S-protein (Tables 5.1 and 5.2), and, indeed, both residues form part of the major hydrophobic nucleus in RNase.[59] The contribution made by Phe-8 may actually exceed that of Met-13: RNase 1–15[Gly-13] still binds to S-protein to a significant degree (Table 5.1), whereas neither RNase 1–20[Gly-8, Orn-10] nor RNase 1–20[Ala-8, Orn-10] is detectably bound (Table 5.2). Analysis of the major hydrophobic nucleus in RNase reveals that Phe-8 makes six nonpolar contacts while Met-13 makes only three.[59] Given the structural importance of Phe-8, a study of the type made by the Richards and Sturtevant laboratories for Met-13 seems warranted.

5.2.1.3   The Role of the Helical Propensity of the S-Peptide in Generating the Binding Energy between S-Peptide and S-Protein

That the source of the binding energy between S-peptide and S-protein may not reside exclusively or even sufficiently in the interactions of Phe-8 and Met-13 with the rest of the molecule is dramatically demonstrated by the finding that RNase 8–13 at a mole ratio of 2000 reactivates S-protein only to the extent of 4% and, at a mole ratio of 8000, only to the extent of 15% (50). It appears that additional structural features are required to achieve the strength of binding seen in RNase S.

In aqueous solution at room temperature, isolated S-peptide has little or no helical content (60–63), whereas, in RNase S, residues 3–13 are found to be in an a-helix.[37] Thus, a strong candidate for an important structural feature is the helical propensity of the S-peptide. A considerable number of analogs have been prepared and characterized that address the question of the relationship between the helical propensity of the S-peptide and the strength of binding between it and the S-protein.

The finding that guanidinated RNase 2–20[Orn-10] caused full activation of S-protein against RNA at a mole ratio of 1 demonstrated that Lys-1 is not necessary for binding or activity[64] (Table 5.3). Earlier measurements with RNase 2–13 had indicated this was the case[50] (Table 5.3), as had the discovery by Eaker et al. of a Des-Lys-1 RNase A in nature that had full activity.[65] In contrast, guanidinated RNase 3–20[Orn-10], which lacks Glu-2, caused a maximal activation of only 50%, and that was achieved only at a mole ratio of 7.[64] A mole ratio of RNase 3–13 to S-protein of 2000 was required to achieve 50% activation; additional shortening of the chain by the removal of Thr-3 further weakened binding only marginally[50] (Table 5.3). Eaker et al. also found reduced activity in a naturally-occurring Des-Lys-1, pyroglutamyl-2 RNase A.[65] On the basis of these findings, Hofmann et al. proposed that an ionic interaction between Glu-2 and Arg-10 stabilizes RNase S.[66] They also noted that the shorter side chain of Orn-10 probably could not interact in the same way with Glu-2, with the result that only upon the conversion of Orn-10 to an Arg by guanidination did RNase 1–20[Orn-10] and RNase 2–20[Orn-10] cause full activation of S-protein.[66] Small conformational differences not attributable to the Orn-Arg structural difference were seen by Valle and co-workers in the regions corresponding to the side chains of Glu-2 and Arg-10 in their difference Fourier analysis at 2.8 Å of crystals of RNase 1–20[Orn-10]:21–124 versus crystals of RNase S.[67] Subsequently, refined X-ray diffraction analysis of crystals of RNase A at 2.0 Å has confirmed that when residues 3–13 form an a-helix, the side chain oxygens of Glu-2 are optimally placed with respect to two of the side chain nitrogens of Arg-10 (Glu-2 OE1 to Arg-10 NH1, 2.87 Å; Glu-2 OE2 to Arg-10 NE, 2.75 Å; PDB file 5RSA).[68,69]

Dunn and Chaiken tested the idea that the helical propensity of individual residues in the 3–13 sequence of S-peptide were a factor in establishing binding energy by examining peptides in which Glu-9 had been replaced by Leu, a relatively good helix former, and by Gly, a poor helix former.[70] While the affinity of the Leu-9 analog for S-protein was decreased 3-fold, that of the Gly-9 analog was reduced 22-fold, as judged by activity measurements (Table 5.3). These decreases were not a reflection of a disturbed interface between S-peptide and S-protein as Glu-9 is on the solvent-exposed side of the helix.[37] At sufficiently high mole ratios of either analog, S-protein could be fully activated.

The replacement of Ala-6, a very good helix former, by Pro, a potential helix breaker, reduced affinity 27-seven fold with respect to S-peptide itself[64] (Table 5.3).

The degree of helicity observed in 97% trifluoroethanol for a series of Phe-8 analogs of RNase 1–20[Orn-10] agreed quite well with the ranking of helical propensity of amino acid side chains that had been made by Chou and Fasman,[71] but there was no correlation between this ranking and the relative binding affinities for S-protein.[57] Filippi et al. concluded that the hydrophobic character of the residue at position 8, as well as "its side-chain interactions with surrounding residues," were of paramount importance in establishing binding and that these overwhelmed any possible effects of helical propensity at this position.[57]

In a highly simplified S-peptide analog where 6 of the 15 residues had been replaced by Ala, Komoriya and Chaiken found the affinity for S-protein only an order of magnitude less than that of S-peptide itself (in the presence of substrate), with the resulting complex exhibiting 36% of the activity of RNase S against cytidine

TABLE 5.3

Synthetic S-Peptide Analogs that Examine the Role of Helical Propensity in the Binding between S-Peptide and S-Protein

Analog	K_D with Substrate	Ref.	K_D without Substrate	Ref.
RNase 2–13	a	1	-	-
RNase 3–13	b	1	-	-
RNase 4–13	c	1	-	-
Guanidinated RNase 2–20[Orn-10]	d	2	-	-
Guanidinated RNase 3–20[Orn-10]	e	2	-	-
RNase 1–15[Leu-9]	1.1×10^{-7} M; f; g	3	-	-
RNase 1–15[Gly-9]	7.7×10^{-7} M; f; g	3	-	-
Guanidinated RNase 1–20[Pro-6,Orn-10]	8×10^{-8} M; h	2	-	-
RNase 1–15[Ala-1, Ala-3, Ala-9, Ala-11, Ala-14, Ala-15]	1.1×10^{-6} M; i	4	-	-
RNase 1–15-NH$_2$[Leu-9, Glu-11]	-	-	2.0×10^{-7} M; j	5
RNase 1–15-NH$_2$[Ala-1, Leu-9, Glu-11]	-	-	1.6×10^{-7} M; j	5
N$^\alpha$-acetyl-RNase 1–15-NH$_2$ [Ala-1, Leu-9, Glu-11]	-	-	1.5×10^{-7} M; j	5
N$^\alpha$-succinyl-RNase 1–15-NH$_2$ [Ala-1, Leu-9, Glu-11]	-	-	1.0×10^{-7} M; j	5
RNase 1–15-NH$_2$[Leu-9]	-	-	4.6×10^{-7} M; j	5
N$^\alpha$–succinyl-RNase 1–15-NH$_2$[Ala-1, Leu-9]	-	-	2.5×10^{-7} M; j	5

[a]A mole ratio of analog:S-protein of 41 provided 50% activation of S-protein against RNA. Under the same conditions, a mole ratio of RNase 1–15:S-protein of 3 provided 50% activation.

[b]A mole ratio of analog:S-protein of 2000 provided 50% activation of S-protein against RNA.

[c]A mole ratio of analog:S-protein of 3500 provided 50% activation of S-protein against RNA.

[d]The RNase 2–20[Orn-10] analog was guanidinated to provide an Arg residue at position 10. Full activation of S-protein against RNA, which occurred at a mole ratio of 1, was indistinguishable from that afforded by guanidinated RNase 1–20[Orn-10].

[e]The RNase 3–20[Orn-10] analog was guanidinated to provide an Arg residue at position 10. Maximal activation of S-protein against RNA was only 50% of that found with guanidinated 2-20[Orn-10] and was achieved only at a mole ratio of 7.

[f]Compared to a value of 0.36×10^{-7} M for S-peptide under the same conditions.

[g]The complex of the analog and S-protein was fully active against cytidine 2', 3'-cyclic phosphate.

[h]Compared to a value of 0.3×10^{-8} M for S-peptide under the same conditions. Maximal activation of S-protein against RNA was only 50% of that found with S-peptide.

[i]Compared to a value of 9×10^{-8} M for S-peptide under the same conditions. At a mole ratio of analog:S-protein of 12, a maximal activation of 36% was obtained.

[j]Compared to a value of 1.0×10^{-6} M for RNase 1–19 measured under the same conditions from the increase in ellipticity at 222 nm as a function of an increasing ratio of RNase 1–19 or analog to S-protein; conditions: 0.1 M NaCl, 1 mM phosphate, 1 mM borate, 1 mM citrate, pH 5.3, 35°.

[1]Finn, F.M., and Hofmann, K. (1965). *J. Am. Chem. Soc.* 87, 645-651.

[2]Marchiori, F., Borin, G., Moroder, L., Rocchi, R., and Scoffone, E. (1972) *Biochim. Biophys. Acta* 257, 210-221.

[3]Dunn, B.M., and Chaiken, I.M. (1975). *J. Mol. Biol.* 95, 497-511.

[4]Komoriya, A., and Chaiken, I.M. (1982). *J. Biol. Chem.* 257, 2599-2604.

[5]Mitchinson, C., and Baldwin, R.L. (1986). *Proteins* 1, 23-33.

TABLE 5.4
Synthetic S-Peptide Analogs that Examine the Role of His-12 in the Activity of RNase S

Analog	Activity	Ref.	K_D in the Absence of Substrate	Ref.	K_i or K_D in the Presence of Substrate	Ref.
RNase 1–12[3-pyrazolyl-Ala-12]	a	1	—	—	—	—
RNase 1–12[1-pyrazolyl-Ala-12]	b	1	—	—	—	—
RNase 1–14[3-pyrazolyl-Ala-12]	c	2, 3	19.5 µM; d	4	e	2
RNase 1–14[3-CM-His-12]	f	3	10.6 µM; d	4	g	3
RNase 1–14[Ser-12]	h	5	17.1 µM; d	4	i	5
RNase 1–15[Fhis-12]	j	6	—	—	4.5×10^{-8} M; k	6
RNase 1–14[β(pyrid-3-yl)-Ala-12]	l	7,8	—	—	7×10^{-9} M; m	7
RNase 1–14[β(pyrid-2-yl)-Ala-12]	n	8	—	—	—	—
RNase 1–14[3-methyl-His-12]	o	7	—	—	2×10^{-8} M; m	7
Guanidinated RNase 1–20[Orn-10, Orn-12]	p	9	—	—	q	9
RNase 1–14[1-CM-His-12]	r	3	—	—	—	—
RNase 1–14[1-methyl-His-12]	s	7	—	—	2×10^{-5} M; m	7
RNase 1–14[β(pyrid-4-yl)-Ala-12]	t	8	—	—	2.6×10^{-8} M; u	10
RNase 1–14[D,L-4-imidazolyl-Gly-12]	v	7	—	—	2×10^{-4} M; m	7
RNase 1–14[homohis-12]	v	7	—	—	2×10^{-7} M; m	7

[a]No activity detectable at mole ratios of analog:S-protein as high as 1500.

[b]No activity detectable at mole ratios of analog:S-protein as high as 5500.

[c]No activity detectable at mole ratios of analog:S-protein as high as 1000.

[d]From UV difference spectra; the corresponding value for S-peptide = 9.6 µM and for RNase 1–14 = 13.3 µM.

[e]Able to inhibit RNase S 50% at a mole ratio of analog:S-peptide of 1.

[f]No activity detectable at mole ratios of analog:S-protein as high as 1400.

[g]Able to inhibit RNase S 50% at a mole ratio of analog:S-peptide of 0.8.

[h]"...fails to activate S-protein..." (Ref. 5).

[i]Able to inhibit RNase S 50% at a mole ratio of analog:S-peptide of 8.

[j]No activity detectable at a mole ratio of analog:S-protein of 10.

[k]From the inhibition of hydrolysis of cytidine 2', 3'-cyclic phosphate by RNase S at pH 7.13. The corresponding dissociation constant for S-peptide under these conditions is 1.8×10^{-8} M.

[l]Not detectable. Mole ratio of analog:S-protein not given.

[m]From the inhibition of RNase S against RNA at pH 5.0. The corresponding dissociation constant for S-peptide under these conditions is 2×10^{-9} M.

[n]No detectable activity against RNA, cytidylyl 3',5'-cytidine or cytidine 2',3'-cyclic phosphate. Activity of 0.2% against 5'-phosphouridine 2',3'-cyclic phosphate.

[o]Less than 1% activity against RNA at pH 5.0. Mole ratio of analog:S-protein not given.

[p]Failed to activate S-protein against RNA, cytidine 2',3'-cyclic phosphate or cytidylyl 3', 5'-cytidine.

[q]No inhibition of RNase S activity at an analog:S-peptide ratio of 500.

[r]At an analog:S-protein ratio of 1000, a 60% activation against RNA was achieved.

[s]A maximum of 20% activation against RNA at pH 5.0. Mole ratio of analog:S-protein not given.

[t]Activities of 34% against RNA, 4% against cytidine 2',3'-cyclic phosphate, and 11% against 5'-phospho-uridine 2',3'-cyclic phosphate.

[u]From the activation of S-protein against RNA at pH 5.0. The corresponding dissociation constant for S-peptide under these conditions is 2×10^{-9} M.

[v]A maximum of 80% activity against RNA at pH 5.0. Mole ratio of analog:S-protein not given.

[1]Hofmann, K., and Bohn, H. (1966). *J. Am. Chem. Soc.* 88, 5914–5919.

[2]Finn, F.M., and Hofmann, K. (1967). *J. Am. Chem. Soc.* 89, 5298–5300.

[3]Hofmann, K., Visser, J.P., and Finn, F.M. (1970). *J. Am. Chem. Soc.* 92, 2900–2909.

[4]Finn, F.M. (1972). *Biochemistry* 11, 1474–1478.

[5]Hofmann, K., Andreatta, R., Finn, F.M., Montibeller, J., Porcelli, G., and Quattrone, A.J. (1971). *Bioorganic Chem.* 1, 66–83.

[6]Dunn, B.M., DiBello, C., Kirk, K.L., Cohen, L.A., and Chaiken, I.M. (1974). *J. Biol. Chem.* 249, 6295–6301.

[7]vanBatenburg, O.D., Voskuyl-Holtkamp, I., Schattenkerk, C., Hoes, K., Kerling, K.E.T., and Havinga, E. (1977). *Biochem. J.* 163, 385–387.

[8]Hoes, C., Kerling, K.E.T., and Havinga, E. (1983). *Recl. Trav. Chim. Pays-Bas* 102, 140–147.

[9]Borin, G., Toniolo, C., Moroder, L., Marchiori, F., Rocchi, R., and Scoffone, R. (1972). *Int'l J. Peptide Prot. Res.* 4, 37–45.

[10]Hoes, C., Hoogerhout, P., Bloemhoff, W., and Kerling, K.E.T. (1979). *Recl. Trav. Chim. Pays-Bas* 98, 137–139.

2',3'-cyclic phosphate.[72] From CD measurements in water and trifluoroethanol, the helical content of the analog and S-peptide were judged to be comparable. A 3.0-Å electron density map of this complex revealed that Met-13 was essentially unchanged in its position; the side chain of Phe-8 had rotated almost 90° around its C_α-C_β bond, but was still deeply buried and well-packed against hydrophobic surfaces.[73] The side chain of Glu-2 now pointed away from the side chain of Arg-10, however, rendering any interaction between these two residues highly unlikely. Thus, the structural basis for the ten-fold lower binding affinity observed with this analog may have been pinpointed. In addition, active site histidines 12 and 119 were found to be approximately 1 Å farther apart in this complex than in RNase S, and the authors suggest this may be the basis for the reduced enzymatic activity.[73] Another possibility is a shift in the pH optimum occasioned by the removal of three charged residues from the peptide.[74] The activity of the complex against cytidine 2',3'-cyclic phosphate was determined at one pH value only (a practice that did not make this study in any way exceptional).

Mitchinson and Baldwin, who reasoned that substitutions which stabilized the helix dipole of residues 3–13 should increase the helical propensity of S-peptide analogs, prepared and characterized six 15-residue structures in which the negative charge was increased at the N-terminus and decreased at the C-terminus[75] (Table 5.3). The helical content of the isolated peptides was determined from measurements of ellipticity at 222 nm at pH 5.3 and 3°, and binding constants for complex formation with S-protein were calculated from the increase in ellipticity at 222 nm as a function of an increasing ratio of analog to S-protein at pH 5.3, 35°. For five of the six peptides, the higher the helical content of the isolated peptide at 3°, the higher was the association constant with S-protein at 35° and the higher was the thermostability of the resulting semisynthetic RNase S. The best helix former, N^α-succinyl-RNase 1–15-NH$_2$[Ala-1, Leu-9, Glu-11], had a helix content of 63% at pH 5.3 and 3°, conditions where RNase 1–19 was found to be only 15% helical. Its binding constant with S-protein was 10-fold greater than that of RNase 1–19, and it generated a semisynthetic RNase S with a thermostability 6.0° higher than that formed with RNase 1–19 and S-protein. The ability of the two analogs containing Gln-11 (rather than the inactivating Glu replacement) to generate full activity upon combination with S-protein was taken as evidence that the peptides assumed a conformation similar to that of RNase 1–19 or S-peptide upon combination with S-protein. Unfortunately, X-ray structures for these analogs are not available, as the set of peptides examined by Mitchinson and Baldwin represents a unique instance of being able to enhance the properties of a semisynthetic RNase using coded amino acids as replacements.

From the experiments discussed above, we see that three distinct factors that influence the helical propensity of S-peptide may be discerned: the ionic interaction between Glu-2 and Arg-10 when residues 3–13 are in an α-helical conformation, the high intrinsic helical propensity of several of the residues in the 3–13 sequence (also, the absence of helical-disrupting residues), and the charge stabilization of the 3–13 helical dipole. Increased helical propensity was correlated in every instance with increased binding to S-protein. Much has been learned here that is of interest to the protein engineer.

In his study of the acid-induced dissociation of RNase S, Labhardt was unable to distinguish whether dissociation of the complex occurred before or after unfolding of the S-peptide helix.[76] If the helical propensity of the S-peptide does make a *kinetic* contribution to the formation of the complex, a more thorough study of this aspect of the binding between S-peptide and S-protein seems warranted, as it would appear to be a good model for the study of protein folding. Cohen and Chaiken and co-workers have demonstrated that shifts in the ^{13}C NMR spectra of isotopically enriched residues in RNase 1–15 allow monitoring of both α-helix formation and complex-ation with S-protein.[40,77–80]

5.2.1.4 The Role of Aspartic Acid-14

Potts et al. found that the removal of residues 16–20 from S-peptide by digestion with carboxypeptidase A did not noticeably diminish its strength of binding to S-protein, at least in the presence of substrate, or its ability to form a completely active complex.[38] On the other hand, Hofmann et al. found that synthetic RNase 1–13 exhibited lowered affinity toward S-protein (also measured in the presence of sub-strate) and provided a complex with a maximal activity only 80% of that of RNase S (22). Thus, a role for the asp[14]-ser[15] sequence in the binding between S-peptide and S-protein and possibly also in the generation of enzymatic activity in the resulting complex was indicated.

The situation was clarified further by a comparison of the properties of three additional semisynthetic analogs. Hofmann and co-workers found that preparations of synthetic RNase 1–15 and RNase 1–14 (in which the sulfoxide forms of Met-13 had been completely reduced) bound to S-protein in the presence of RNA as tightly as did S-peptide and formed complexes with full enzymatic activity (81). It was concluded that Asp-14, but not Ser-15, plays a significant role in binding, and possibly also in activation. The finding that an Asn-14 analog, namely, guanidinated RNase 1–20[Orn-10, Asn-14], generated a RNase S which was 100% active against RNA indicated that Asp-14 does not play a direct role in the activity of the enzyme.[82] The requirement for a mole ratio of analog to S-protein of 20 to achieve full activity confirmed the involvement of Asp-14 in generating binding energy between the two chains, however. The lowered affinity of the Asn analog was also found in the absence of substrate from ellipticity measurements at 222 nm and pH 6.8 of solutions containing a progressively increasing ratio of guanidinated RNase 1–20[Orn-10, Asn-14] to S-protein. A K_D of 1.7×10^{-6} M could be calculated from these mea-surements. The authors did not provide corresponding K_D values for S-peptide or guanidinated RNase 1–20[Orn-10, Asp-14] under comparable conditions, but did state that "A sensibly lower S-protein affinity derives from substitution of the side-chain carboxyl function in position 14 with a carboxamide one."[82]

The addition of S-peptide to S-protein at neutral pH values generates a difference spectrum in the near UV that is believed to reflect the perturbation of both phenylalanyl and tyrosyl residues upon complex formation.[83] From structural studies with RNase A, we now appreciate that one of these interactions involves the formation of a charge-stabilized hydrogen bond between Asp-14 and the phenolic hydroxyl of Tyr-25 (distance from OD2 of Asp-14 to the OH oxygen of Tyr-25 =

2.67 Å; PDB file 5RSA).[68,69] When Filippi et al. compared the UV difference spectra generated upon the addition to S-protein of guanidinated S-peptide or of guanidinated RNase 1–20[Orn-10, Asn-14], they saw differences which they attributed to the weaker, uncharged hydrogen bond formed between Tyr-25 and Asn-14.[82] These workers also noticed that the value of K_D for guanidinated RNase 1–20[Orn-10, Asn-14] was pH-dependent over the pH range of 5.0 to 7.0, becoming weaker at lower pH values. From a measurement of the quantitative dependence of K_D upon pH, an apparent pK for this phenomenon of 6.0 ± 0.2 was determined. They suggested that, upon protonation, the imidazole ring of nearby His-48 might then effectively compete with Asn-14 in forming a hydrogen bond with the phenolic hydroxyl of Tyr-25. Subsequently, however, Niu et al., using an RNase 1–15 preparation with a $^{13}C^\gamma$-enriched Asp-14, established a pK value of 2.4 for the β-carboxyl group of this residue and concluded that Asp-14 must function in the complex as a hydrogen bond acceptor, not a donor.[80] Therefore, a direct competition with the protonated ring of His-48, which can only function as a hydrogen donor, seems to be ruled out.

5.2.2 ACTIVE SITE RESIDUES IN S-PEPTIDE

5.2.2.1 Histidine-12

The observation that S-protein alone exhibited no activity toward the substrates of RNase indicated that elements of the active site were contained in the S-peptide portion of the molecule.[84] Soon after, and prior to the preparation and characterization of any semisynthetic analogs of RNase S, carboxymethylation studies of RNase A by Moore and Stein and co-workers[25,26,85,86] and Barnard and Stein and co-workers[87-90] pinpointed His-12 as a critical element of the active site.

The synthesis and characterization of a semisynthetic RNase S where the S-peptide was replaced by a synthetic RNase 1–14 in which 3-pyrazolyl-Ala had been substituted for His-12 provided a notable test of this hypothesis[91,92] (Table 5.4). That the 3-pyrazolyl-Ala-containing tetradecapeptide bound to S-protein as tightly as S-peptide was evidenced by its ability to cause a 50% inhibition of the activity of RNase S against RNA at a mole ratio of analog:S-peptide of 1. Nevertheless, no enzymatic activity was detected, even at a mole ratio of analog to S-protein of 1000.[91,92] Hoffman appreciated the probable widespread usefulness of pyrazolylalanines for synthetic structure-function studies involving histidine residues and of the potential they presented for the elaboration of powerful and specific synthetic peptide inhibitors of many systems.[93] Here is his description of the virtues of these structures: "Among conceivable histidine substitutes the β-pyrazolylalanines command considerable interest. Pyrazole, like imidazole, is a five-membered planar aromatic ring system which contains two nitrogens. Its molecular dimensions are very similar, if not identical, with those of imidazole, and the geometry of histidine and the pyrazolylalanines is the same. However, the different spacing of the nitrogens endows the two ring systems with remarkably different acid-base properties. The pK of the imidazole portion of histidine is 5.97 in contrast to the pKs of the pyrazole portion of Pyr(1)Ala and Pyr(3)Ala which are 2.2 and approximately 2.1, respectively."[93]

Hofmann and co-workers also prepared RNase 1–14 analogs in which His-12 was replaced with 3-CM-His (also denoted N^{tele}-CM-His or Nε2-CM-His) or with Ser (Table 5.4).[92] The carboxymethylated analog bound very well to S-protein, even slightly better than S-peptide itself, but no detectable activity ensued, even at a mole ratio of analog to S-protein of 1400.[92] The Ser-12-containing analog did not bind quite so tightly: a mole ratio of this analog to S-peptide of 8 was required to cause 50% inhibition of RNase S against RNA. The relatively high degree of binding by this analog, which itself could generate no enzymatic activity upon combination with S-protein,[66] indicated that the aromatic ring of His-12 contributes little to the binding energy between S-peptide and S-protein. This finding was in marked contrast to the conclusion of Borin et al., who found that guanidinated RNase 1–20[Orn-10, Orn-12] failed to generate enzymatic activity upon combination with S-protein, and also failed to inhibit the activity of RNase S at mole ratios of analog to S-peptide as high as 500.[94] From these data, they concluded that His-12 made a significant contribution to the binding energy between S-peptide and S-protein. The replacement of His-12 with a bulkier, highly charged Arg residue does not appear to have been the best choice to test this hypothesis, however.

Four additional His-12 analogs which bound well to S-protein, but generated little or no activity, were prepared using β-(pyrid-3-yl)-Ala,[95,96] β-(pyrid-2-yl)-Ala,[96] 3-methyl-His (also denoted N^{tele}-methyl-His or Nε2-methyl-His),[84] or 4-fluoro-His.[97] In all cases, inhibition studies indicated a level of binding to S-protein only slightly lower than that seen with S-peptide itself (Table 5.4).

The complete inactivity of the first analog is a bit puzzling. The pK value of the pyridine moiety should allow it to function both as an acid and a base during catalysis over the pH range of 5–7,[4,5] while the substantial activity exhibited by a number of modified histidine structures (*vide infra*) suggests that a certain amount of structural play can be accommodated at position 12. Perhaps the environment stabilizes a conformation of the ring in which the single nitrogen is pointed away from the active site: if such a conformation is modeled into position 12 in RNase A (PDB file 5RSA), the ring nitrogen is positioned 2.6 Å from Oδ1 of Asn-44 (P.D. Martin, unpublished). Hydrogen bond formation between these two atoms would require that the nitrogen be protonated.

The failure of the β-(pyrid-2-yl)-Ala analog to generate a complex exhibiting detectable enzymatic activity (except perhaps a very low level (0.2%) with the unusual substrate, **pU > p**) is more understandable as neither orientation of the ring would place the nitrogen atom coincident with the Nε2 nitrogen of imidazole. Nonetheless, the equally noncoincident positioning of the nitrogen atom in the β-(pyrid-4-yl)-Ala analog allowed the attainment of significant levels of activity toward both RNA and **pU > p** (*vide infra*).[96]

The very low ("less than 1%") or perhaps nil activity of the complex of RNase 1–14[3-methyl-His-12] with S-protein is consistent with the results of the alkyation studies with RNase A[25,26,85–90] and with the properties observed for the complex of RNase 1–14[3-CM-His-12] and S-protein:[92] enzymatic activity is lost if the Nε2 nitrogen of His-12 is substituted.

The inactivity of the 4-fluoro-His analog was also anticipated, as the pK of the fluorinated ring is only 2.5, thereby precluding the participation of the acid form of this structure during catalysis over the normal pH range.[97] Nevertheless, the complex of this analog with S-protein binds a variety of active site ligands, including cytidine 2',3'-cyclic phosphate and uridylyl-3',5'-adenosine.[98,99] Moreover, it was possible to grow diffraction-quality crystals of the complex. These crystals were in the same space group and had essentially the same unit cell dimensions as crystals of the complex between RNase 1–15 and S-protein,[100] or of that between synthetic RNase 1–20 and S-protein,[36] or of RNase S itself.[37] A comparison at 2.6-Å resolution of the electron density maps from crystals of RNase 1–15:21–124 and RNase 1–15[Fhis-12]:21–124 revealed there were no significant conformational differences between the two structures.[39] It is unfortunate that no further structural studies of the complex of this analog in the presence of good substrates have been forthcoming.

Substitution of the N–1 (N^{pros} or Nδ1) nitrogen of His-12 with either a carboxymethyl[92] or a methyl group[95] is tolerated moderately well, however (Table 5.4). In both cases, the dissociation constant between the analog and S-protein has increased at least two orders of magnitude; nevertheless, the resulting complexes exhibit very substantial activity against RNA (60% in the former case and 20% in the latter case). Part of these activity changes may reflect shifts in pH-activity profiles as the pK values of these substituted histidines are higher than that of histidine itself (pK of the imidazole in His = 6.13; in 1-carboxymethyl-His = 6.33; in 1-methyl-His = 6.77).[101] Aside from steric and electrostatic considerations, the drastic decrease in binding upon substitution of the Nδ1 nitrogen of His-12 almost certainly reflects the loss of the strong hydrogen bond between this atom and the carbonyl oxygen of Thr-45 that is seen in both RNase S[102] and RNase A.[68]

Perhaps the most surprising finding reported for semisynthetic His-12 analogs is the very high enzymatic activity exhibited by S-protein complexes with 14-residue peptides containing β-(pyrid-4-yl)-Ala, D,L-4-imidazolyl-Gly or homohis at position 12[95,96,103,104] (Table 5.4). In the first case, the imidazole ring has been replaced by a 4-pyridyl moiety; in the second case, the methylene group in the histidine side chain has been removed, and, in the third case, a second methylene group has been added to the side chain. Given our general understanding of the precision with which atoms must be positioned at enzyme active sites, none of these structures would have been expected to generate significant activity.

The complex formed between S-protein and the 14-residue peptide containing β-(pyrid-4-yl)-Ala at position 12 had 34% activity against RNA at pH 5.0 and hydrolytic activities against cytidine 2',3'-cyclic phosphate and 5'-phospho-uridine 2',3'-cyclic phosphate (pU > p) of 4% and 11%, respectively, at pH 6.0.[96] Interestingly, transphosphorylation of the small substrate cytidylyl-3',5'-cytidine (CpC) was drastically reduced to 0.4% of that catalyzed by RNase S. Some insight into the basis for this low activity was afforded by the measurement of the K_M and k_{cat} values for the hydrolysis of pU > p by this complex. With respect to RNase S, the value of k_{cat} had fallen only four-fold (to 0.18 sec^{-1}), but the value of K_M had increased 41-fold (to 4.5 mM). On the basis of these results, the authors noted that the occurrence of a comparable collapse in substrate binding in the transphosphorylation step could explain the great difference in activity toward RNA and CpC. They also speculated that His-12 might have a structural role in orienting substrate, presumably either in addition to or instead of, playing a role as an acid/base in the catalytic steps.

The strength of binding to S-protein of the peptide containing D,L-4-imidazolyl-Gly-12 was drastically reduced (K_D in the presence of substrate = 2×10^{-4} M versus 2×10^{-9} M for S-peptide), but the complex, when formed, exhibited 80% activity against RNA.[95,103] (The chiral α-carbon in 4-imidazolyl-Gly spontaneously racemizes within several hours under acid as well as alkaline conditions, hence, the use of the racemate.) The observed activity is all the more remarkable when it is appreciated that the pK of the imidazole ring in this amino acid is 4.6.[105]

The binding energy of the homohis-12 analog was much greater (K_D in the presence of substrate = 2×10^{-7} M) and its complex with S-protein also exhibited 80% activity against RNA.[104] A considerably greater deterioration in catalytic efficiency was observed when the homohis-12-containing analog was tested against cytidine 2',3'-cyclic phosphate: the combined increase in K_M and decrease in k_{cat} resulted in a 30-fold decrease in k_{cat}/K_M.[104] In general, hydrolytic activity with small substrates declines to a greater extent than does depolymerase activity toward RNA.[4,5] In fact, only two examples of semisynthetic RNases are known where the loss of depolymerase activity exceeds the loss in hydrolytic activity: the complex of RNase 1–20[Orn-10, Lys-11] with S-protein exhibits 24% activity against RNA and 35% activity against cytidine 2',3'-cyclic phosphate at pH 6.0[106] while the complex of RNase 111–124[Tyr-118] with RNase 1–118 is 3% active against RNA and 24% active against cytidine 2',3'-cyclic phosphate at pH 6.0 (M.L. Ram and M.S. Doscher, unpublished).

Given the implications of the properties of the complexes of RNase 1–14[β-(pyrid-4-yl)-Ala-12], RNase 1–14[D,L-4-pyrazolyl-Gly-12], and RNase 1–14[homohis-12] for a increased understanding of the mechanism of action of RNase, as well as the potential usefulness of these amino acids, particularly the last, in the fields of semisynthetic and synthetic protein engineering, it is unfortunate that further characterization of these materials has not been carried out.

5.2.2.2   Glutamine-11

Glutamine-11 is an invariant residue in the pancreatic RNases from over 40 mammalian species (106) and it is positioned so intimately with respect to the active site of the enzyme[67,68] that many consider it to be an active site residue. Investigations with semisynthetic analogs containing Leu, Glu, or Lys at position 11 have indeed demonstrated that deleterious effects ensue if Gln-11 is replaced with any of these three residues[50,107–110] (Table 5.5).

At very high mole ratios of peptide, the complex of RNase 1–13[Glu-11] with S-protein was found to have a maximum of 26% of the activity of RNase S against RNA[50] (Table 5.5). Thus, the introduction of a negative charge at this position lowers both the binding energy between the peptide and S-protein and the maximal depolymerase activity of the resulting complex. These findings were confirmed by Marchiori et al., who utilized guanidinated RNase 1–20[Orn-10, Glu-11] in their studies.[107,108] The maximal activation achievable against RNA was 33%, but the affinity of this latter peptide for S-protein appeared to be only marginally reduced. With cytidine 2,'3'-cyclic phosphate as substrate, the complex was only 5% active and this degree of activation required a mole ratio of peptide to S-protein of 15–20[107] (Table 5.5).

TABLE 5.5
Synthetic S-Peptide Analogs that Examine the Role of Gln-11 in the Activity of RNase S

Analog	Activity	Ref.	K_D in the Presence of Substrate
RNase 1–13[Glu-11]	26%; a	1	b
Guanidinated RNase 1–20[Orn-10, Glu-11]	33%; c; 5%; d	2	e
Guanidinated RNase 1–20[Orn-10, Lys-11]	18%; c; 12%; d	2	f
Guanidinated RNase 1–20[Orn-10, Leu-11]	56%; c; 14%; d	2	g

[a]Against RNA at pH 5.0, 25°.

[b]The activity of 26% was achieved at a mole ratio of analog: S-protein of 700. At a mole ratio of analog:S-protein of 100, the activity was 10%.

[c]Against RNA at pH 6.0, 25°.

[d]Against cytidine 2',3'-cyclic phosphate at pH 6.0, 25°.

[e]With RNA, a mole ratio of analog:S-protein of <5 gave maximal activity; with cytidine 2',3'-cyclic phosphate, a mole ratio of 15–20 was necessary.

[f]A mole ratio of analog:S-protein of <2 was adequate to achieve maximal activity with both RNA and cytidine 2',3'-cyclic phosphate.

[g]A mole ratio of analog:S-protein of <2 was adequate to achieve maximal activity with both RNA and cytidine 2',3'-cyclic phosphate.

[1]Finn, F.M., and Hofmann, K. (1965). *J. Am. Chem. Soc.* 87, 645–651.

[2]Marchiori, F., Borin, G., Moroder, L., Rocchi, R., and Scoffone, E. (1974). *Intl. J. Peptide Prot. Res.* 6, 337–345.

The guanidinated RNase 1–20[Orn-10, Lys-11] analog (where Lys-11 has been converted to a homoarg residue) bound very well to S-protein, whether this binding was judged by the generation of activity against RNA or cytidine 2',3'-cyclic phosphate.[107] The respective maximal activities were only 18% and 12%, however (Table 5.5). The pH-activity profile against cytidine 2',3'-cyclic phosphate of the complex was also determined, an undertaking as laudable as it is rare for studies of semisynthetic RNases, and it was found that the midpoint of the ascending limb had shifted upward from pH 5.6 to pH 5.9, while that for the descending limb had shifted from pH 7.1 to pH 7.3 (Figure 2 in Ref. 107). The replacement of a neutral side chain by a positively charged one will depress the pK values of nearby dissociating residues, as it destabilizes the positively charged (or neutral) protonated species relative to the neutral (or negatively charged) basic forms. If the ascending limb of the bell-shaped RNase pH-activity profile reflects the deprotonation of a histidine residue and the descending limb the deprotonation of a second such residue,[4] *downward* shifts of both limbs should occur upon the introduction of a positively charged residue. The observed pK shifts for the Lys-11 analog are in the direction opposite to those which would be predicted by this simple scheme, however. Interestingly, an earlier determination of the pH-activity profile against RNA for the complex of RNase 1–20[Orn-10, Glu-11] (not guanidinated) had also revealed paradoxical shifts, in this case, downward upon the introduction of a negatively charged residue (Figure 2 in Ref. 109).

The replacement of Gln-11 by the neutral but hydrophobic leucine did not reduce the binding energy between the peptide and S-protein to the extent seen in the case of the glutamate replacement, and the resulting complex exhibited the highest activity of the three analogs[107] (Table 5.5).

Somewhat later, Filippi and co-workers examined the binding to S-protein in the absence of substrate of RNase 1–20[Orn-10, Glu-11], RNase 1–20[Orn-10, Lys-11], and RNase 1–20[Orn-10, Leu-11], peptides which had not been guanidinated and therefore lacked the stabilization of the Glu-2, Arg-10 interaction when in an α-helical conformation *(vide supra).*[110] From the increase in ellipticity at 222 nm that occurs upon complex formation, they found the Glu-11 analog bound slightly better than the Gln-11 structure (which, nevertheless, does not bind as well as its guanidinated derivative), the Leu-11 had an intermediate affinity and the Lys-11 analog bound least well. This complete reversal of the relative affinities observed previously in the presence of substrate illustrates dramatically the potential pitfalls associated with the measurement of apparent dissociation constants from the generation of enzymatic activity, perhaps particularly when highly charged substrates and products are used to evaluate a series of analogs in which the net charge at the active site is being varied.

From their thorough ^{15}N-NMR study of the complex between S-protein and a RNase 1–15 containing ^{15}N enrichment at the epsilon amino group of Lys-7, the amide nitrogen of Gln-11, and the two imidazole nitrogen atoms of His-12, Knoblauch et al.[111] concluded that neither Gln-11 nor Lys-7 is directly involved in the interaction between the peptide and the protein.

The invariant nature of Gln-11 and the results of the above experiments are consistent with a pertinent role for this residue in the mechanism of action of the enzyme, but its exact nature remains obscure. Perhaps Gln-11 functions as part of the scaffolding that positions and steadies other active site residues that participate more directly in the binding of substrate and its conversion to product. Alternatively, del Cardayré et al. have concluded, from their recent study of RNase A mutants containing Ala, Asn or His at position 11, that Gln-11 makes a negligible contribution to stabilizing the transition state, but instead functions to prevent non-productive binding of substrate.[14]

5.2.2.3 Lysine-7

Lysine-7 is another invariant residue in the pancreatic RNases from over 40 mammalian species.[106] A functional role was suggested for this residue when the complex between RNase 1–20[Orn-10, Ala-7] and S-protein was found to exhibit only 18% activity against RNA and 8% activity against cytidine 2',3'-cyclic phosphate.[112] Binding by this peptide was somewhat reduced: maximal activity was achieved at mole ratios of peptide to S-protein of 10 and 5, respectively. In the absence of substrate, binding as measured by the increase in molar ellipticity at 222 nm was equal to that of RNase 1–20[Orn-10] (0.6 and 0.5 mM, respectively), and upon guanidination, activities of the complex with S-protein rose to 30% for RNA and

20% for cytidine 2',3'-cyclic phosphate.[112] The null effect of the Ala-7 replacement upon binding between the peptide and S-protein is in agreement with the results of the ^{15}N-NMR study of Knoblauch et al.[111]

The determination of the kinetic parameters for complexes of both the Ala-7 analog and its guanidinated derivative against cytidine 2',3'-cyclic phosphate at pH 6.0 revealed that the diminished activity reflected principally a decrease in k_{cat} (0.31 sec^{-1} and 1.0 sec^{-1}, respectively, vs. 3.0 sec^{-1} for RNase S) with only marginal increases in K_M (1.35 mM and 1.5 mM vs. 1.0 mM for RNase S).[112] These authors concluded that Lys-7 contributes in some way to optimize the catalytic process, but did not elaborate further.

From their study of the activity of the complex between RNase 1–20[Nle-7] and S-protein, Irie and co-workers[113] concluded that Lys-7 functions as part of the binding site for the 3'-phosphate in **UpU > p**, that is, as part of the P_2 site, as designated by Richards and Wyckoff.[4] The isosteric, but uncharged, Nle replacement had no effect on k_{cat} and almost no effect on K_M for the transphosphorylation of UpU (0.20 mM and 231 min^{-1} vs. 0.17 mM and 231 min^{-1} for RNase S), but for UpU > p the values changed significantly to 0.54 mM (vs. 0.16 mM) and 185 min^{-1} (vs. 272 min^{-1}).

In their model of the active site cleft of RNase A complexed with a pentanucleotide substrate, deLlorens et al. found that Lys-7 was correctly positioned to form part of subsite P_2.[114]

5.3 THE COMPLEX OF RNASE 1–118 AND SYNTHETIC RNASE 111–124

The creation by Merrifield and co-workers of a second semisynthetic system with which to study RNase, consisting in this case of a large N-terminal segment, RNase 1–118, and a small C-terminal fragment, RNase 111–124, has allowed an examination of the role played by several additional residues in generating the characteristic specificity and catalytic efficiency of the enzyme[27,102] (Figure 5.4). This fully active noncovalent complex, denoted RNase 1–118:111–124, also provides an additional example of a strong and specific interaction between a relatively small peptide and a protein.

Prior to the determination of a high-resolution crystal structure for RNase 1–118:111–124,[30] a worrisome aspect of the complex, one not found in RNase S, was the redundancy of residues 111–118, which are present in each chain. (As will become evident from the discussion below, it was not possible to deal with this problem by shortening the peptide.) However, the crystal structure revealed that the backbone hydrogen bonding very closely resembled that of RNase A, even in the C-terminal β-strand switchback[30] (Figure 5.6). Moreover, no electron density was seen for redundant residues, which must be disordered in the crystal (and therefore are not circled in Figure 5.6). Good electron density was found for residues 1–113 of RNase 1–118 and residues 114–124 of RNase 111–124. Hence, the ordering of the complex is reminiscent of the hybridization of nucleic acids wherein only properly paired elements interact.

The assumption that studies with RNase 1–118:111–124 would be of direct pertinence to questions about the mechanism of action of RNase A received support

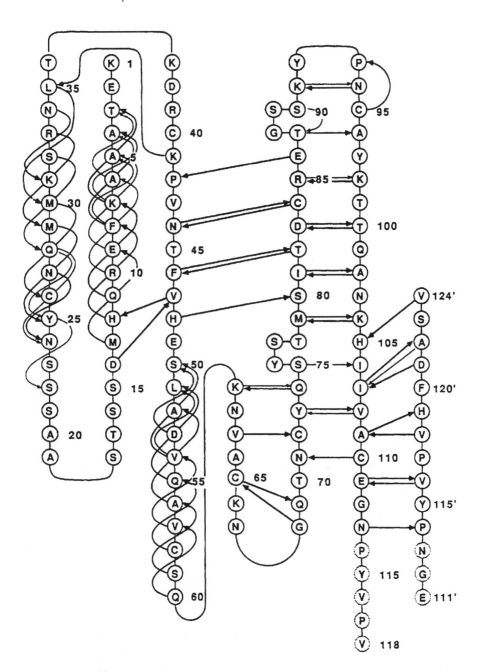

FIGURE 5.6 Backbone hydrogen bonding scheme for RNase 1–118:111–124. Backbone hydrogen bonds with a donor-acceptor distance less than 3.35 Å and an angle greater than 110° are shown for RNase 1–118:111–124. The arrows go from donor to acceptor. The thicker lines represent short hydrogen bonds between 2.5 and 3.15 Å; the thinner lines represent long hydrogen bonds between 3.15 and 3.35 Å. There is no appreciable electron density for residues 114–118 of RNase 1–118 or residues 111–113 of RNase 111–124.

from the observation that the pK values measured from proton NMR spectra of active site histidines 12 and 119, as well as that of His-105, were, for the most part, identical within experimental error, whether measured at 10°[116] or 30°.[117] The only exception was the pK value of His-119 at 10°, which was found to be 0.15 pH unit lower in RNase 1–118:111–124 than in RNase A.

5.3.1 STRUCTURAL SOURCES OF THE BINDING ENERGY BETWEEN RNASE 1–118 AND RNASE 111–124

5.3.1.1 Effects of the Number of Residues in the Peptide

The seven-residue peptide containing Val-118, as well as the six residues that are removed from RNase A to form RNase 1–118, did not reactivate RNase 1–118 to an extent greater than 1%, even at very high mole ratios of peptide to RNase 1–118[28] (Table 5.6). The addition of Pro-117 to form RNase 117–124 also had very little effect upon reactivation, but a striking increase in activity occurred upon the addition of Val-116[28] (Table 5.6). Val-116 is an invariant residue across 40 mammalian pancreatic RNases,[106] but there is no evidence that it participates in substrate binding or in the catalytic machinery. As to its structural role, it forms two backbone hydrogen bonds with Glu-111 in RNase 1–118:111–124,[30] as it also does in both RNase S[37] and RNase A,[68] but it participates only marginally in the formation of a hydrophobic nucleus, forming a single nonpolar contact with Pro-117.[59] Pro-117 plays a major role in the structure of this hydrophobic nucleus, however, forming nonpolar contacts with four additional residues, including Phe-8.[59] Perhaps a significant stabilizing contribution of Val-116 is the neutralization of the charged, hydrophilic α-amino group of Pro-117, which would be expected to destabilize a hydrophobic region.

The successive addition of Tyr-115, Pro-114, Asn-113, and the Gly-112/Glu-111 segment each causes further modest increases in maximal reactivation, until essentially full activity is exhibited by the complex between RNase 1–118 and the 14-residue peptide composed of residues 111–124[28] (Table 5.6). Residue 110 is a half-cystine in RNase 1–118, so no further additions to the synthetic peptide were made. From the crystal structure of the complex, one might have predicted that residues 111–113 would not make a contribution to binding energy, as they are disordered[30] (Figure 5.6). Perhaps the solution structure is different in this region, or perhaps these residues make an electrostatic contribution to binding energy in spite of their mobility.

The strength of binding between RNase 1–118 and RNase 111–124 in the presence of substrate appears to be one to two orders of magnitude weaker than that found for S-protein and S-peptide under comparable conditions (200 vs. 10 nM).[28,115] However, the increased strength of binding caused by substrate may not be as great for RNase 1–118:111–124 as it is for RNase S: preliminary experiments using tritiated RNase 111–124 indicate binding to RNase 1–118 is 50-fold weaker in the absence of substrate, whereas the reduction in binding is 2,800-fold for S-peptide and S-protein.[115]

TABLE 5.6
Effect of Peptide Length upon the Apparent Strength of Binding to RNase 1–118 and upon the Activity of the Resulting Complex

Peptide	K_D[a]	Ref.	Activity[b]	Ref.
RNase 111–124	0.2 μM	1	98%	1
RNase 113–124	—	—	90%	1
RNase 114–124	—	—	80%	1
RNase 115–124	1.0 μM	2	70%	1
RNase 116–124	2.5 μM	1	60%	1
RNase 117–124	—	—	1.5%	1
RNase 118–124	—	—	1%; c	1

[a]All K_D values were measured from the increase in activity against cytidine 2',3'-cyclic phosphate at pH 6.0 observed at successively higher mole ratios of peptide to RNase 1–118.
[b]Maximal regenerable activity relative to an equimolar amount of RNase A.
[c]The mole ratio of peptide to RNase 1–118 was 40.

[1]Gutte, B., Lin, M.C., Caldi, D.G., and Merrifield, R.B. (1972). *J. Biol. Chem.* 247, 4763–4767.
[2]Merrifield, R.B., and Hodges, R.S. (1975) in *Proc. Intl. Symp. on Macromolecules* (Mano, E.B., ed.), pp. 417–431, Elsevier, Amsterdam.

5.3.1.2 Contributions of Several Individual Residues to Binding Energy

The effect on apparent K_D values of structural changes made in several individual residues in RNase 111–124, including His-119, Phe-120, Asp-121, and Ser-123 has also been examined. As these measurements were also all made in the presence of substrate, the actual thermodynamic contributions of individual residues in the peptide to the binding energy of the complex remain to be determined.

The Role of Histidine-119

The detectable, but moderate, reduction in binding to RNase 1–118 that occurs upon the replacement of His-119 in RNase 111–124 with the structurally similar 3-methyl-His, 1-methyl-His, and homohis analogs is consistent with a relatively marginal contribution of this residue to binding energy, but given the conservative nature of these structural changes, these experiments do not very rigorously test this hypothesis[118,119] (Table 5.7). The analog containing the completely isosteric 3-(β-pyrazolyl)-Ala at position 119 binds almost as well as RNase 111–124 itself[119] (Table 5.7).

Merrifield and Hodges reported that high levels of RNase 111–124[Ala-119] caused no reactivation whatsoever of RNase 1–118, but they did not test whether this analog could inhibit reactivation by RNase 111–124.[115] Even if RNase 111–124[Ala-119] bound well to RNase 1–118, no reactivation would be expected, given the critical role played by His-119 in the activity of the enzyme.[25,26]

TABLE 5.7
Effect of Histidine-119 Replacements upon the Apparent Strength of Binding to RNase 1–118 and upon the Activity of the Resulting Complex

Peptide	$K_D{}^a$	Activity[b]	$K_M{}^c$	$k_{cat}(sec^{-1})$	Ref.
RNase 111–124	0.5 µM	—	1.6mM; d	3.2; e	1
RNase 111–124[3-methyl-His-119]	7.7 µM	20%; f	1.7 mM; d	0.25; e	1
RNase 111–124[1-methyl-His-119]	2.0 µM; g	Not detectable	—	—	1
RNase 111–124[homohis-119]	4.3 µM; g	Not detectable	—	—	2
RNase 111–124[3-β-pyrazolyl-Ala-119]	0.8 µM; g	<0.4%	—	—	1

[a]All K_D values were measured from the increase in activity against cytidine 2',3'-cyclic phosphate at pH 6.0 observed at successively higher mole ratios of peptide to RNase 1–118.
[b]Maximal regenerable activity against cytidine 2',3'-cyclic phosphate at pH 6.0 relative to an equimolar amount of RNase A unless otherwise noted.
[c]For cytidine 2',3'-cyclic phosphate at pH 6.0, 25°.
[d]The K_M value for RNase A under these conditions was 2.4 mM.
[e]The k_{cat} value for RNase A under these conditions was 5.4 sec⁻¹.
[f]Maximal regenerable activity against RNA at pH 5.0 relative to an equimolar amount of RNase 1–118:111–124.
[g]From the inhibition of the activity of RNase 1–118:111–124 against cytidine 2',3'-cyclic phosphate.

[1]Serdijn, J., Bloemhoff, W., Kerling, K.E.T., and Havinga, E. (1984). *Recl. Trav. Chim. Pays-Bas* 103, 50–54.
[2]Serdijn, J., Hoes, C., Raap, J., and Kerling, K.E.T. (1980). *Recl. Trav. Chim. Pays-Bas* 99, 349–352.

The Role of Phenylalanine-120

The replacement of Phe-120 by Ala reduced the apparent affinity between RNase 1–118 and RNase 111–124 50-fold to a value of 10 µM[120] (Table 5.8), while replacement with Leu or Ile reduced the apparent K_D 12-fold and 20-fold, respectively[121] (Table 5.8). Thus, the important structural role played by Phe-120 in RNase A first discovered by Lin[24] manifests itself also in the strength of binding between the two chains. Phe-120 is one of the residues comprising the major hydrophobic nucleus of RNase, where it makes nonpolar contacts with Phe-8, His-12, Thr-45, Val-47, and Ile-106.[59] Nevertheless, its replacement by the bulkier, more hydrophilic Tyr is well tolerated and K_D does not change significantly[120] (Table 5.8). This outcome differs significantly from that seen when Tyr replaces Phe-8, where the affinity of the RNase 1–20[Orn-10] analog for S-protein is reduced 13-fold[47,57] (Table 5.2). The difference may reflect the different energetic cost of burying a tyrosine with or without concomitant hydrogen bond formation by the phenolic hydroxyl group.[122] The phenolic hydroxyl of Tyr-8 does not form any hydrogen bonds when guanidinated RNase 1–20[Orn-10] combines with S-protein, according to UV spectroscopic evidence,[57] whereas the oxygen of the phenolic hydroxyl of Tyr-120 accepts the proton from the hydroxyl group of Ser-123 (2.92 Å), and the proton of the phenolic hydroxyl forms a good hydrogen bond with a water molecule,

TABLE 5.8
Effect of Phenylalanine-120 Replacements upon the Apparent Strength of Binding to RNase 1–118 and upon the Activity of the Resulting Complex

Peptide	K_D[a]	Activity[b]	K_M[c]	k_{cat}(sec^{-1})	Ref.
RNase 111–124	0.2 μM	98%	0.7 mM; d	—	1
RNase 111–124[Leu-120]	2.5 μM	13%	1.0 mM	—	1
	9.4 μM; e	—	1.4 mM; f	0.07; f	2
RNase 111–124[Ile-120]	4.0 μM;	12%	1.2 mM	—	1
RNase 111–124[trp-120]	35 μM;	0.5%	—	—	1
RNase 111–124[Tyr-120]	0.06 μM; g	97%	—	—	3
	1.4 μM; h	—	3.1 mM; i	4.0; i	4
	0.2 μM; j	176%; k	—	—	3
	0.2 μM; j	192%; 1	—	—	3
	0.6 μM; e	—	1.1 mM; f	0.4; f	2
	—	100%; m	—	—	3
RNase 111–124[Ala-120]	10 μM	0.8%; n	—	—	3

[a]All K_D values were measured from the increase in activity against cytidine 2",3'-cyclic phosphate at pH 6.0 observed at successively higher mole ratios of peptide to RNase 1–118.

[b]Maximal regenerable activity against cytidine 2',3'-cyclic phosphate at pH 6.0 relative to an equimolar amount of RNase A, unless otherwise indicated.

[c]For cytidine 2',3'-cyclic phosphate at pH 6.0, unless otherwise noted.

[d]The value for RNase A under these conditions was also 0.7 mM.

[e]K_D value measured from the increase in activity against uridine 2',3'-cyclic phosphate at pH 6.0 observed at successively higher mole ratios of peptide to RNase 1–118. The corresponding value for RNase 111–124 is 5.5 μM.

[f]For uridine 2',3'-cyclic phosphate at pH 6.0. The corresponding values for RNase 111–124 are 1.0 mM and 0.35 sec^{-1}.

[g]The corresponding value for RNase 111–124 obtained by these investigators is 0.08 μM.

[h]The corresponding value for RNase 111–124 obtained by these investigators is 1.1 μM.

[i]The corresponding values for RNase 111–124 are 1.1 mM and 2.1 sec^{-1}.

[j]The corresponding value for RNase 111–124 obtained by these investigators is 0.2 μM.

[k]Maximal regenerable activity against uridine 2',3'-cyclic phosphate at pH 6.0 relative to an equimolar amount of RNase 1–118:111–124.

[l]Maximal regenerable activity against uridine 2',3'-cyclic phosphate at pH 7.0 relative to an equimolar amount of RNase 1–118:111–124.

[m]Activity against 0.05% w/v yeast RNA relative to an equimolar amount of RNase 1–118:111–124.

[n]Activity against cytidine 2',3'-cyclic phosphate at pH 6.0 relative to an equimolar amount of RNase 1–118:111–124; mole ratio of peptide to RNase 1–118 was 65.

[1]Lin, M.C., Gutte, B., Caldi, D.G., Moore, S., and Merrifield, R.B. (1972). *J. Biol. Chem.* 247, 4768–4774.

[2]M.L. Ram and M.S. Doscher, unpublished results

[3]Hodges, R.S., and Merrifield, R.B. (1974). *Intl. J. Peptide Prot. Res.* 6, 397–405.

[4]de Mel, V.S.J., Doscher, M.S., Glinn, M.A., Martin, P.D., Ram, M.L., and Edwards, B.F.P. (1994). *Protein Sci.* 3, 39–50.

as revealed by X-ray diffraction analysis of crystals of RNase 1–118:111–124[Tyr-120] (123; PDB File 1SSB).

The trp side chain proved to be too bulky a structure, however, as its presence in place of Phe-120 reduced the apparent K_D 105-fold[121] (Table 5.8).

The Role of Aspartic Acid-121

The replacement of aspartic acid-121, even by a negatively charged Glu or cysteic acid residue, reduces the apparent K_D 6-fold[115] and 3-fold,[124] respectively, whereas the loss of the negative charge that accompanies its replacement by Asn or Ala reduces the apparent K_D 30-fold and 33-fold, respectively[124,125] (Table 5.9). Evidently, electrostatic effects make a significant contribution to the binding energy between RNase 111–124 and RNase 1–118. The positioning of the oxygens of the β-carboxyl of Asp-121 within 3-4 Å of the positively charged ε-nitrogen of Lys-66 in RNase 1–118:111–124 may be the structural basis for this effect (PDB File 1SRN).[30]

The Role of Serine-123

The hydroxyl group of Ser-123 does not appear to play a role in the binding of RNase 111–124 to RNase 1–118 as its replacement by Ala causes no change in the value of K_D[126] (Table 5.10). The slight increase in apparent K_D observed with the Ser(OMe) analog is presumably a steric effect[126] (Table 5.10). Ser-123 has been replaced by either Thr or Tyr in some mammalian pancreatic RNases.[106]

5.3.2 Active Site Residues in RNase 111–124

Rather remarkably, of the six non-redundant C-terminal residues found in RNase 111–124, four and possibly five play a role in generating the full enzymatic activity of RNase 1–118:111–124 and, presumably, of RNase A as well. Most replacements of His-119 abolish activity completely, an outcome consistent with the major contribution to the catalytic machinery that has been proposed for this residue.[4,5,10,127–129] All analogs with structural changes in Phe-120, Asp-121, or Ser-123 exhibit modified values of K_M, k_{cat}, or both, but in no case is activity completely lost. The synthetic peptide lacking Val-124, that is, RNase 111–123, generated only 7% activity upon combination with RNase 1–118, while the replacement of Val-124 with Ala to provide RNase 111–124[Ala-124] gave a complex with RNase 1–118 that was 53% active.[115] These observations are in conflict with the results of experiments using RNase A, which indicate no involvement of Val-124 in generating enzymatic activity.[130]

Histidine-119

In contrast to the case with His-12[95,104] (Table 5.4), the replacement of His-119 with homohis abolishes activity completely[118] (Table 5.7). The high degree to which the tetradecapeptide containing this analog could inhibit the activity of RNase 1–118:111–124 demonstrated that the lack of activity did not reflect an inability to bind to RNase 1–118.[118] The apparent K_D was increased from 0.5 μM to 4.3 μM, however. The pK value of His-12 in this analog has dropped from 6.22 to 5.99, and the pK value of the homohis-119 is 6.98 (as compared to a pK value of 6.42 for His-119 in the parent complex).[116] However, this increased gap between the pK values of the two active site histidine residues should not completely eliminate the

TABLE 5.9
Effect of Asp-121 Replacements upon the Apparent Strength of Binding to RNase 1–118 and upon the Activity of the Resulting Complex

Peptide	$K_D{}^a$	Activity	$K_M{}^b$	$k_{cat}(sec^{-1})$	Ref.
RNase 111–124	0.2 μM	98%; c	0.7 mM; d	—	1
	1.1 μM	—	1.1 mM	2.1	2,3
RNase 111–124[Glu-121]	2.1 μM; e	21%; f	—	—	4
RNase 111–124[Asn-121]	—	4%; f	—	—	4
	—	4.5%; g	—	—	5
	33 μM; h	—	4.3 mM; i	0.24; i	2,3
	—	6%; j	—	—	6
RNase 111–124[Ala-121]	—	7%; f	—	—	4
	36 μM; h	—	1.9 mM; i	0.47; i	2,3
	—	15%; j	—	—	6
RNase 111–124[cysteic acid-121]	2.9 μM; h	—	2.8 mM; i	0.76; i	3
	—	19%; j	—	—	6

[a]All K_D values were measured from the increase in activity against cytidine 2',3'-cyclic phosphate at pH 6.0 observed at successively higher mole ratios of peptide to RNase 1–118.

[b]For cytidine 2',3'-cyclic phosphate at pH 6.0.

[c]Maximal regenerable activity against cytidine 2',3'-cyclic phosphate at pH 6.0 relative to an equimolar amount of RNase A.

[d]The value for RNase A under these conditions was 0.7 mM.

[e]The corresponding value for RNase 111–124 obtained by these investigators is 0.34 μM.

[f]Maximal regenerable activity against cytidine 2',3'-cyclic phosphate at pH 6.0 relative to an equimolar amount of RNase 1–118:111–124.

[g]Maximal regenerable activity against cytidine 2',3'-cyclic phosphate at pH 7.0 relative to an equimolar amount of RNase 1–118:111–124.

[h]The corresponding value for RNase 111–124 is 1.1 μM.

[i]The corresponding values for RNase 111–124 are 1.1 mM and 2.1 sec⁻¹.

[j]Maximal regenerable activity against RNA at pH 6.0 relative to an equimolar amount of RNase A.

[1]Gutte, B., Lin, M.C., Caldi, D.G., and Merrifield, R.B. (1972). *J. Biol. Chem.* 247, 4763–4767.

[2]de Mel, V.S.J., Martin, P.D., Doscher, M.S., and Edwards, B.F.P. (1992). *J. Biol. Chem.* 267, 247–256.

[3]Kelly, C.R., Ram, M.L., and Doscher, M.S. (1993). *Frontiers in Macromolecular Structure and Function,* Toronto, June 7–8, Abstract M10.

[4]Merrifield, R.B., and Hodges, R.S. (1975) in *Proc. Intl. Symp. on Macromolecules* (Mano, E.B., ed.), pp. 417–431, Elsevier, Amsterdam.

[5]Stern, M.S., and Doscher, M.S. (1984) *FEBS Lett.* 171, 253–256.

[6]Kelly, C.R., and Doscher, M.S., unpublished.

combination of one unprotonated and one protonated histidine believed necessary to achieve catalysis.[10,127–129]

Similarly, the 1-methyl-His (also denoted N^pros-methyl-His or Nδ1-methyl-His) substitution at position 119 resulted in a peptide that bound well to RNase 1–118, but generated no detectable activity[119] (Table 5.7). This finding gave strong support

TABLE 5.10

Effect of Ser-123 Replacements upon the Apparent Strength of Binding to RNase 1–118 and upon the Activity of the Resulting Complex[a]

Peptide	K_D[b]	Activity[c]	K_M	$k_{cat}(sec^{-1})$
RNase 111–124	0.1 μM; d	100%; d	1.5 mM; d	6.6; d
	—	100%; e	1.6 mM; e	1.4; e
	0.2 μM; f	100%; f	11 mM; f	4.1; f
RNase 111–124[Ser(O-Me)-123]	0.2 μM; d	48%; d	2.0 mM; d	4.3; d
	—	52%; e	3.3 mM; e	1.1; e
	0.3 μM; f	38%; f	20 mM; f	3.4; f
	—	46%; g	—	—
	—	55%; h	—	—
	—	38%; i	—	—
RNase 111–124[Ala-123]	0.05 μM; d	100%; d	—	—
	—	42%; e	—	—
	0.1 μM; f	25%; f	62 mM; f	5.8; f
	—	50%; g	—	—
	—	110%; h	—	—
	—	101%; i	—	—

[a]All data are from Hodges, R.S., and Merrifield, R.B. (1975). *J. Biol. Chem.* 250, 1231–1241.

[b]All K_D values were measured from the increase in activity against the indicated substrate as observed at successively higher mole ratios of peptide to RNase 1–118.

[c]Maximal regenerable activity relative to an equimolar amount of RNase 1–118:111–124

[d]For cytidine 2',3'-cyclic phosphate at pH 6.0.

[e]for uridine 2',3"-cyclic phosphate at pH 6.0.

[f]For uridine 2',3'-cyclic phosphate at pH 7.0.

[g]Activity against polyuridylate at pH 7.0.

[h]Activity against polycytidylate at pH 7.0.

[i]Activity against copolyformycin/adenosine at pH 6.0.

to the hypothesis that the N^{pros}-nitrogen of His-119 successively functions during the course of catalysis as a proton donor and acceptor,[11,127-129] although the greater steric bulk of the methyl group does introduce the possibility of conformational distortion. Both the homohis and 1-methyl-His semisynthetic complexes bind 3'-cytidylate, the product of the hydrolysis of cytidine 2',3'-cyclic phosphate, with only slightly diminished affinity.[116,119]

A third analog in which there is good binding between the peptide and RNase 1–118, but extremely low or nil enzymatic activity, is formed by substituting His-119 with 3-(β-pyrazolyl)-Ala.[119] The side chain of this amino acid is a planar aromatic ring that is isosteric with the imidazole side chain of His, but the pK is 2.1 rather than 6.0.[93] Consequently, the conjugate acid that is formed by protonation of the second nitrogen will exist only in a vanishingly small amount at neutral pH values. The substitution of His-12 in the S-peptide by this structure also resulted in an analog that bound well (to S-protein), but generated no enzymatic activity *(vide supra).* [91,92]

The substitution of His-119 with 3-methyl-His (also denoted N^{tele}-methyl-His or Nε2-methyl-His) results in a complex with very substantial enzymatic activity in both the transphosphorylation (depolymerase) and hydrolytic reactions[119] (Table 5.7). Although the pK value of the 3-methyl-His side chain in RNase 111–124 is lower than that of His-119 (6.60 vs. 6.79), a smaller downward shift occurs upon binding to RNase 1–118 with the result that the respective pK values in the complex are 6.41 and 6.42.[116] The pK value of His-12 in the complex of the analog is 6.06, as compared to 6.22 in the parent structure.[116]

In the 3-methyl-His analog, there is no possibility of forming a hydrogen bond between the N^{tele}-nitrogen of the imidazole ring of residue-119 and the β-carboxyl group of Asp-121. The existence of such a hydrogen bond has been postulated in RNase A on the basis of the 2.74-Å distance found between the N^{tele}-nitrogen of His-119 and one of the oxygens of the β-carboxyl of Asp-121 (PDB File 5RSA).[68] Brooks et al. have proposed that the formation of this bond serves to modulate the acid/base properties of His-119 and is present only when His-119 is functioning as a base during the hydrolytic reaction, not when it is functioning as an acid during the transphosphorylation reaction.[10] On the basis of this hypothesis, a differential loss in the catalytic efficiency of the transphosphorylation and hydrolysis reactions upon the elimination of this bond can be predicted. Such a differential loss may indeed be present in the 3-methyl-His-119 analog as k_{cat} for the hydrolysis reaction is reduced 13-fold while activity against RNA is reduced only 5-fold[119] (Table 5.7). Nevertheless, as noted previously, hydrolytic activity is almost always more severely reduced than is depolymerase activity, regardless of the site of the modification.

The side chain of His-119 exhibits a conformational mobility that is unique among the active site amino acid residues in RNase. By rotation about its C_α-C_β bond, the imidazole side chain can occupy two distinct conformations, denoted position A ($\chi_1 = +149°$ to $+168°$) and position B ($\chi_1 = -43°$ to $-60°$).[29-31,69,131,132] The relative occupancy of the two conformations in the crystal appears to be modulated by the crystallization solvent, with position A being favored in aqueous organic solvents[31,69,131] and position B being favored in salt solutions.[29–31,132] Baudet-Nessler et al. found His-119 to be in position A in a fluorescent His-12-alkylated derivative of RNase A crystallized from salt solutions, but, in this case, the imidazole ring of His-119 is stacked on the naphthyl moiety of the His-12 substituent.[133] In addition, Harris et al. have reported that a non-bonded potential energy map for His-119 reveals two discrete side chain positions possessing energy minima; these positions correspond to the crystallographically observed A and B positions and are linked by a low energy pathway.[134]

A high-resolution 2D-NMR study of RNase A has revealed that position B also predominates in solution at pH 4.0.[135] The equilibrium between the two conformations in solution appears to be a function of pH, with position B being favored at low pH values and position A being favored at high pH values.[136]

Working with crystals of RNase 1–118:111–124, de Mel et al. have now shown that the conformation of His-119 is also modulated by pH in the crystalline state.[137]

As in solution, position B predominates at low pH values with conversion to position A occurring as the pH is raised.

Given the central role ascribed to His-119 in the catalytic mechanism of RNase,[10,127–129] the possibility that its dual positioning has mechanistic significance is an intriguing idea. In their studies of the active site dynamics of RNase A, Brünger et al. noted that His-119 underwent dihedral-angle transitions in some of the simulations and suggested that this freedom of movement might be important for the catalytic mechanism.[138] Modelling studies have revealed, however, that His-119 is positioned almost equally well to participate in both the transphosphorylation and hydrolytic reactions catalyzed by the enzyme whether it is in position A or position B.[125] As yet, no experimental verification or disproof of a linkage between the dual conformations of His-119 and the mechanism of action of the enzyme has been forthcoming.

Phenylalanine-120

The initial X-ray structural analysis of the complex of RNase S with 3'-cytidylate strongly suggested that this residue might play a direct role in binding substrate by interacting with the pyrimidine moiety in the B_1 base binding site.[4,37] The fact that this residue appeared at that time to be invariant across mammalian pancreatic RNases reinforced this idea.[106] However, its replacement by Leu or Ile did not significantly modify the value of K_M for cytidine 2',3'-cyclic phosphate[121] or for uridine 2',3'-cyclic phosphate (M.L. Ram and M.S. Doscher, unpublished) (Table 5.8). Instead, contrary to expectations, the rate of turnover with both substrates was found to be reduced an order of magnitude (Table 5.8). The Ala and trp analogs bound poorly and generated only very low activity[120,121] (Table 5.8).

A 2.0-Å resolution structure of RNase 1–118:111–124[Leu-120] has now been determined and compared with the 1.8-Å structure of the fully active parent molecule, RNase 1–118:111–124.[123] As revealed by a difference distance matrix for the C_α positions in the two molecules, a number of structural changes have occurred, the majority of which are at a considerable distance from the point of modification itself. These changes include the movement of the loop containing residues 65–72 away from the active site, as well as the rearrangement of hydrogen bonding networks and solvent molecules at the active site. Nevertheless, the overall conformation and the major elements of secondary structure remain undisturbed. Such findings are neither unique to this analog (*see* discussion of Aspartic acid-121, below) nor to structural alterations in semisynthetic RNases: a comparable case would be that of staphylo-coccal nuclease, where the 1400-fold reduction in catalytic efficiency caused by replacing active site Glu-43 with an Asp is accompanied by a multiplicity of significant structural changes throughout the molecule.[139] The working hypothesis developed by deMel et al.[123] for the structural basis of the inactivating Leu-120 replacement is discussed below, under Aspartic Acid-121, as replacements of this residue have also caused comparable structural changes and reductions in turnover rates.[125]

The analog containing Tyr at position 120 is fully active against cytidine 2',3'-cyclic phosphate and exhibits enhanced activity toward uridine 2',3'-cyclic phosphate[120] (Table 5.8). The determination of K_M and k_{cat} for the former substrate has now disclosed, however, that these two parameters have, in fact, been modified in a countervailing manner, with an increase in k_{cat} compensating for an increase in K_M[123](Table 5.8). The 2.0-Å crystal structure of this analog reveals a possible basis for these changes: the phenolic hydroxyl of Tyr-120 is involved in a hydrogen bonding network with Ser-123 (as an acceptor) and a water molecule (as a donor), a network not possible in the Phe-containing parent structure.[123] This same water molecule is seen to be hydrogen bonded to the O4 of the uridine ring in the complex between RNase A and the transition-state analog, uridine vanadate (PDB File 6RSA).[140] By accepting the hydrogen of the phenolic hydroxyl of Tyr-120 in the tyrosine-containing analog, this water would become a better hydrogen donor to the O4 oxygen of the uracil ring, with a consequent increase in catalytic efficiency of the enzyme. In the case of cytidine 2',3'-cyclic phosphate, on the other hand, the cytosine ring requires a hydrogen *acceptor;* interaction with Tyr-120 would therefore diminish the suitability of the water and could lead to the increased K_M experimentally observed for this substrate. However, it is not evident from structural considerations why k_{cat} should also be increased.

While Hodges and Merrifield were carrying out their studies of the Tyr-120 analog, it was reported that, in fact, Phe-120 was not a completely invariant residue as giraffe RNase contained Tyr at position 120.[141] Upon testing a sample of this RNase, Hodges and Merrifield found that it was indeed 2.6-fold more selective than bovine pancreatic RNase for uridine 2',3'-cyclic phosphate over cytidine 2',3'-cyclic phosphate.[120] They noted that the Phe to Tyr replacement was not the only structural difference between bovine and giraffe RNase, however. Tyrosine has subsequently been found at position 120 in four other mammalian pancreatic RNases.[106]

Aspartic Acid-121

In crystals of RNase A, one of the oxygen atoms of the β-carboxyl group of Asp-121, which is invariant across mammalian pancreatic RNases,[106] is found to be within hydrogen bonding distance (2.74 Å) of the Nε2 nitrogen of His-119, a confirmed active site residue (PDB File 5RSA).[68] The probable functional relevance of this interaction is supported by the finding that replacement of Asp-121 even by a negatively charged residue such as Glu or cysteic acid reduces activity toward cytidine 2',3'-cyclic phosphate 5-fold[115,124] (Table 5.9). If the negative charge is also removed, activity is reduced still farther: in the Ala-121 analog, K_M for cytidine 2',3'-cyclic phosphate is increased from 1.1 mM to 1.9 mM and k_{cat} is reduced from 2.1 sec-1 to 0.47 sec^{-1}, while the corresponding values are 4.3 mM and 0.24 sec^{-1} for the Asn-121 analog,[123] (Table 5.9). Significant reductions in transphosphorylation (depolymerase) rates at pH 6.0 against RNA are also seen with the cysteic, Asn, and Ala analogs (C.R. Kelly and M.S. Doscher, unpublished) (Table 5.9).

Using a somewhat different semisynthetic RNase system (*see* Other Semisynthetic RNases, below), Irie and co-workers found that significant reductions occurred in both depolymerase and hydrolytic activities upon the replacement of Asp-121 with either Ala or Asn.[142,143]

Structures at 2.0-Å resolution have been determined for both RNase 1–118:111–124[Ala-121] and RNase 1–118:111–124[Asn-121] and have led to hypotheses concerning the basis for the activity loss seen in these analogs as well as that seen in the Leu-120 analog.[123,125] In all three structures, the interactions between Lys-66 and Asp-121 that are seen in both the parent complex, that is, RNase 1–118:111–124, and in the fully active Tyr-120 analog have been modified or entirely disrupted, with the result that the loop containing residues 65–72 and, in particular, the positively charged N^ε of Lys-66, has moved away from the active site. de Mel et al. propose that the positive charge on the N^ε of Lys-66, which is an invariant residue,[106] is a factor stabilizing the negatively charged pentacoordinate transition states in both the transphosphorylation and hydrolysis reactions,[128] and its movement away from the active site destabilizes these transition states, reducing the magnitude of k_{cat}.[125]

In all three analogs, the side chain of His-119 is found predominantly in the conformation known as position B, which places the Nε2 nitrogen of the imidazole ring too distant from the oxygens of the β-carboxyl group of Asp-121 to allow the hydrogen bond formation seen when the side chain is in position A (see the discussion of the dual conformation of His-119 in Section 5.2.2.1). Consequently, for the analog structures in hand, it is not possible to see what the modification of residue 120 or 121 may have done to the environment of His-119 when it occupies position A. In the Asn-121 analog, the imidazole ring of His-119 has also flipped 180° in a majority of the molecules, carrying the critical Nδ1 atom out of the active site.[125] As this conformation would not be expected to be catalytically active, its existence very likely accounts for the greater reduction in activity seen in this analog.

The replacement of a negatively charged residue by a neutral one should destabilize the positively charged forms of nearby residues, thereby causing depressed pK values. Such electrostatic effects have been shown experimentally to persist over considerable distances and at moderate ionic strengths; in an extracellular subtilisin from *B. amyloliquefaciens,* Russell and co-workers observed that an Asp to Ser replacement at a distance of 12–13 Å reduced the pK of the active site His by 0.29 pH unit at an ionic strength of 0.1 M.[144,145] Given the relatively short distances between His-119 and Asp-121, whether the former residue is in position A (2.74 Å) or position B (9.9 Å), a large decrease in its pK value would be predicted upon making the Asp-121 to Asn replacement in RNase 1–118:111–124. When the pK value of His-119 in the Asn-121 analog was measured using 1H NMR spectroscopy and compared to the corresponding value in the parent complex, an increase in pK value of only 0.05 pH unit was found.[146] The predicted change, calculated using the 2.0-Å coordinate set for the crystal structure of the Asn-121 analog[125] (PDB file 3SRN) and a finite difference algorithm to solve the Poisson-Boltzmann equation,[147] was a decrease in pK value of 0.41 pH unit.[146] Discrepant pK values were also found for His-12 (–0.09 pH unit experimentally vs. –0.36 pH unit theoretically), but the values for His-105 were in agreement (–0.12 pH unit experimentally vs. –0.07 pH unit theoretically). The most likely basis for the differences between the observed and predicted pK values for His-12 and His-119 are departures in local dielectric constant from the umbrella value of 2 that was used to model crystallographically bound water:[148] a comparison of the structures of RNase 1–118:111–124 and the Asn-121 analog reveals numerous differences in the location and structure of crystallographically bound water networks.[125]

Serine-123

Examination of the structures of the complexes of RNase S with 3'-cytidylate and 3'-uridylate[37] led Hodges and Merrifield to propose that Ser-123 might function in binding the substrates corresponding to these products by interacting with the various substituents on C4 of the pyrimidine ring.[126] With uridine-containing substrates, the oxygen on C4 would accept the proton of the serine hydroxyl group while, with cytidine-containing substrates, the amino group on C4 would donate a proton to the oxygen of the serine hydroxyl group. To test this hypothesis they prepared and characterized the Ser(OMe)-123 analog, which can accept, but not donate a proton, and the Ala-123 analog, which can neither accept nor donate a proton.[126]

The pertinence of the proposed interaction for cytidine-containing substrates was not evident as the Ala-123 analog exhibited full activity in both the hydrolytic and depolymerase reactions involving this base (Table 5.10). The possibility of countervailing changes in K_M and k_{cat} of the type seen with the Tyr-120 analog (*vide supra*) has not been ruled out, however. In the face of the results with the Ala-123 analog, it seems likely that the reduced activity seen with the Ser(OMe)-123 analog and cytidine-containing substrates is the manifestation of a steric effect.

The predicted sensitivity of uridine-containing substrates to the two structural changes was seen experimentally with two-fold reductions in activity occurring in both depolymerase and hydrolytic reactions (Table 5.10). These observations are also consistent, for the most part, with the water-mediated interaction of Ser-123 seen with the transition state analog, uridine vanadate[140] and with the kinetic changes observed with the Tyr-120 analog[120,123] (*vide supra*). However, if a mediating water is present in the interaction of Ser(OMe)-123 with uridine-containing substrates, one might predict no loss in activity in this analog as the OG oxygen can still accept a proton from the water. Perhaps the observed activity loss should be attributed to steric effects of the methyl group in this case as well.

Valine-124

The complex of RNase 111–123 and RNase 1–118 was found to have only 7% of the activity of the Val-124-containing parent complex, while the analog containing Ala at position 124 was only 53% active.[115] These findings are at variance with observations of the effect of removing Val-124 from RNase A: here, no reduction in activity occurred.[130] Val-124 participates in forming one of the minor hydrophobic clusters in RNase, but only to the extent of making one nonpolar interaction, with Ile-107.[59] It is an invariant residue, however, even in those mammalian pancreatic RNases where it is not the C-terminal residue.[106]

5.4 OTHER SEMISYNTHETIC RNASES

As part of their early characterization of the RNase 1–118:111–124 system, Gutte et al. examined the capacity of RNase 111–124 to reactivate RNase 1–119 and RNase 1–120.[28] Complexes highly active against cytidine 2',3'-cyclic phosphate were formed in both cases, to levels of 90% and 85%, respectively, but the apparent K_D values for the interaction between the peptide and the proteins were increased correspondingly to values of 0.5 μM and 4.0 μM, as compared to 0.2 μM for the parent complex. With two His-119 residues present in each of these complexes, it was not evident which was participating in the active site. Using the criterion that an enhanced rate of alkylation by haloacetates correlates with a catalytically functional His-119,[26, 86, 89, 90] Merrifield and co-workers made the interesting discovery that the two His-119 residues in the RNase 1–120 complex both became alkylated at an accelerated rate, with the His-119 of the 1–120 chain being favored 4 to 1 over the His-119 in the peptide.[28] In the 1–119 complex, both histidines also become alkylated at an accelerated, but equal, rate. The existence of two alternative enzymatically active conformations has also been seen with overlapping fragments of staphylococcal nuclease.[149]

In spite of the ambiguity imposed by the presence of two potentially reactive His-119 residues, Irie and co-workers prepared and characterized a set of analogs of RNase 1–120:115–124 in which Asp-121 was replaced by Glu, Asn, Gln, or Ala.[142,143] The activity of the glutamate-containing analog against cytidine 3',5'-adenosine was 13% of that of RNase A, in close agreement with the 21% activity toward cytidine 2',3-cyclic phosphate found for RNase 1–118:111–124[Glu-121].[115] The reduction of V_{max} to 7% for the asparagine-containing analog compared closely as well to the reduction of initial velocity toward cytidine 2',3-cyclic phosphate to 4%[115] and of k_{cat} to 10% for RNase 1–118:111–124[Asn-121].[125] The activities of the Gln-121 and Ala-121 analogs prepared by Irie and co-workers were also sharply reduced.[142,143]

Hayashi et al. found that RNase 1–118 could be further shortened by successive digestion with carboxypeptidase Y, which removed residues 116–118, and by carboxypeptidase A, which removed residue 115 (residue 114 is a Pro).[150] The complexes which these shortened chains formed with the nonapeptide containing the nine C-terminal residues of RNase, namely, RNase 1–115:116–124, and RNase 1–114:116–124, were found to have high binding affinities, possessing apparent K_D values of 80 and 30 nM, respectively, but they were less active than RNase 1–118:111–124, exhibiting maximal activities of 50% and 54%, respectively, against cytidine 2',3'-cyclic phosphate.[150] Nevertheless, Chaiken and co-workers realized that the RNase 1–115:116–124 system might be susceptible to an enzyme-catalyzed resynthesis of the peptide bond between residues 115 and 116 in a manner analogous to the subtilisin-catalyzed reformation of RNase A from RNase S.[34] Their attempts to use chymotrypsin in 90% aqueous glycerol to effect peptide bond formation between residues 115 and 116 were not successful, however.[151]

Prior to the demonstration that RNase A can be reformed in 50% yield from RNase S by a subtilisin-catalyzed reaction in glycerol,[34] Hoogerhout and Kerling examined the possibility of forming a RNase A analog from the reaction of the lactone of RNase 1–20[Ile-13, hser-20] and S-protein.[152] This approach, which had been proposed earlier,[153] and utilized with success in the semisynthesis of cytochrome c[154] and basic trypsin inhibitor,[155] provided only a 10% yield of product in this case, however.

Neumann and Hofsteenge have successfully used subtilisin-catalyzed peptide bond formation to prepare RNase A analogs modified at Lys-1 or Lys-7 starting from S-protein and appropriately modified synthetic S-peptide analogs.[156] The properties of these analogs revealed that the major portion of the binding energy between

RNase A and the RNase inhibitor from porcine liver is derived from electrostatic interactions of Lys-7 with a more minor contribution coming from Lys-1.

Imperiali and co-workers have used the RNase S system to construct coenzyme-amino acid chimeras that exhibit novel chemical activities.[157,158] Substitution of the Phe-8 moiety in RNase 1–14-NH$_2$ with a pyridoxal analog provided a structure which, when complexed with S-protein, effected the single turnover conversion of L-alanine to pyruvate at a rate 14 times greater than that found with 5'-deoxypyridoxal alone. The dependence of the rate upon complexation with S-protein as well as upon the nature of the residue at position 7 (Gly or Nle replacing a Lys) demonstrated the involvement of the peptidyl component in the reaction.

Two three-chain semisynthetic RNase systems have been prepared and characterized. The first of these is a partially overlapping system consisting of a combination of S-peptide (RNase 1–20), carboxypeptidase A-digested S-protein (RNase 21–118), and synthetic RNase 111–124.[27] It was found to exhibit 30% activity against cytidine 2',3'-cyclic phosphate at pH 6.0.[27] The second system, which is non-overlapping, is composed of synthetic RNase 1–15, carboxypeptidase Y-digested S-protein (RNase 21–111), and synthetic RNase 116–124; the three-fragment complex was 4% active against cytidine 2',3'-cyclic phosphate.[159] No structural analogs have been reported for either of the three-component systems.

5.5 CONCLUDING REMARKS

More than 75 semisynthetic analogs of RNase have been prepared up to this time, and their characterization has revealed much about the structural basis for the specificity and catalytic efficiency of this enzyme as well as the structural basis for the high affinity binding that can occur between a relatively small peptide of approximately 15 residues and a larger polypeptide. The results of many of these latter studies have a generality that extends well beyond the enzymology of RNase itself.

On the other hand, it is notable that in no instance was the full range of information that might have been forthcoming from a thorough examination of the properties of an analog actually obtained and, in many cases, characterization was very limited indeed. In part, this situation reflects the unavailability to early workers of some of the powerful techniques such as single crystal X-ray diffraction analysis, NMR spectrometry, molecular dynamics calculations, etc., that can be applied today. Yet, it may also reflect a legacy from classical medicinal chemistry, wherein the aim often was to measure a given physiological property for the largest possible number of structures in order to find the one in which this property was optimized. From this viewpoint, a structure that did not show promise with respect to the particular feature of immediate interest would not be subjected to further study.

With but one exception,[75] the replacement of a *single* residue by another *coded* amino acid, of which there are 25 examples tabulated here, resulted in unchanged or reduced binding between the two chains and unchanged or reduced catalytic efficiency, thus lending credence to those who assert that nature has already done most, if not all, of this type of experiment. Only the replacement of Glu-9 by Leu in RNase 1–15-NH$_2$ resulted in improved binding of the peptide to S-protein.[75] Improved binding was also seen when Fphe replaced Phe-8 in RNase 1–20[Orn-

10][47] or when two or more residues in RNase 1–15-NH$_2$ were replaced by coded alternatives chosen to stabilize the helix dipole that is formed when the peptide binds to S-protein.[75] In no case did a peptide analog, whether it incorporated non-naturally-occurring or alternative coded residues, provide a complex with S-protein or with RNase 1–118 that exhibited improved catalytic efficiency.

5.6 ABBREVIATIONS

RNase A	Bovine pancreatic ribonuclease A.
RNase S	RNase A in which the 20–21 peptide bond has been hydrolyzed by the action of subtilisin.
S-peptide	Peptide consisting of residues 1–20 of RNase A, also denoted RNase 1–20.
S-protein	Polypeptide consisting of residues 21–124 of RNase A, also denoted RNase 21–124. Shortened synthetic analogs are designated by RNase followed by the sequence number of the N- and C-terminal residues. For example, RNase 1–15 is the peptide consisting of residues 1–15 of RNase A. The type and location of a replacement in a synthetic peptide analog is given within square brackets that follow the abbreviation for the parent structure. For example, the RNase 1–15 analog containing a tyrosine residue at position 8 is designated RNase 1–15[Tyr-8]. Its non-covalent complex with S-protein is denoted RNase 1–15[Tyr-8]:21–124.
RNase 1–120	Polypeptide consisting of residues 1–120 of RNase A.
RNase 1–119	Polypeptide consisting of residues 1–119 of RNase A.
RNase 1–118	Polypeptide consisting of residues 1–118 of RNase A.
Aba	α-amino-n-butyric acid.
CM-	carboxymethyl.
cha	cyclohexylalanine.
cpg	cyclopentylglycine.
Fhis	4-fluorohis.
Fphe	p-fluorophenylalanine.
homoarg	homoarginine.
homohis	homohistidine.
1-methyl-His	N^{pros}-methyl-histidine.
3-methyl-His	N^{tele}-methyl histidine.
Nle	norleucine.
Orn	ornithine.
pU > p	5'-phosphouridine 2',3'-cyclic phosphate.
UpU	uridylyl-3',5'-uridine.
UpU > p	uridyly-3',5'-uridine 2',3'-cyclic phosphate.

PDB Protein Data Bank (Bernstein, F.C., et al. [1977] J. Mol. Biol. 112, 535–542).

5.7 ACKNOWLEDGMENTS

I thank Dr. Phil Martin for the modelling study of RNase 1–14[β-(pyrid-3-yl)-Ala-12]:21-124. Work done in the author's laboratory was supported in part by NIH grant GM 40630.

REFERENCES

1. Kunitz, M. (1939). Isolation from beef pancreas of a crystalline protein possessing ribonuclease activity. *Science* 90, 112-113.
2. Barnard, E.A. (1969). Biological function of pancreatic ribonuclease. *Nature* 221, 340-344.
3. Confalone, E., Beintema, J.J., Sasso, M.P., Carsana, A., Palmieri, M., Vento, M.T., and Furia, A. (1995). Molecular evolution of genes encoding ribonucleases in ruminant species. *J. Mol. Evol.* 41, 850-858.
4. Brooks, C., III, Brünger, A., Francl, M., Haydock, K., Allen, L.C., and Karplus, M. (1986). Role of active site residues and solvation in RNase A. *Ann. NY Acad. Sci.* 471, 295-298.
5. Richards, F.M., and Wyckoff, H.W. (1971). Bovine pancreatic ribonuclease. In *The Enzymes* (Boyer, P.D., ed.) 3rd Ed.,Vol. 4, pp. 647-806, Academic Press, New York.
6. Blackburn, P., and Moore, S. (1982). Pancreatic ribonuclease. In *The Enzymes* (Boyer, P.D., ed.) 3rd Ed.,Vol. 15, pp. 317-433, Academic Press, New York.
7. Wlodawer, A. (1985). Structure of bovine pancreatic ribonuclease by X-ray and neutron diffraction. In *Biological Macromolecules and Assemblies* (Jurnak, F., and McPherson, A., eds.) Vol. 2, pp. 393-439, John Wiley & Sons, Inc.
8. Eftink, M. R., and Biltonen, R. L. (1987). Pancreatic ribonuclease A: the most studied endoribonuclease. In *Hydrolytic Enzymes* (Neuberger, A., and Brocklehurst, K., eds.) pp. 333-376, Elsevier Science Publishers B.V. (Biomedical Division).
9. Fersht, A. (1985). *Enzyme Structure and Mechanism,* 2nd Ed., pp. 426-431, W.H. Freeman & Co., New York.
10. Stryer, L. (1995). *Biochemistry,* 4th Ed., pp. 216-218, W.H. Freeman & Co., New York.
11. Zegers, I., Maes, D., DaoThi, M.H., Poortmans, F., Palmer, R., and Wyns, L. (1994). The structures of RNase A complexed with 3'-CMP and d(CpA): Active site conformation and conserved water molecules. *Protein Sci.* 3, 2322-2339.
12. Thompson, J.E., Venegas, F.D., and Raines, R.T. (1994). Energetics of catalysis by ribonucleases: Fate of the 2',3'-cyclic phosphodiester intermediate. *Biochemistry* 33, 7408-7414.
13. delCardayré, S.B., and Raines, R.T. (1995). A residue to residue hydrogen bond mediates the nucleotide specificity of ribonuclease A. *J. Mol. Biol.* 252, 328-336.
14. delCardayré, S.B., Ribo, M., Yokel, E.M., Quirk, D.J., Rutter, W.J., and Raines, R.T. (1995). Engineering ribonuclease A: Production, purification and characterization of wild-type enzyme and mutants at Gln 11. *Protein Eng.* 8, 261-273.
15. Stowell, J.K., Widlanski, T.S., Kutateladze, T.G., and Raines, R.T. (1995). Mechanism-based inactivation of ribonuclease A. *J. Org. Chem.* 60, 6930-6936.

16. Moussaoui, M., Guasch, A., Boix, E., Cuchillo, C.M., and Nogues, M.V. (1996). The role of non-catalytic bonding subsites in the endonuclease activity of bovine pancreatic ribonuclease A. *J. Biol. Chem.* 271, 4687-4692.

17. Baker, W.R., and Kintanar, A. (1996). Characterization of the pH titration shifts of ribonuclease A by one- and two-dimensional nuclear magnetic resonance spectroscopy. *Arch. Biochem. Biophys.* 327, 189-199.

18. Straub, J.E., Lim, C., and Karplus. M. (1994). Simulation analysis of the binding interactions in the RNase-A 3'-UMP enzyme-product complex as a function of pH. *J. Am. Chem. Soc.* 116, 2591-2599.

19. Wladkowski, B.D., Krauss, M., and Stevens, W.J. (1995). Transphosphorylation catalyzed by ribonuclease A: Computational study using *ab initio* effective fragment potentials. *J. Am. Chem. Soc.* 117, 10537-10545.

20. Richards, F.M. (1955). On an active intermediate produced during the digestion of ribonuclease by subtilisin. *Compt. rend. trav. lab. Carlsberg, Ser. chim.* 29, 329-346.

21. Richards, F.M., and Vithayathil, P.J. (1959). The preparation of subtilisin-modified ribonuclease and the separation of the peptide and protein components. *J. Biol. Chem.* 234, 1459-1465.

22. Hofmann, K., Finn, F.M., Haas, W., Smithers, M.J., Wolman, Y., and Yanaihara, N. (1963). Studies on polypeptides. XXVI. Partial synthesis of an enzyme possessing high RNase activity. *J. Am. Chem. Soc.* 85, 833-834.

23. Anfinsen, C.B. (1956). The limited digestion of ribonuclease with pepsin. *J. Biol. Chem.* 221, 405-412.

24. Lin, M.C. (1970). The structural roles of amino acid residues near the carboxyl terminus of bovine pancreatic ribonuclease. *J. Biol. Chem.* 245, 6726-6731.

25. Crestfield, A.M., Stein, W.H., and Moore, S. (1963). Alkylation and identification of the histidine residues at the active site of ribonuclease. *J. Biol. Chem.* 238, 2413-2420.

26. Crestfield, A.M., Stein, W.H., and Moore, S. (1963) Properties and conformation of the histidine residues at the active site of ribonuclease. *J. Biol. Chem.* 238, 2421-2428.

27. Lin, M.C., Gutte, B., Moore, S., and Merrifield, R.B. (1970). Regeneration of activity by mixture of ribonuclease enzymically degraded from the COOH terminal and a synthetic COOH-terminal tetradecapeptide. *J. Biol. Chem.* 245, 5169-5170.

28. Gutte, B. Lin, M.C., Caldi, D.G., and Merrifield, R.B. (1972). Reactivation of des (119-, 120-, or 121-124) ribonuclease A by mixture with synthetic COOH-terminal peptides of varying lengths. *J. Biol. Chem.* 247, 4763-4767.

29. Kim, E.E., Varadarajan, R., Wyckoff, H.W., and Richards, F.M. (1992). Refinement of the crystal structure of ribonuclease S. Comparison with and between various ribonuclease A structures. *Biochemistry* 31, 12304-12314.

30. Martin, P.D., Doscher, M.S., and Edwards, B.F.P. (1987). The refined crystal structure of a fully active semisynthetic ribonuclease at 1.8 Å resolution. *J. Biol. Chem.* 262, 15930-15938.

31. de Mel, V.S.J., Doscher, M.S., Martin, P.D., Rodier, F., and Edwards, B.F.P. (1995). The 1.6-Å structure of a semisynthetic ribonuclease crystallized from aqueous ethanol. Comparison with crystals from salt solutions and with ribonuclease A from aqueous alcohol solutions. *Acta Cryst.* D51, 1003-1012.

32. Doscher, M.S., and Hirs, C.H.W. (1967). The heterogeneity of bovine pancreatic ribonuclease S. *Biochemistry* 6, 304-312.

33. Gross, E., and Witkop, B. (1967). The heterogeneity of the S-peptide of bovine pancreatic ribonuclease A. *Biochemistry* 6, 745-748.

34. Homandberg, G.A., and Laskowski, M., Jr. (1979). Enzymatic resynthesis of the hydrolyzed peptide bond(s) in ribonuclease S'. *Biochemistry* 18, 586-592.

35. Hofmann, K., Visser, J.P., and Finn, F.M. (1969). Studies on polypeptides. XLIII. Synthesis of S-peptide$_{1-20}$ by two routes. *J. Am. Chem. Soc.* 91, 4883-4887.

36. Pandin, M., Padlan, E.A., DiBello, C., and Chaiken, I.M. (1976). Crystalline semisynthetic ribonuclease-S'. *Proc. Natl. Acad. Sci.* 73, 1844-1847.

37. Wyckoff, H.W., Tsernoglou, D., Hanson, A.W., Knox, J.R., Lee, B., and Richards, F. M. (1970). The three-dimensional structure of ribonuclease-S. *J. Biol. Chem.* 245, 305-328.

38. Potts, J.T., Young, D.M., and Anfinsen, C.B. (1963). Reconstitution of fully active RNase S by carboxypeptidase-degraded RNase S-peptide. *J. Biol. Chem.* 238, 2593-2594.

39. Taylor, H.C., Richardson, D.C., Richardson, J.S., Wlodawer, A., Komoriya, A., and Chaiken, I.M. (1981). "Active" conformation of an inactive semi-synthetic ribonuclease-S. *J. Mol. Biol.* 149, 313-317.

40. Niu, C., Shindo, H., Matsuura, S., and Cohen, J.S. (1980). Direct observation of peptide exchange by stable isotope enrichment. *J. Biol. Chem.* 255, 2036-2037.

41. Hearn, R.P., Richards, F.M., Sturtevant, J.M., and Watt, G.D. (1971). Thermodynamics of the binding of S-peptide to S-protein to form ribonuclease S'. *Biochemistry* 10, 806-817.

42. Connelly, P.R., Varadarajan, R., Sturtevant, J.M, and Richards, F.M. (1990). Thermodynamics of protein-peptide interactions in the ribonuclease S system studied by titration calorimetry. *Biochemistry* 29, 6108-6114.

43. Woodfin, B.M., and Massey, V. (1968). Spectrophotometric determination of the dissociation constant of ribonuclease S'. *J. Biol. Chem.* 243, 889-892.

44. Rocchi, G., Borin, G., Marchiori, F., Moroder, L., Peggion, E., Scoffone, E., Crescenzi, V., and Quadrifoglio, F. (1972). Interaction of S-protein with S-peptide and with synthetic S-peptide analogs. A spectroscopic and calorimetric investigation. *Biochemistry* 11, 50-57.

45. Levit, S., and Berger, A. (1976). Ribonuclease S-peptide. A model for molecular recognition. *J. Biol. Chem.* 251, 1333-1339.

46. Finn, F.M. (1972). Spectrophotometric measurement of binding of S-peptide analogs to S-protein. *Biochemistry* 11, 1474-1478.

47. Borin, G., Filippi, B., Moroder, L., Santoni, C., and Marchiori, F. (1977). Kinetic and conformational studies on some partially synthetic ribonuclease S' analogues modified in position 8. *Int'l. J. Peptide Prot. Res.* 10, 27-38.

48. Vithayathil, P.J., and Richards, F.M. (1960). Modification of the methionine residue in the peptide component of ribonuclease-S. *J. Biol. Chem.* 235, 2343-2351.

49. Kauzmann, W. (1959). Some factors in the interpretation of protein denaturation. *Adv. Prot. Chem.* 14, 1-63.

50. Finn, F.M., and Hofmann, K. (1965). Studies on polypeptides. XXXIII. Enzymic properties of partially synthetic ribonucleases. *J. Am. Chem. Soc.* 87, 645-651.

51. Rocchi, R., Scatturin, A., Moroder, L., Marchiori, F., Tamburro, H.M., and Scoffone, E. (1969). Synthesis of peptide analogs of the N-terminal eicosapeptide sequence of ribonuclease A. XI. Synthesis and conformational studies of [orn[10], nle[13]]-S-peptide. *J. Am. Chem. Soc.* 91, 492-496.

52. Thomson, J., Ratnaparkhi, G.S., Varadarajan, R., Sturtevant, J.M., and Richards, F.M. (1994). Thermodynamic and structural consequences of changing a sulfur atom to a methylene group in the M13Nle mutation in ribonuclease-S. *Biochemistry* 33, 8587-8593.

53. Varadarajan, R., and Richards, F.M. (1992). Crystallographic structures of ribonuclease S variants with nonpolar substitution at position 13: packing and cavities. *Biochemistry* 31, 12315-12327.

54. Varadarajan, R., Connelly, P.R., Sturtevant, J.M., and Richards, F.M. (1992). Heat capacity changes for protein-peptide interactions in the ribonuclease S system. *Biochemistry* 31, 1421-1426.

55. Sturtevant, J.M. (1977). Heat capacity changes in processes involving proteins. *Proc. Natl. Acad. Sci.* 74, 2236-2240.

56. Simonson, T., and Brünger, A.T. (1992). Thermodynamics of protein-peptide interactions in the ribonuclease-S system studied by molecular dynamics and free energy calculations. *Biochemistry* 31, 8661-8674.

57. Filippi, B., Borin, G., and Marchiori, F. (1976). The influence of amino acid side-chains on alpha-helix stability: S-peptide analogues and related ribonucleases S'. *J. Mol. Biol.* 106, 315-324.

58. Filiipi, B., Borin, G., Moroder, L., and Marchiori, F. (1976). Near-ultraviolet difference absorption and circular dichroism studies on partially synthetic ribonucleases S'. *Biochim. Biophys. Acta* 454, 514-523.

59. Kolbanovskaya, E.Y., Sathyanarayana, B.K., Wlodawer, A., and Karpeisky, M.Y. (1992). Intramolecular interactions in pancreatic ribonucleases. *Protein Sci.* 1, 1050-1060.

60. Scatturin, A., Tamburro, A.M., Rocchi, R., and Scoffone, E. (1967). The conformation of bovine pancreatic ribonuclease S-peptide. *Chem. Commun.* 1273-1274.

61. Tamburro, A.M., Scatturin, A., Rocchi, R., Marchiori, F., Borin, G., and Scoffone, E. (1968). Conformational-transitions of bovine pancreatric ribonuclease S-peptide. *FEBS Lett.* 1, 298-300.

62. Klee, W.A. (1968). Studies on the conformation of ribonuclease S-peptide. *Biochemistry* 7, 2731-2736.

63. Brown, J.E., and Klee, W.A. (1969). Conformational studies of a series of overlapping peptides from ribonuclease and their relationship to the protein structure. *Biochemistry* 8, 2876-2879.

64. Marchiori, F., Borin, G., Moroder, L., Rocchi, R., and Scoffone, E. (1972). Relation between structure and function in some partially synthetic ribonucleases S'. I. Kinetic determinations. *Biochim. Biophys. Acta* 257, 210-221.

65. Eaker, D.L., King, T.P., and Craig, L.C. (1965). Des-lysyl glutamyl and des-lysyl pyroglutamyl ribonucleases. III. Enzymatic activities and conformational stabilities. *Biochemistry* 4, 1486-1490.

66. Hofmann, K., Andreatta, R., Finn, F.M., Montibeller, J., Porcelli, G., and Quattrone, A.J. (1971). Studies with the S-peptide-S-protein system: the role of glutamic acid-2, lysine-7, and methionine-13 in S-peptide$_{1-14}$ for binding to and activation of S-protein. *Bioorganic Chem.* 1, 66-83.

67. Valle, G., Zanotti, G., Filippi, B., and Del Pra, A. (1977). X-ray analysis of bovine pancreatic ribonuclease analogs: a difference Fourier at 2.8 Å resolution of (orn^{10})-ribonuclease S'. *Biopolymers* 16, 1371-1376.

68. Wlodawer, A., Bott, R., and Sjölin, L. (1982). The refined crystal structure of ribonuclease A at 2.0 Å resolution. *J. Biol. Chem.* 257, 1325-1332.

69. Wlodawer, A., and Sjölin, L. (1983). Structure of ribonuclease A: results of joint neutron and X-ray refinement at 2.0-Å resolution. *Biochemistry* 22, 2720-2728.

70. Dunn, B.M., and Chaiken, I.M. (1975). Relationship between α-helical propensity and formation of the ribonuclease-S complex. *J. Mol. Biol.* 95, 497-511.

71. Chou, P.Y., and Fasman, G.D. (1974). Conformational parameters for amino acids in helical, β-sheet, and random coil regions calculated for proteins. *Biochemistry* 13, 211-222.

72. Komoriya, A., and Chaiken, I.M. (1982). Sequence modeling using semisynthetic ribonuclease S. *J. Biol. Chem.* 257, 2599-2604.

73. Taylor, H.C., Komoriya, A., and Chaiken, I.M. (1985). Crystallographic structure of an active, sequence-engineered ribonuclease. *Proc. Nat'l Acad. Sci.* 82, 6423-6426.

74. Russell, A.J., and Fersht, A.R. (1987). Rational modification of enzyme catalysis by engineering surface charge. *Nature* 328, 496-500.

75. Mitchinson, C., and Baldwin, R.L. (1986). The design and production of ribonucleases with increased thermostability by incorporation of S-peptide analogues with enhanced helical stability. *Proteins* 1, 23-33.

76. Labhardt, A.M. (1984). Kinetic circular dichroism shows that the S-peptide α-helix of ribonuclease S unfolds fast and refolds slowly. *Proc. Nat'l. Acad. Sci.* 81, 7674-7678.

77. Chaiken, I.M., Freedman, M.H., Lyerla, J.R., Jr., and Cohen, J.S. (1973). Preparation and studies of ^{19}F-labeled and enriched ^{13}C-labeled semisynthetic ribonuclease-S' analogues. *J. Biol. Chem.* 248, 884-891.

78. Chaiken, I.M. (1974). Carbon 13 as a probe of helix formation in semisynthetic ribonuclease-S'. *J. Biol. Chem.* 249, 1247-1250.

79. Chaiken, I.M., Cohen, J.S., and Sokoloski, E.A. (1974). Microenvironment of histidine 12 in ribonuclease-S as detected by ^{13}C nuclear magnetic resonance. *J. Am. Chem. Soc.* 96, 4703-4705.

80. Niu, C., Matsuura, S., Shindo, H., and Cohen, J.S. (1979). Specific peptide-protein interactions in the ribonuclease S' system studied by ^{13}C nuclear magnetic resonance spectroscopy with selectively ^{13}C-enriched peptides. *J. Biol. Chem.* 254, 3788-3796.

81. Hofmann, K., Finn, F.M., Limetti, M., Montibeller, J., and Zanetti, G. (1966). Studies on polypeptides. XXXIV. Enzymic properties of partially synthetic de (16-20)- and de (15-20)-ribonucleases S'. *J. Am. Chem. Soc.* 88, 3633-3639.

82. Filippi, B., Moroder, L., Borin, G., Samartsev, M., and Marchiori, F. (1975). Relation between structure and function in some partially synthetic ribonucleases S'. Enzymic and spectroscopic investigation on [orn^{10}, asn^{14}]-RNase S' and 1$^\varepsilon$, 7$^\varepsilon$, 10$^\delta$-triguanidino-[orn^{10}, asn^{14}]-RNase S'. *Eur. J. Biochem.* 52, 65-76.

83. Richards, F.M., and Logue, A.D. (1962). Changes in absorption spectra in the ribonuclease-S system. *J. Biol. Chem.* 237, 3693-3697.

84. Richards, F.M. (1958). On the enzymic activity of subtilisin-modified ribonuclease. *Proc. Natl. Acad. Sci.* 44, 162-166.

85. Gundlach, H.G., Stein, W.H., and Moore, S. (1959). The nature of the amino acid residues involved in the inactivation of ribonuclease by iodoacetate. *J. Biol. Chem.* 234, 1754-1760.

86. Stark, G.R., Stein, W.H., and Moore, S. (1961). Relationships between the conformation of ribonuclease and its reactivity toward iodoacetate. *J. Biol. Chem.* 236, 436-442.

87. Barnard, E.A., and Stein, W.D. (1959). The histidine residue in the active centre of ribonuclease. I. A specific reaction with bromoacetic acid. *J. Mol. Biol.* 1, 339-349.

88. Barnard, E.A., and Stein, W.D. (1959). The histidine residue in the active centre of ribonuclease. II. The position of this residue in the primary protein chain. *J. Mol. Biol.* 1, 350-358.

89. Goren, H.J., and Barnard, E.A. (1970). Relations of reactivity to structure in pancreatic ribonuclease. I. An analysis of the various reactions with bromoacetate in the pH range 2-7. *Biochemistry* 9, 959-973.

90. Goren, H.J., and Barnard, E. A. (1970). Relations of reactivity to structure in pancreatic ribonuclease. II. Positions of residues alkylated in certain conditions by bromoacetate. *Biochemistry* 9, 974-983.

91. Finn, F.M., and Hofmann, K. (1967). Studies on polypeptides. XXXVII. Competitive inhibition in the S-peptide-S-protein system. *J. Am. Chem. Soc.* 89, 5298-5300.

92. Hofmann, K., Visser, J.P., and Finn, F.M. (1970). Studies on polypeptides. XLIV. Potent synthetic S-peptide antagonists. *J. Am. Chem. Soc.* 92, 2900-2909.

93. Hofmann, K., and Bohn, H. (1966). Studies on polypeptides. XXXVI. The effect of pyrazole-imidazole replacements on the S-protein activating potency of an S-peptide fragment. *J. Am. Chem. Soc.* 88, 5914-5919.

94. Borin, G., Toniolo, C., Moroder, L., Marchiori, F., Rocchi, R., and Scoffone, E. (1972). Synthesis of peptide analogs of the N-terminal eicosapeptide sequence of ribonuclease A. Part XV. Synthesis and guanidination of [orn^{10}, orn^{12}]-S-peptide. *Int'l. J. Peptide Prot. Res.* 4, 37-45.

95. van Batenburg, O.D., Voskuyl-Holtkamp, I., Schattenkerk, C., Hoes, K., Kerling, K.E.T., and Havinga, E. (1977). The role of the imidazolyl nitrogen atoms of histidine-12 in ribonuclease S. *Biochem. J.* 163, 385-387.

96. Hoes, C., Kerling, K.E.T., and Havinga, E. (1983). Studies on polypeptides XXXIX. The role of the imidazole *tele*-nitrogen atom of histidine-12 in the catalytic action of RNase S'. *Recl. Trav. Chim. Pays-Bas* 102, 140-147.

97. Dunn, B.M., DiBello, C., Kirk, K.L., Cohen, L.A., and Chaiken, I.M. (1974). Synthesis, purification, and properties of a semisynthetic ribonuclease S incorporating 4-fluoro-L-histidine at position 12. *J. Biol. Chem.* 249, 6295-6301.

98. Chaiken, I.M., and Taylor, H.C. (1976). Analysis of ribonuclease-nucleotide interactions by quantitative affinity chromatography. J. Biol. Chem. 251, 2044-2048.

99. Taylor, H.C., and Chaiken, I.M. (1977). Active site ligand binding by an inactive ribonuclease S analogue. *J. Biol. Chem.* 252, 6991-6994.

100. Chaiken, I.M., Taylor, H.C., and Ammon, H.L. (1977). Crystal properties of [des 16-20] semisynthetic sequence variants of ribonuclease S'. *J. Biol. Chem.* 252, 5599-5601.

101. Hapner, K.D., Carboxymethyl and azo derivatives of histidine. Ph.D. Thesis, Indiana University, 1966.

102. Richards, F.M., and Wyckoff, H.W. (1973). in *Atlas of Molecular Structures in Biology. I. Ribonuclease S'.* (Phillips, D.C., and Richards, F.M., eds) p. 16, Clarendon, Oxford.

103. van Batenberg, O.D., Kerling, K.E.T., and Havinga, E. (1976). Synthesis and enzymatic activity of an RNase S' analogue in which the 4-imidazolylglycyl residue takes the position and the role of histidine-12. *FEBS Lett.* 68, 228-230.

104. Hoes, C., van Batenberg, O.D., Kerling, K.E.T., and Havinga, E. (1977). Enzymatic hydrolysis of 2',3'-cyclic CMP by homohistidine-12-ribonuclease S'. *Biochem. Biophys. Res. Commun.* 77, 1074-1077.

105. Schneider, F. (1961). Synthese des DL-imidazolyl-glycins und einiger derivate. *Z. physiol. Chem.* 324, 206-210.

106. Beintema, J.J., Schuller, C., Irie, M., and Carsana, A. (1988). Molecular evolution of the ribonuclease superfamily. *Prog. Biophys. Mol. Biol.* 51, 165-192.

107. Marchiori, F., Borin, G., Moroder, L., Rocchi, R. and Scoffone, E. (1974). Relation between structure and function in some partially synthetic ribonucleases S'. II. Kinetic determinations on [orn^{10},glu^{11}]-, [orn^{10},leu^{11}]- and [orn^{10}, lys^{11}]-RNase S'. *Int'l. J. Peptide Protein. Res.* 6, 337-345.

108. Borin, G., Marchiori, F., Moroder, L., Rocchi, R., and Scoffone, E. (1974). Synthesis of analogs of the N-terminal eicosapeptide sequence of ribonuclease A. Part XVI. Synthesis and guanidination of [orn^{10}, leu^{11}]- and [orn^{10}, lys^{11}]-S-peptide. *Int'l. J. Peptide Protein Res.* 6, 329-335.

109. Marzotto, A., Marchiori, F., Moroder, L., Boni, R., and Galzigna, L. (1967). Enzymatic activity of partially synthetic ribonucleases. *Biochim. Biophys. Acta* 147, 26-31.

110. Filippi, B., Chessa, G., and Borin, G. (1981). Semisynthetic ribonucleases S'. The role of glutamine 11. *J. Mol. Biol.* 147, 597-600.

111. Knoblauch, H., Rüterjans, H., Bloemhoff, W., and Kerling, K.E.T. (1988). ^{15}N- and ^{1}H-NMR investigations of the active-site amino acids in semisynthetic RNase S' and RNase A. *Eur. J. Biochem.* 172, 485-497.

112. Marchiori, F., Borin, G., and Moroder, L. (1974). Studies on ribonuclease S': the role of lysine-7 for the activation of S-protein. *Int'l. J. Protein Peptide Res.* 6, 419-434.

113. Irie, M., Ohgi, K., Yoshinaga, M., Yanagida, T., Okada, Y., and Teno, N. (1986). Roles of lysine$_1$ and lysine$_7$ residues of bovine pancreatic ribonuclease in the enzymatic activity. *J. Biochem.* 100, 1057-1063.

114. deLlorens, R., Arus, C., Pares, X., and Cuchillo, C.M (1989). Chemical and computer graphics studies on the topography of the ribonuclease A active site cleft. A model of the enzyme-pentanucleotide substrate complex. *Prot. Eng.* 2, 417-429.

115. Merrifield, R.B., and Hodges, R.S. (1975). Synthetic approaches to the study of proteins. In *Proc. Int'l. Symp. on Macromolecules* (Mano, E.B., ed.), pp. 417-431, Elsevier, Amsterdam.

116. Serdijn, J., Bloemhoff, W., Kerling, K.E.T., and Havinga, E. (1984). Studies on polypeptides. XLI. ^{1}H NMR studies of non-covalent, semisynthetic ribonuclease A analogues in which histidine-119 has been replaced by L-homohistidine, N-tele-methyl-L-histidine, N-pros-methyl-L-histidine and 3-(3-pyrazolyl)-L-alanine, respectively. *Recl. Trav. Chim. Pays-Bas* 103, 351-360.

117. Doscher, M.S., Martin, P.D., and Edwards, B.F.P. (1983). Characterization of the histidine ^{1}H nuclear magnetic resonances of a semisynthetic ribonuclease. *Biochemistry* 22, 4125-4131.

118. Serdijn, J., Hoes, C., Raap, J., and Kerling, K.E.T., (1980). Studies on polypeptides. XXXIII. Semisynthetic ribonuclease analogues. The role of histidine-119. *Recl. Trav. Chim. Pays-Bas* 99, 349-352.

119. Serdijn, J., Bloemhoff, W., Kerling, K.E.T., and Havinga, E. (1984). Studies on polypeptides. XL. The role of histidine-119 in non-covalent semisynthetic ribonuclease; its replacement by 3-(3-pyrazolyl)-L-alanine, N-pros-methyl-L-histidine and N-tele-methyl-L-histidine. *Recl. Trav. Chim. Pays-Bas* 103, 50-54.

120. Hodges, R.S., and Merrifield, R.B. (1974). A synthetic study of the effect of tyrosine at position 120 of ribonuclease. *Int'l. J. Peptide Prot. Res.* 6, 397-405.

121. Lin, M.C., Gutte, B., Caldi, D.G., Moore, S., and Merrifield, R.B. (1972). Reactivation of des (119-124) ribonuclease A by mixture with synthetic COOH-terminal peptides; the role of phenylalanine-120. *J. Biol. Chem.* 247, 4768-4774.

122. Chignell, D.A., and Gratzer, W.B. (1968). Solvent effects on aromatic chromophores and their relation to ultraviolet difference spectra of proteins. *J. Phys. Chem.* 72, 2934-2941.

123. de Mel, V.S.J., Doscher, M.S., Glinn, M.A., Martin, P.D., Ram, M.L., and Edwards, B.F.P. (1994). A structural investigation of catalytically modified F120L and F120Y semisynthetic ribonucleases. *Protein Sci.* 3, 39-50.

124. Kelly, C.R., Ram, M.L., and Doscher, M.S. (1993). Kinetic characterization of catalytically modified semisynthetic ribonucleases. *Frontiers in Macromolecular Structure and Function,* Toronto, June 7-8, (Abstract) M10.

125. de Mel, V.S.J., Martin, P.D., Doscher, M.S., and Edwards, B.F.P. (1992). Structural changes that accompany the reduced catalytic efficiency of two semisynthetic ribonuclease analogs. *J. Biol. Chem.* 267, 247-256.

126. Hodges, R.S., and Merrifield, R.B. (1975) The role of serine-123 in the activity and specificity of ribonuclease. *J. Biol. Chem.* 250, 1231-1241.

127. Findlay, D., Herries, D.G., Mathias, A.P., Rabin, B.R., and Ross, C.A. (1961). The active site and mechanism of action of bovine pancreatic ribonuclease. *Nature* 190, 781-784.

128. Roberts, G.C.K., Dennis, E.A., Meadows, D.H., Cohen, J.S., and Jardetzky, O. (1969). The mechanism of action of ribonuclease. *Proc. Natl. Acad. Sci.* 62, 1151-1158.

129. Lim, C., and Tole, P. (1992). Endocyclic and exocyclic cleavage of phosphorane monoanion: a detailed mechanism of the RNase A transphosphorylation step. *J. Am. Chem. Soc.* 114, 7245-7252.

130. Sela, M., Anfinsen, C.B., and Harrington, W.F. (1957). The correlation of ribonuclease activity with specific aspects of tertiary structure. *Biochim. Biophys. Acta* 26, 502-512.

131. Borkakoti, N., Moss, D.S., and Palmer, R.E. (1982). Ribonuclease-A: least squares refinement of the structure at 1.45 Å resolution. *Acta Cryst.* B38, 2210-2217.

132. Nachman, J., Miller, M., Gilliland, G.L., Carty, R., Pincus, M., and Wlodawer, A. (1990). Crystal structure of two covalent nucleoside derivatives of ribonuclease A. *Biochemistry* 29, 928-937.

133. Baudet-Nessler, S., Jullien, M., Crosio, M.-P., and Janin, J. (1993). Crystal structure of a fluorescent derivative of RNase A. *Biochemistry* 32, 8457-8464.

134. Harris, G.W., Borkakoti, N., Moss, D.S., Palmer, R.A., and Howlin, B. (1987). Ribonuclease A. Analysis of the hydrogen bond geometry, and spatial accessibility at the active site. *Biochim. Biophys. Acta* 912, 348-356.

135. Santoro, J., Gonzalez, C., Bruix, M., Neira, J.L., Nieto, J. L., Herranz, J. and Rico, M. (1993). High-resolution three-dimensional structure of ribonuclease A in solution by nuclear magnetic resonance spectroscopy. *J. Mol. Biol.* 229, 722-734.

136. Rico, M., Santoro, J., Gonzalez, C., Bruix, M., Neira, J.L., Nieto, J.L., and Herranz, J. (1991). 3D structure of bovine pancreatic ribonuclease A in aqueous solution: an approach to tertiary structure determination from a small basis of ^{1}H NMR NOE correlations. *J. Biomolecular NMR* 1, 283-298.

137. de Mel, V.S.J., Doscher, M. S., Martin, P.D., and Edwards, B.F.P. (1994). The occupancy of two distinct conformations by active-site histidine 119 in crystals of ribonuclease is modulated by pH. *FEBS Lett.* 349, 155-160.

138. Brünger, A.T., Brooks III, C.L., and Karplus, M. (1985) Active site dynamics of ribonuclease. *Proc. Natl. Acad. Sci.* 82, 8458-8462.

139. Loll, P.J., and Lattman, E.E. (1990). Active site mutant glu-43 → asp in staphylococcal nuclease displays nonlocal structural changes. *Biochemistry* 29, 6866-6873.

140. Borah, B., Chen, C., Egan, W., Miller, M., Wlodawer, A., and Cohen, J.S. (1985). Nuclear magnetic resonance and neutron diffraction studies of the complex of ribonuclease A with uridine vanadate, a transition-state analogue. *Biochemistry* 24, 2058-2067.

141. Gaastra, W., Groen, G., Welling, G.W., and Beintema, J.J (1974). The primary structure of giraffe pancreatic ribonuclease. *FEBS Lett.* 41, 227-232.

142. Okada, Y., Teno, N., Miyao, J., Mori, Y., and Irie, M. (1984). Amino acids and peptides. XII. Synthesis of C-terminal decapeptide of bovine pancreatic ribonuclease A (RNase A) and its analogs and determination of their ability to reactivate des (121-124) RNase A. *Chem. Pharm. Bull.* (Tokyo) 32, 4585-4592.

143. Irie, M., Miyoa, J., Mori, Y., Okada, Y., and Teno, N. (1988). The role of the carboxyl group of asp_{121} of bovine pancreatic ribonuclease A in enzymatic activity. *Agric. Biol. Chem.* 52, 1291-1292.

144. Russell, A.J., Thomas, P.G., and Fersht, A.R. (1987). Electrostatic effects on modification of charged groups in the active site cleft of subtilisin by protein engineering. *J. Mol. Biol.* 193, 803-813.

145. Sternberg, M.J.E., Hayes, F.R.F., Russell, A.J., Thomas, P.G., and Fersht, A.R. (1987). Prediction of electrostatic effects of engineering of protein charges. *Nature* 330, 86-88.

146. Cederholm, M.T., Stuckey, J. A., Doscher, M.S., and Lee, L. (1991). Histidine pK_a shifts accompanying the inactivating $asp^{121} \rightarrow$ asn substitution in a semisynthetic bovine pancreatic ribonuclease. *Proc. Natl. Acad., Sci.* 88, 8116-8120.

147. Sharp, K.A., and Honig, B. Electrostatic interactions in macromolecules: theory and application. (1990). *Annu. Rev. Biophys. Biophys. Chem.* 19, 301-332.

148. Gilson, M.K., and Honig, B. Energetics of charge-charge interactions in proteins. (1988). *Proteins Struct. Funct. Genet.* 3, 32-52.

149. Taniuchi, H., and Anfinsen, C.B. (1971). Simultaneous formation of two alternative enzymically active structures by complementation of two overlapping fragments of staphylococcal nuclease. *J. Biol. Chem.* 246, 2291-2301.

150. Hayashi, R., Moore, S., and Merrifield. R.B. (1973). Preparation of pancreatic ribonucleases 1-114 and 1-115 and their reactivation by mixture with synthetic COOH-terminal peptides. *J. Biol. Chem.* 248, 3889-3892.

151. Chaiken, I.M., Komoriya, A., and Homandberg, G.A. (1979). Protein semisynthesis and the chemical basis of folding and function. In *Peptides, Structure and Biological Function* (Gross, E., and Meienhofer, J., eds.), *Proc., Sixth Amer. Peptide Symp.*, Pierce Chemical Co., Publishers, Illinois, pp. 587-595.

152. Hoogerhout, P., and Kerling, K.E.T. (1982). Studies on polypeptides. XXXVII. Synthesis of [isoleucine-13, homoserine-20]-S-peptide lactone and semisynthesis of [isoleucine-13, homoserine-20]-RNase A. *Recl. Trav. Chim. Pays-Bas* 101, 246-253.

153. Offord, R.E. (1972). The possible use of cyanogen bromide fragments in the semisynthesis of proteins and polypeptides. *Biochem.* J. 129, 499-501.

154. Corradin, G., and Harbury, H.A. (1974). Reconstitution of horse heart cytochrome c: reformation of the peptide bond linking residues 65 and 66. *Biochem. Biophys. Res. Commun.* 61, 1400-1406.

155. Dyckes, D.F., Creighton, T., and Sheppard, R.C. (1974). Spontaneous re-formation of a broken peptide chain. *Nature* 247, 202-204.

156. Neumann, U., and Hofsteenge, J. (1994). Interaction of semisynthetic variants of RNase A with ribonuclease inhibitor. *Protein Sci.* 3, 248-256.

157. Imperiali, B., and Sinha Roy, R. (1994). Coenzyme-amino acid chimeras: New residues for the assembly of functional proteins. *J. Am. Chem. Soc.* 116, 12083-12084.

158. Imperiali, B., Sinha Roy, R., Walkup, G.K., and Wang, L. (1996). Unnatural amino acids for the design of functional proteins: Biomimetic catalysis using coenzyme-amino acids. In *Molecular Design and Bioorganic Catalysis* (Wilcox, C.S., and Hamilton, A.D., eds.) pp. 35-52, Kluwer Academic Publishers, Netherlands.

159. Kanmera, T., Homandberg, G.A., Komoriya, A., and Chaiken, I.M. (1983). Minimum information content and formation of interacting ribonuclease fragment complexes. *Int'l. J. Peptide Prot. Res.* 21, 74-83.

6 Hemoglobin Semisynthesis

Parimala Nacharaju and Seetharama A. Acharya

CONTENTS

6.1   INTRODUCTION

Structure-function correlation studies of proteins involve the determination of the amino acid sequence and the elucidation of their three dimensional structure by X-ray crystallography[1,2] and/or multidimensional NMR spectroscopy[3–5] that provides an atomic resolution to facilitate a systematic alteration of the side-chain interactions of the proteins. The latter is achieved by site-specific replacement of amino acid residues in the protein sequence. In carrying out these studies, one needs to conserve the basic structural motif of the protein and ascertain that the alterations introduced by site-specific events are very limited. Only under these circumstances will the interpretations of the results be straightforward.

The proteins serve very diverse functions (enzymic, immunological, transportation of biologically important materials, binding to ligands, and service as the first or second messengers), and the basic structural motif (units) selected for a function appears to be very distinct and is generally conserved beyond the species boundaries. Besides, the basic structural principles used for the assembly of the desired motifs also appear to be common. The early attempts to correlate the role of amino acid side chains of proteins with the function have depended heavily on the development of chemical modification reactions specific for functional groups of proteins.[6–9] Only the side chains of the amino acid residues with *reactive* functional groups are amenable to this approach, and the presence of many reactive groups of the same class in a given protein is a major limitation of this approach. Attempts to develop direct approaches to engineer structural changes into proteins have been evolving over the years.

The early attempts to engineer structural changes into proteins were based on the chemical synthesis of peptides and proteins. These attempts have culminated in the total chemical synthesis of bovine pancreatic RNase A by Merrifield and his associates using solid phase methods of peptide synthesis,[10–12] and also by Hofmann and his colleagues using solution methods of peptide synthesis.[13–14] The Merrifield solid phase peptide synthesis is now a conventional technique for the synthesis of medium-sized peptides. The exponential decrease in the yields as a function of the increase in the chain length of the polypeptide was an early limitation for the widespread use of this technology, particularly for the total synthesis of proteins. The recent improvements in the coupling chemistry and automation have made it possible to synthesize proteins with about 100 amino acid residues by chemical synthesis.[15] The synthesis of HIV protease and its mutants by Kent and his associates authenticates the improved technologies for the total chemical synthesis.[16–20] However, a much wider application of peptide synthesis is only in the preparation of medium sized peptides, particularly for immunological studies.

The advent of recombinant DNA technology and site-directed mutagenesis has simplified the design and the preparation of molecular variants of proteins. Proteins differing from the wild type by a single amino acid at a specific site[21–23] can be generated easily. Such a capability has revolutionized the approaches available for the structure function correlation studies of proteins on one hand and the development of new Biotechnology Industries[24–30] on the other. This technology has provided a new dimension to the development of enzymes with improved activity and/or stability and the design of novel therapies for diseases.

One limitation with recombinant DNA methods of protein engineering is its restriction to the naturally occurring amino acid residues. The desire of the protein chemists to introduce noncoded amino acid residues into the proteins demanded a higher flexibility in the approaches chosen for the design and fabrication of new versions of proteins. The chemical synthesis of peptides affords the flexibility for incorporating noncoded amino acids,[31–33] as well as amino acids with spectroscopic probes[34,35] (NMR, EPR, or fluorophores) or with radionucleides,[36–37] and for the introduction of unnatural (backbone) bond functionalities into proteins and/or peptides. An *in vitro* translation system has also been developed by Schultz and his colleagues, which also gives the flexibility to insert noncoded amino acid residues at the desired sites of the proteins. The chemical synthesis of peptides and proteins still provides a higher degree of flexibility in assembling proteins and peptides with noncoded amino acids.

6.2 SEMISYNTHESIS

Protein semisynthesis is a valuable alternative when one desires to conserve the flexibility of chemical synthesis of peptides and to avoid or minimize difficulties encountered in total chemical synthesis of proteins. This approach, as the name itself implies, involves the chemical synthesis of a polypeptide segment of a protein with the desired alterations at the sites of choice. This chemically synthesized polypeptide segment is then assembled with the complementary fragment(s) derived from the wild type protein to generate a functional protein.[38–43] The current approaches of protein semisynthesis for assembling the functional protein analog can be broadly grouped as non-covalent and covalent semisynthesis.

6.2.1 Non-covalent Semisynthesis

This approach utilizes the *principle of complementation* and involves mixing of synthetic mutant fragment with the complementary fragment prepared from the parent protein. Thus, these semisynthetic proteins are the mutant versions of the parent protein except that these do not contain the covalent continuity present in the parent protein. The complementary fragments are held together by strong non-covalent interactions, maintaining nearly the same overall conformation of the parent protein in spite of the presence of the discontinuity in the polypeptide chain. The noncovalent interacting system is referred as a *fragment complementing* system. The fragment complementing system of a single-chain protein could be considered as the new version of the protein with two subunits. As in the multimeric (dimeric in this case) proteins, the noncovalent interactions are strong enough to permit their assembly from the purified segments (artificial subunits).

The region of the protein wherein a discontinuity in the polypeptide chain is tolerated is referred to as *permissible discontinuity region* of the protein. Such regions of permissible discontinuity have been identified in a number of proteins such as RNase A,[44] staphylococcal nuclease-T,[45] cytochrome c,[46,47] human somatotropin,[48] prolactin,[49] thermolysin,[50] and rhodopsin.[51,52] The noncovalent complexes of these proteins are reported to have nearly the same structural properties as their respective native forms.

These complexes have been used to explore the structure-function relationships of the respective proteins. The first step to develop a fragment complementing system of a protein is the identification of *permissible discontinuity site* in that protein and represents the most difficult challenge. Much work is needed in this area to appreciate the structural basis and the significance of the permissible discontinuity sites in proteins and the potential use of such discontinuity sites of proteins in the semisynthesis and structure-function correlation studies of proteins.

6.2.1.1 Subunit Exchange of Multimeric Proteins

The noncovalent semisynthesis of single-chain peptide can be considered as a special case of the exchange of one subunit of a fragment complementing system with a chemically synthesized subunit (fragment). The chemical synthesis of RNAse S-peptide as well as its analogs and the study of their assembly to generate RNAse S mutants (these are referred to as RNAse S' to distinguish them from the parent material, RNAse S) is the best studied system of this class.[53-55] In the case of RNAse S, Beintima and his colleagues have crossed the species boundary in their attempts to assemble chimeric RNAse S'. They have carried out the subtilisin modification of pancreatic RNase from various mammals that exhibit multiple sequence differences compared to bovine pancreatic RNase A. Besides, they have also exchanged S-peptide and S-protein over the species boundaries to generate novel RNAse S mutants[56,57] or chimeric species of RNAse S. Chemically synthesized S-peptide of rat pancreatic RNAse has also been used to generate chimeric hybrids.[58]

Exchange of subunits of proteins consisting of more than one type of subunit has been a common practice for assembling new structural variants. Early attempts of this kind involved the exchange of the subunits of proteins (carrying out the same function) by crossing over species boundaries. Riggs and his colleagues assembled interspecies hybrid Hb (donkey, elephant, and mouse), and this represents one such pioneering attempt.[59] These interspecies hybrids are therefore chimeric hemoglobins and could be considered as semisynthetic proteins in that these attempt to introduce a defined set of sequence differences into an evolutionarily conserved structural motif to establish the influence on the structure and function by the sequence difference(s) in one of the chains. Preparation of hybrid immunoglobulin by exchanging the heavy and light chains of antibodies specific for haptens is another example of an early study that contributed to the recognition of the basic immunoglobulin fold.[60,61]

The subunit exchange approach has been used to generate Hb with two sequence differences to map the intermolecular contact sites of deoxy HbS polymer. In HbS, Glu-6(β) has been mutated to Val, which decreases the solubility of deoxy protein and results in its polymerization. Nature has provided the hematologists a large repertoire of Hb variants with single mutations, either in the α or in the β–chains. The Benesch group has used the available naturally occurring mutant α chains to hybridize with β^s-chains to generate Hb with two sequence differences per $\alpha\beta$ dimer. They have used such double mutants extensively to map the intermolecular contact regions of the deoxy HbS polymer.[62–66] Nagel and Bookchin[67–69] have generated a new class of mutant Hb *in situ*, namely asymmetric Hb, using two mutant Hbs. Their investigations have taken advantage of the high stability of the $\alpha_1\beta_1$ interface of $\alpha\beta$ dimers of Hb and stabilization of the $\alpha_1\beta_2$ interface of Hb in the deoxy conformation, thereby limiting the dissociation of the tetramers into $\alpha\beta$ dimers. The *in situ* formation of such asymmetric hybrids, where one β-chain of one dimer contains a mutation at one site and the other β-chain in the other dimer of the same tetramer contains the mutation at another site, has proved powerful in establishing that only one of the Val-6(β) of HbS tetramer participate in the polymerization of the deoxy protein.[70]

6.2.2 Covalent Semisynthesis

Despite the utility of non-covalent approaches, the reformation of the missing peptide bond between the peptide fragments is sometimes necessary or desirable. As discussed in Chapter 3, the complementing systems can greatly simplify the religation as exemplified in the semisynthesis of cytochrome *c* and its variants.[79–83] The proximity of the amino and the carboxyl groups of the discontinuity sites in the fragment complementing systems, which is provided by the *native-like* conformation of such systems, has been ingeniously exploited by Laskowski and his colleagues to reestablish the chain continuity using proteases as the splicing enzymes.[71–74] The synthetic activity of proteases (formation of peptide bonds) has been demonstrated by Fruton and his colleagues.[75–78] However, these were mostly transpeptidation reactions. The ligation of complementary fragments of a noncovalent complex of a protein is distinct from these transpeptidation reactions in that the α-carboxyl is present as the free carboxyl group and not as an amide. The pioneering work of Laskowski and his colleagues has established the generalized concepts and protocols for developing new systems of protease mediated covalent semisynthesis of proteins. They demonstrated the propensity of proteases to catalyze the reaction in a reverse direction, namely synthesis of the peptide bonds from free α-carboxyl and α-amino groups. The organic cosolvents like glycerol were used to shift the equilibrium of the reaction from the hydrolysis of peptide bond toward the synthetic direction.[84] The propensity of these solvents to facilitate the protonation of the carboxyl groups appears to be related to the decreased activity of water as well as the hydrophobic medium that they generate.

As noted earlier, subtilisin catalyzed splicing of the discontinuity sites of RNase S to RNase A is the first example of this class of splicing reactions.[73] The trypsin catalyzed splicing of staphylococcal nuclease-T[85,86] and thrombin catalyzed semi-synthesis of human somatotropin[88] represent the translation of the original principles derived from the studies of Laskowski and represent some of the best known examples for the protease catalyzed semisynthetic reactions. A wider application of the enzyme catalyzed semisynthetic reactions for the structural-functional studies has been advanced by Offord and his colleagues.[89–91]

6.3 SEMISYNTHESIS OF α-GLOBIN

For the studies on Hb, our interest has been to develop covalent semisynthesis to introduce sequence difference at desired sites of α and β chains. The efforts to develop semisynthetic approaches for the α and β chains of HbA was initiated to determine whether the principles learned from subtilisin catalyzed splicing of RNAse S could be translated to other protein systems. However, although a semisynthetic approach has been developed, the approach that finally evolved turned out to be very distinct from the previously studied protease catalyzed covalent semisynthetic reactions of proteins. The α-globin semisynthetic reaction represents the V8 protease catalyzed ligation of the complementary fragments of α-globin at the Glu-30 Arg-31 peptide bond[92] (Figure 6.1). This selective splicing reaction is apparently dictated by the organic-cosolvent induced secondary structure of the nascent peptide. The secondary structure is determined by the amino acid sequence of the contiguous segment. For the α-globin semisynthetic reaction, it represents the amino acid

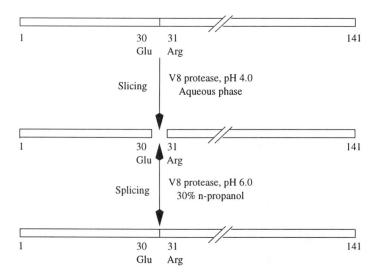

FIGURE 6.1 Diagramatic representation of the α-globin semisynthetic reaction.

sequence around Glu-30 Arg-31 peptide bond of human and/or other mammalian α-globins. The details of the development of this splicing reaction and the elucidation of the mechanistic aspects, as well as the application of the α- globin semisynthetic reaction to delineate new molecular aspects of polymerization of deoxy HbS, is discussed in this chapter. This semisynthetic reaction has many potential applications to other structure function correlation studies of Hb. The splicing reaction, along with approaches to the generation of chimeric HbS (symmetric interspecies hybrids) as well as the *in situ* generation of asymmetric hybrids, has helped us to deduce new structural information on the α-chain contact sites of the deoxy HbS polymer, which are also discussed in this chapter.

6.3.1 LIMITED PROTEOLYSIS OF α-CHAIN AND α-GLOBIN BY STAPHYLOCOCCAL V8 PROTEASE

The initial step in the development of protease catalyzed covalent semisynthetic reactions is the identification of the permissible discontinuity site in the polypeptide chain. Limited proteolysis of the protein is the only approach available at this stage to pursue this line of investigation. In view of the relatively high selectivity of V8 protease for the Glu and Asp residues of proteins, and higher specificity for Glu residues in buffers containing ammonium acetate or chloride,[93] as well as the presence of limited number of Glu residues in the α-chain, staphylococcal V8 protease was chosen for the initial studies of limited proteolysis. This protease exhibits two pH optima, at 7.8 and 4.0. Intact HbA is resistant to V_8 protease digestion.[94]

On the other hand, the isolated α- and β-chains are digested readily by this enzyme. Glu-27, Glu-30, and Asp-47 are the major sites of digestion in the α-chain at neutral pH, but the peptide bond of Glu-30 Arg-31 is the one that is most readily hydrolyzed by V8 protease.[94] If the reaction is carried out at pH 6.0, the digestion could be restricted to Glu-30 Arg-31 peptide bond, though the cleavage at this site is still not quantitative. Under physiological conditions, the complementary fragments α_{1-30} and α_{31-141} (with the bound heme) conserve enough of the *native-like* noncovalent interactions between them to prevent the separation of the α_{1-30} by centricon filtration into the filtrate. However, if the pH of the V8 protease digest is adjusted to 2.0 before the centricon separation, the segment α_{1-30} accumulates in the filtrate. Thus, the noncovalent interactions present between the segment α_{1-30} and α_{31-141} at neutral pH facilitating the retention of α_{1-30} during the concentration using centricon, is lost when the pH is lowered to 2.0 (when all the ionizable groups of the protein are protonated).

On the other hand, at pH 4.0 the digestion of α-chain is restricted to the Glu-30 Arg-31 bond, and the proteolytic cleavage is quantitative. This facilitates the isolation of the complementary fragments, α_{1-30} and α_{31-141}. The separation of the complementary fragments from the V8 protease digest is readily achieved by RPHPLC[95] or by the gel filtration on Sephadex G-25 equilibrated with 0.1% trifluoroacetic acid, or by urea CM-cellulose chromatography.[96] The heme of the α-chain does not seem to play any role in restricting the digestion to Glu-30 Arg-31 peptide bond when the digestion is carried out at pH 4.0. The same selective and quantitative digestion can be obtained when α-globin is used as the substrate instead of α-chain.[97]

6.3.2 V8 PROTEASE CATALYZED SPLICING OF THE COMPLEMENTARY
FRAGMENTS OF α-GLOBIN

The splicing of the complementary fragments is also catalyzed by V_8 protease. The splicing of the complementary fragments of α-globin,[95] namely, α_{1-30} and α_{31-141}, can be carried out over a wide range of pH, from 5.5 to 8.0 and 4° C. The ligation reaction is facilitated by incorporating 30% n-propanol as the organic cosolvent in the buffer. The ligation reaction is generally carried out at a protein concentration of 1 mM; the amino component (α_{31-141}) is generally present at 1.2 equivalent to that of the carboxyl component (α_{1-30}). If the solubility of the components permit, a higher concentration of the fragments could also be used. Equilibrium yields of 40 to 50% are generally obtained in 24 to 48 h. Other organic solvents such as propane diol, butane diol, and trifloro ethanol have also been found to be suitable for the splicing reaction.[95] However, glycerol, which has almost become a universal organic cosolvent to carry out the protease catalyzed semisynthetic reaction, was not a good solvent if heme free globin fragments are used. On the other hand, if the splicing reaction is tried with the heme containing α_{31-141}, 50% glycerol serves as a useful organic cosolvent.

6.4 MOLECULAR ASPECTS OF α-GLOBIN
SEMISYNTHETIC REACTION

Addition of organic cosolvents to a mixture of two small noninteracting peptides (one of which is generally an amide/ester) favors the synthesis (condensation) of the two peptides forming a new peptide bond.[76, 98–101] The organic cosolvent increases the pKa of the α-carboxyl group of the peptide. This aspect makes it readily susceptible for the splicing reaction, and the use of a large excess of the amino component, the nucleophile, facilitates the synthetic reaction by favoring the microscopic reversibility toward the synthetic direction. A number of organic solvents have been shown to facilitate the protease catalyzed ligation reactions of peptides, and this reaction does not exhibit any unique preference for one organic solvent over the other. The protease catalyzed semisynthesis of proteins is distinct from the protease catalyzed splicing reactions of small peptides in that the protein semisynthetic reactions are generally carried out with an equimolar mixture of complementary fragments, namely the amino and the carboxyl component. Besides, all the organic cosolvents that facilitated the subtilisin catalyzed splicing of the noninteracting peptides did not work in the protease catalyzed semisynthesis of RNAse A from RNAse S. Glycerol (90%) served as a good organic cosolvent, and this has remained a versatile solvent for the protease catalyzed semisynthesis of proteins. This unusual specificity aspect of the organic cosolvent in the protease catalyzed semisynthetic reaction is reflective of the distinct differences that exist in the molecular basis of the protease catalyzed covalent semisynthesis of proteins and in the splicing of smaller noninteracting peptides. The trypsin catalyzed addition[102] of Gly-NH_2 to Arg-141(α) of hemoglobin as well as the trypsin catalyzed formation of humulin[103,104] are examples of the latter category.

The noncovalent interactions of the fragment complementing systems that help to maintain a native like structure of the protein also ensures the proximity of the α-amino and the α-carboxyl groups at the permissible discontinuity site. This is one of the essential, as well as unique, structural aspects of the protease catalyzed covalent semisynthetic reactions of proteins. Accordingly, the organic cosolvent used to facilitate the protease catalyzed splicing reaction should be compatible with the conservation of the noncovalent interactions of the fragment complementing system. Glycerol is one such compatible organic cosolvent, and hence its widespread use in the protease catalyzed semisynthesis of proteins.

6.4.1 Unique Aspects of α-Globin Semisynthetic Reaction

V8 protease catalyzed α-globin semisynthetic reaction of the complementary fragments, α_{1-30} and α_{31-141}, in the presence of n-propanol is distinct from other fragment complementation mediated protease catalyzed protein semisynthetic reactions in that it proceeds well in the presence of diverse organic cosolvents like butane diol, propane diol, trifluroethanol,[95] etc. As noted above, the α-globin semisynthetic reaction exhibits a very high degree of flexibility both in terms of the compatibility of the organic cosolvent and the apparent pH optimum of the splicing reaction. The semisynthetic reaction is generally carried out in the presence of 30% n-propanol at pH 6.0 but could also be carried out over a wide range of pH from 5.0 to 8.5. Interestingly, even at pH 7.5, where the hydrolytic activity of the V8 protease is high, the splicing reaction proceeds with an equilibrium yields nearly comparable to that at pH 6.0.

As noted above, one of the common factors among the various protease catalyzed semisynthesis is the noncovalent interactions between the complementary fragments that are to be spliced by the protease. However, such noncovalent interactions between the two complementary fragments of α-globin, $\alpha(1-30)$ and $\alpha(31-141)$ could not be detected in the presence of 30% n-propanol.[105] This observation clearly demonstrates that the mechanistic aspects of this splicing reaction are distinct from the other protease catalyzed splicing of the discontinuity sites of fragment complementing systems. Besides, glycerol, the versatile organic cosolvent for the protease catalyzed proteosynthetic reactions, is not a suitable organic cosolvent for the semisynthesis of α-globin.[93] These unique and distinguishing features of the α-globin semisynthesis suggest that the organic cosolvent used in this reaction probably performs roles in addition to its generally implicated role for the protonation of the α-carboxyl group. One unifying property of the various organic cosolvents that are compatible with the α-globin semisynthetic reaction is that all of these are helix inducing solvents.[106-110] Therefore, it is conceivable that the induction of an α-helical conformation into either the complementary fragments or to the protease spliced nascent segment could play a role in the splicing reaction.

Another unusual feature of the α-globin semisynthetic reaction (as compared with other protease catalyzed splicing reaction) is that the truncated globin fragments also participate in this reaction.[105,111] For example, the segments α_{31-47} and α_{31-40} can be spliced (Figure 6.2) with α_{1-30} with nearly the same overall equilibrium yields obtained when α_{31-141} is used as the amino component (nucleophile). Similarly, α_{17-30}

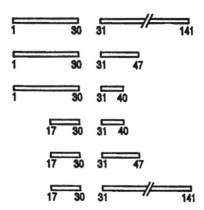

FIGURE 6.2 Schematic representation of the propensity of V8 protease to catalyze the splicing of the human α-globin fragments at Glu-30 and Arg-31. α_{1-30} gets spliced with α_{31-141} or α_{31-47} or α_{31-40} with nearly the same yield. Similarly, α_{17-30} gets spliced with α_{31-40} or α_{31-47} or α_{31-141} with comparable yields.

could be spliced with α_{31-40} or α_{31-47} or with α_{31-141}, with an overall equilibrium yield obtained when α_{1-30} is used as the carboxyl component. On the other hand, the truncation of α_{1-30} on its carboxyl side to generate α_{1-27} or α_{1-23} completely abolished the propensity of the truncated fragments to splice with α_{31-40} or α_{31-47} or α_{31-141}. This implies a critical role for the residues α_{28-30} in the splicing reaction.

6.4.2 INDUCTION OF α-HELICAL CONFORMATION INTO THE NASCENT PEPTIDE BY THE ORGANIC COSOLVENT

The unusual features of the α-globin semisynthetic reactions led us to speculate on the unique role for the organic cosolvent in this splicing reaction that was either absent in other protein semisynthetic reactions or has not been recognized previously. We have hypothesized that the organic cosolvent induced α-helical conformation of the nascent contiguous segment (generated in the synthetic reaction) plays a dominant role in the facilitation of synthesis. In support of this concept, it was observed that the segment α_{17-40} exhibits a considerably higher amount of α-helical conformation in the presence of n-propanol than the equimolar mixture of α_{17-30} and α_{31-40}. Establishing the chain contiguity facilitated the α-helical conformation induction process (Figure 6.3). Thus, the higher propensity of the contiguous segment to assume an α-helical conformation in the presence of n-propanol, as compared to the equimolar mixture of the complementary fragments, could serve as the *molecular trap* of the α-globin semisynthetic reaction. Conceivably, the α-helical conformation induced into the nascent polypeptide chain with the newly synthesized Glu-30 Arg-31 peptide bond endows this peptide a degree of resistance to V8 protease digestion. Establishing the contiguity between the complementary fragments *(similar to that present in the native protein)* is the event that facilitates the *in situ* induction of the α-helical conformation. Accordingly, this could also be the molecular event that triggers the generation of the molecular trap *in situ*.

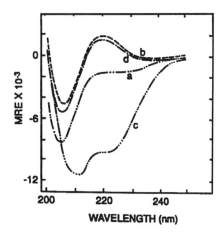

FIGURE 6.3   Influence of n-propanol on the helical conformation of α_{17-40}. CD spectra of contiguous and discontiguous α_{17-40} in the presence and absence of n-propanol at pH 6.0 is shown. Curves a and c represent the contiguous α_{17-40} in the absence and presence of n-propanol, respectively. Curves b and d represent the discontiguous α_{17-40} (equimolar mixture of α_{17-30} and α_{31-40}) in the absence and presence of n-propanol (30%).

6.4.3   Propensity of the Organic Cosolvent Induced α-Helical Conformation of the Nascent Polypeptide to Endow a Resistance against Protease Digestion

If the induction of the α-helical conformation into the nascent polypeptide chain is the *molecular trap* of the semisynthetic reaction, it should be possible to splice the complementary fragments of the α-helical segments using proteases. This aspect has been studied using complementary fragments of RNAse S-peptide, namely RNAse A_{1-9} and RNAse A_{10-20}. A significant level of the α-helical conformation could be induced into RNAse S-peptide in the presence of n-propanol. The two complementary fragments of S-peptide failed to undergo V8 protease catalyzed segment condensation reaction in the presence of 30% n-propanol *(conditions similar to the splicing of α_{1-30} and α_{31-47} of α-globin)* to generate RNAse A_{1-20} *(S-peptide)* even though the splicing of the complementary fragments by forming Glu-9 Arg-10 peptide bond is consistent with specificity requirement of the enzyme. On the other hand, the formation of Glu-9 Arg-10 peptide bond proceeded smoothly if RNAse S protein was also present in the splicing mixture.

The formation of V8 protease catalyzed splicing of Glu9 Arg10 peptide bond in the mixture of the complementing segments RNAse A_{1-9} and RNAse A_{10-20} is not unique to the presence of n-propanol as the organic cosolvent. The splicing reaction also proceeded smoothly in the presence of 50% n-glycerol as long as S-protein was present in the system.[111] These results demonstrated that the organic cosolvent induced α-helical conformation into the enzymically ligated nascent contiguous segment by itself does not serve as the universal *molecular trap* for protease mediated splicing of the noninteracting fragments. Thus, the results suggest that the *quality* of the n-propanol induced α-helical conformation of RNAse S peptide in the presence of organic cosolvent is distinct from that generated by the noncovalent interactions of S-peptide and S-protein wherein the latter acts as a tertiary template for the induction of the helical conformation. It appears that the n-propanol induced α-helical conformation of α_{17-47} and α_{17-40} has some unique features that mimic the role that RNAse S-protein played in facilitating the V8 protease catalyzed splicing of the complementary segments of S-peptide.

6.4.4   THERMODYNAMIC STABILITY OF THE SPLICED PEPTIDE BOND: ROLE OF THE (I + 4) SIDE-CHAIN INTERACTION

In an attempt to determine whether any unique sequence aspects are generated as a consequence of splicing reaction, the amino acid sequence of α_{17-40} has been examined very closely. This has immediately revealed that an (i, i + 4) side-chain interaction involving the γ–carboxyl group of Glu27 and the guanidino group of Arg31 is possible in the α-helical conformation of the contiguous segment, once the peptide chain contiguity is generated *in situ* and the contiguous chain assumes an α-helical conformation (Figure 6.4). Such an interaction should provide a stabilization to the n-propanol induced α-helical conformation of the contiguous peptide. The induction of the α-helical conformation or the (i, i + 4) side-chain interaction independently or a synergy of the two molecular aspects, could impart a resistance to the proteolytic digestion of Glu-30 Arg-31 peptide bond (the thermodynamic stability). Consistent with this, site-specific perturbation of this putative side-chain interaction by replacing Glu27 with Val resulted in nearly 50% reduction in the equilibrium yields of the splicing reaction[112] of α_{17-30} and α_{31-47}. Thus, the propensity to generate an (i, i + 4) side-chain interaction in the enzymically ligated peptide appears to determine the quality of the propanol induced α-helical conformation of α_{17-47}.

In an attempt to determine whether segments shorter than α_{17-30} also retain splicing potential (Figure 6.2), α_{17-30} with Val at α-27 was subjected to V8 digestion

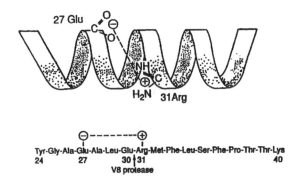

FIGURE 6.4   Diagrammatic presentation of the possible i + 4 interaction between γ-carboxyl group of Glu-27 and guanidino group of Arg[31] in the semisynthetic α-helical fragment that may be anticipated to stabilize the α-helical conformation and act as the *conformational trap* of the semisynthetic reaction. Amino acid sequence of the segment α_{24-40} is also given to show the high content of helix promoting amino acid residues in the peptide segment.

to isolate α_{24-30} containing Val at α-27. (The residue at α-23 is a Glu residue, a V8 protease sensitive site). The mutant α_{24-30} segment generated could also be spliced with α_{31-47} again with an overall efficiency of 25% of the parent peptide (α_{17-47}). The peptide, α_{24-30} with Glu at α-27 has to be prepared by chemical synthesis. This peptide, α_{24-30} (with Glu at α-27) showed that shortening the peptide from α_{17-30} to α_{24-30} has little influence when α_{31-47} is used as the amino component. However, if α_{31-40} is used as the amino component, the equilibrium yield is lowered. Thus, the total chain length of the nascent chain also plays some role in establishing the *quality* of the α-helical conformation induced into the nascent peptide.

6.4.5 γ-CARBOXYLATE-GUANIDINO SIDE-CHAIN INTERACTION IN THE ORGANIC COSOLVENT INDUCED α-HELICAL CONFORMATION OF NASCENT PEPTIDE AS THE PRIMARY DETERMINANT OF THE SPLICING POTENTIAL

Truncated fragment systems involving α_{24-30} and α_{31-40} as well as α_{24-30} and α_{31-47} have been used as the two model peptide systems to establish the influence of the various amino acid residues at α-27 and α-31. The replacement of Glu-27 with Val or Gly at this site retains considerable amount of the splicing reactivity, whereas the replacement with Lys or Pro abolished splicing reaction almost completely. Similarly, substitution of Arg-31 by Val decreased the equilibrium yield of splicing reaction to about 20%. On the other hand, if the positive charge at this site is retained by the replacement of Arg-31 with Lys, the influence on the synthetic activity is very limited. As noted earlier, the truncation of α_{1-30} to fragments α_{1-23} and α_{1-27} completely abolishes the splicing potential. In addition, the fragment α_{28-30} also does not condense with α_{31-47}. These results demonstrate the requirement of the presence of a negatively charged residue at α-27, and a positively charged residue at α-31, to generate an efficient (highly productive) *molecular trap* of the semisynthetic reaction. The (i, i + 4) interaction that can operate in the contiguous segment contributes to the stabilization of the n-propanol induced α-helical conformation (nascent α-helix) and thereby facilitates the splicing reaction.

The absence of the (i, i + 4) side-chain interaction does not completely abolish the splicing potential. Accordingly, it has been interpreted that the *in situ* generated *molecular trap* of α-globin semisynthetic reaction involves two components. The first component is the induction of the α-helical conformation into the contiguous segment *in situ* by the organic cosolvent. The newly established chain contiguity between the complementary segments (formation of the peptide bond Glu-30 Arg-31) facilitates the translation of the *conformational code* of the nascent peptide and, accordingly, the nascent peptide assumes an α-helical conformation in the presence of the organic cosolvent (Figure 6.4). This organic cosolvent induced α-helical conformation itself imparts partial resistance to the Glu30 Arg31 peptide bond against the V8 protease mediated hydrolysis. The second part of the molecular event that imparts further resistance to the V8 protease catalyzed hydrolysis is the stabilization of the n-propanol induced α-helical conformation by the (i, i+ 4) side-chain interaction, discussed above (Figure 6.5). In the case of α-globin semisynthetic reaction, both the molecular components independently contribute to the overall yield of the splicing reaction.

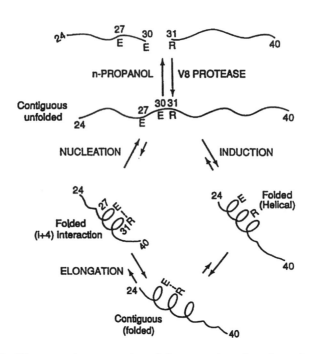

FIGURE 6.5 Diagrammatic presentation of the generation of conformational trap in the enzymically ligated peptides. Once the contiguity is generated, the synthetic peptide could acquire the stable structure in two alternate pathways. In one pathway, the helical conformation is induced at any site (induction) of the peptide and propagated through out the peptide and then stabilized by i + 4 interaction of Glu-27 and Arg-31. Alternatively, the helical conformation is induced in a site-specific fashion (nucleation), simultaneously stabilizing the structure by i + 4 interaction.

6.4.6   PROTECTION OF OTHER PEPTIDE BONDS OF NASCENT α–GLOBIN IN THE PRESENCE OF ORGANIC COSOLVENT

Thus, the *molecular trap* of the α-globin semisynthetic reaction invokes two molecular processes as contributing toward the facilitation of splicing reaction. These elements explain the splicing of the truncated globin fragments. However, with the full-length α-globin, the introduction of α-helical conformation to the contiguous region appears to be coupled to another molecular process, namely the collapse of contiguous α-globin with its organic cosolvent induced α-helical conformation (Figure 6.6) into a native-like fold (format). This second contribution of n-propanol should also be considered here as an equally important molecular process in the overall success of α-globin semisynthetic reaction. As noted earlier, the peptide bond of Asp-47(α) is also susceptible to V8 protease digestion in aqueous medium. However, this peptide bond becomes resistant to proteolysis in the presence of n-propanol. The induction of α-helical conformation into the contiguous B-Helix region of the α-globin chain facilitates the packing of the various helical segments into a *native-like* format with a concomitant resistance of the peptide bond of Asp-

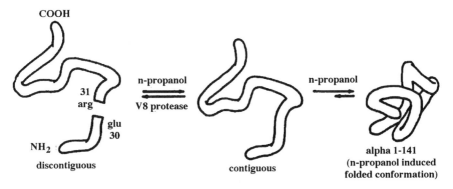

FIGURE 6.6 Schematic representation of the dual role of the organic cosolvent, n-propanol, in the V8 protease catalyzed splicing of Glu-30 Arg-31 peptide bond in the complementary fragments of α-globin. The initial role of n-propanol is to protonate the carboxyl group to facilitate the reverse proteolytic reaction and the second role of the solvent is to induce α-helical conformation into the newly synthesized α-globin. The α-helical conformation induced into the semisynthetic α-globin fragment facilitates the folding of globin chain into a native-like format. This serves as a molecular trap of the semisynthetic reaction.

47(α) toward proteolysis. This induction of native-like format in the presence of organic cosolvents appears to be not unique to α-globin; other apoproteins (that are helical in the native state) as well as the fragments of coiled coil proteins also behave the same way. Therefore, it is conceivable that the mechanistic principles deduced from the α-globin semisynthetic reaction could be translated to other proteins that are stabilized in the presence of organic cosolvents.

6.5   EXPERIMENTAL PROTOCOLS FOR SEMISYNTHESIS OF CHIMERIC α-GLOBIN

The V_8 protease catalyzed semisynthesis of α-globin (as well as the chimeric α–globin) involves at least the following five steps:

1. Purification of α- and β-globin chains of hemoglobin
2. Preparation of complementary fragments of α-globin (α_{1-30} and α_{31-141})
3. Preparation of mutant fragments
4. Splicing of complementary fragments
5. Isolation of semisynthetic α-globin or the chimeric α-globin.

The basic steps are briefly discussed below. For a more detailed account of these methodologies, the reader is advised to refer to a recent article by Roy and Acharya in *Methods in Enzymology.*[96]

6.5.1   PURIFICATION OF α- AND β-GLOBIN OF HEMOGLOBIN

The general procedure that we have used for the separation of α- and β-chains of human hemoglobin is the one originally developed by Bucci and Frontecelli.[113,114]

The PMB-derivatives of the chains are purified by CM-52 cellulose. The globin from the isolated chains is prepared by acid acetone precipitation.[115] Nonetheless, the use of urea CM cellulose chromatography discussed below could also be used.

It is not possible to isolate heme containing chains of animal Hb[116] by modifying the tetramers with hydroxymercuri benzoate and CM cellulose chromatography of the modified Hb. Accordingly, with animal Hbs, the separation has to be done as the heme free globin chains. This is achieved by chromatographing the acid acetone precipitated total globin on CM-cellulose columns in the presence of 8 M urea.[117] The lyophilized total globin, acid-acetone precipitate of animal or human Hb (up to 200 mg), is dissolved in the starting buffer (5 mM phosphate buffer, pH 7.0, containing 50 mM β-mercaptoethanol and 8 M urea) and dialyzed against same buffer for 2 to 3 h. The dialyzed sample is loaded on CM-52 column (2.5 × 10 cm) equilibrated with starting buffer. The column is washed with the starting buffer to elute any material that does not bind strongly to the resin. The globin chains are eluted by applying a linear gradient generated from 250 mL each of starting and final buffers (30 mM phosphate buffer, pH 7.0, containing 50 mM β-mercaptoethanol and 8 M urea). In this gradient system β-globin elutes earlier than the α-globin. The fractions containing α- and β-globin are pooled separately and dialyzed extensively against 0.1% acetic acid and stored as the lyophilized material.

6.5.2 PREPARATION OF COMPLEMENTARY FRAGMENTS OF α_{1-30} AND α_{31-141}

As noted earlier, the slicing of α-globin by V8 Protease at the Glu^{30} Arg^{31} peptide bond is specific and quantitative at pH 4.0 and 37° C. For protease digestion, the lyophilized α-globin is dissolved in 10 mM ammonium acetate buffer, pH 4.0. A protein concentration of 0.5 mg/mL and a V8 protease to α-globin ratio of 1:200 (w/w) is normally used for the digestion. The complete digestion of the Glu-30 Arg-31 peptide bond is achieved in an hour, however digestion period of 2 to 3 h is used to ensure the complete digestion. The digestion of the α-globin could be readily monitored by RPHPLC. After the desired digestion period, the material is lyophilized. However it should be noted that, even though these conditions have proved to be specific for the cleavage of Glu-30 Arg-31 bond by V8 protease with human, mouse, horse, swine, and monkey α-globin, it is advisable to follow the kinetics of the digestion with studies of a new α-globin chain to ascertain whether the conditions are appropriate for quantitative and selective cleavage of the Glu^{30} Arg^{31} peptide bond. RPHPLC analysis of the digestion mixture is a simple and ideal procedure to follow the kinetics of digestion. RPHPLC is also useful for the preparation of the complementary fragments in amounts of 10 to 20 mg.

For a large-scale preparation of the complementary fragments (in 100 to 200 mg range), α_{1-30} and α_{31-141}, we have generally used the chromatography of the V8 protease digest of the α-globin on CM cellulose in the presence of 8 M urea. The lyophilized digest is dissolved in the starting buffer (5 mM phosphate buffer, pH 7.0, containing 50 mM β-mercaptoethanol and 8M urea) and loaded on the column. As α_{1-30} does not bind to the column, it can be eluted out by washing the column with the starting buffer. After the complete elution of α_{1-30} as determined by monitoring the effluent at 280 nm, the column is eluted with a linear gradient, generated from 100 mL each of starting (25 mM phosphate buffer, pH 7.0, 50 mM in β-mercaptoethanol and 8M in urea) and final buffers (50 mM phosphate buffer, pH 7.0, 50 mM in β-mercaptoethanol and 8M in urea). α_{31-141} elutes almost at the end of the gradient. Undigested α-globin, if there is any, elutes before α_{31-141}. The fractions containing α_{31-141} are pooled and dialyzed extensively against 0.1% acetic acid and lyophilized. The fractions containing α_{1-30} are pooled and desalted on a Sephadex G- 25 column (2.5 × 50 cm), equilibrated with 0.1% TFA and lyophilized. The lyophilized fragments are stored at −80° C if not used immediately.

6.5.3 ISOLATION/SYNTHESIS OF MUTANT SEGMENTS OF α_{1-30} AND/OR α_{31-141}

The segment α_{1-30} with desired replacement of amino acids can be readily synthesized by chemical methods. The chemical synthesis allows the incorporation of noncoded amino acids as well as naturally occurring amino acids labeled with desired spectroscopic probes as well. The unique strength of the chemical synthesis and hence of the semisynthetic approaches is the ease with which a spectroscopic probe tagged to a coded or noncoded amino acid residues could be incorporated at the desired sites of the peptide.

Although the larger segment α_{31-141} could also be synthesized by chemical methods, the size of this fragment might make it some what difficult task and may not be a worthwhile effort in view of the ease with a polypeptide chain of this size could be generated by the current recombinant DNA technology, unless the attempt is to introduce noncoded amino acids at one or more sites. Development of additional strategic sites in this segment for the enzymic or chemical condensation of the complementary fragments of α_{31-141} could prove valuable for the chemical synthesis of mutant forms of α_{31-141}.

Mutant complementary fragments can also be prepared from human α-chains containing sequence differences in α_{31-141} region of the chain. As the Glu-30 Arg-31 peptide bond is conserved in most of the non-human α-chains, and the V8 protease is likely to slice the globin to generate the respective complementary fragments, a suitable non-human α-chain with desired mutations can be selected to exchange the complementary fragments with that of human α-chain generating chimeric α-chains. The preparation of the complementary fragments of non-human α-globin can be achieved by the method as described for human α-globin.[96]

6.5.4 SEMISYNTHESIS OF α-GLOBIN

V8 protease catalyzed semisynthesis of α-globin is usually carried out in 50 mM ammonium acetate buffer, pH 6.0 containing 30% n-propanol. But it should be noted that this splicing reaction occurs over a wide range of pH (4.5 to 8.5) and concentration of n-propanol (30 to 70%).

The fragments α_{1-30} and α_{31-141} are mixed at a molar ratio of 1:1.2 (carboxyl to amino component) and lyophilized. The lyophilized sample is dissolved in an appropriate amount of water and then diluted with 100 mM ammonium acetate buffer, pH 6.0, and n-propanol to get a final concentration of 50 mM in ammonium acetate and 30% in n-propanol. The concentration of peptide fragments is kept at 1 mM. The solution is cooled to 4° C in an ice bath for 30 min, and then an aliquot of V8 protease is added so that the ratio of the enzyme to substrate is 1:200 (w/w). The reaction is allowed to proceed at 4° C for 24 to 72 h. The kinetics of the splicing reaction are followed by RPHPLC.

6.5.5 PURIFICATION OF SEMISYNTHETIC α-GLOBIN (CHIMERIC α-GLOBIN)

The purification of semisynthetic α-globin (or chimeric α-globin) from a synthetic mixture that contains unligated complementary fragments is carried out by ion exchange chromatography of the material on CM-cellulose in the presence of 8 M urea. The conditions employed for chromatography are similar to the one used for the separation of the complementary fragments, α_{1-30} and α_{31-141}, from the V8 protease digest of α-globin. However, it should be noted that, in the V8 protease digestion of α-globin, the slicing reaction is quantitative, whereas equilibrium yields in the splicing reaction is generally about 40%. Accordingly, the chromatographic conditions used with a given synthetic mixture may have to be optimized to get a satisfactory separation of all the components present in the mixture. This would depend on the charge and nature of the components in the synthetic mixture, particularly when non-human α-chains are used to generate human-nonhuman or non-human-nonhuman chimera. The fractions containing the semisynthetic α-globin are pooled, dialyzed against 0.1% acetic acid, and lyophilized. The unligated α_{1-30} and α_{31-141} fragments may also be recovered and used again for the next cycle of splicing reaction. The contiguity of the segment, α-globin, can be established by a tryptic map.

6.6 ASSEMBLY OF SEMISYNTHETIC α-GLOBIN INTO THE FUNCTIONAL TETRAMER

The semisynthetic α-globin is reconstituted into tetramers with either HMB-β^A or β^S chains by the *Alloplex intermediate* pathway.[118] This aspect has also been discussed in detail by Roy and Acharya,[96] and a reader interested in the details of the protocol should refer to that chapter in *Methods in Enzymology*. The reconstituted tetramer obtained by this procedure will be in the ferric form. It is reduced to the ferrous state by dithionite and purified on CM-cellulose to get the semisynthetic oxyhemoglobin. To ascertain the appropriate chemical composition of the tetramer, the determination of chain composition and the tryptic peptide mapping of the globin chains may be carried out. The conformational equivalency and/or similarity of the chimeric Hb to the parent Hb can be confirmed by the spectroscopic analysis and oxygen affinity measurements of the assembled tetramer.

In much the same way as with other proteins, site-directed mutagenesis of Hb has become a powerful technique in recent years to correlate the structure with the function. The semisynthetic approach discussed here should not be considered as an alternative to the approaches of recombinant DNA technology. Semisynthetic and the recombinant DNA approaches have unique strengths of their own and, at best, these are only complementary to each another. It is conceivable that a combination approach can be a powerful methodology for the structure-function relationship studies of proteins. In the structural studies of Hb, we anticipate that such a combined approach will facilitate investigations of the structural dynamics of oxygenation-deoxygenation mediated conformational aspects of Hb as well as polymerization studies of deoxy HbS. Of particular interest for the latter is that the semisynthetic approach has the flexibility to introduce noncoded amino acids at a given site or amino acid residues with unique spectroscopic probes.

6.7 QUINTIC STRUCTURE OF THE DEOXY HbS POLYMER

Before we address the application of the semisynthetic reaction to delineate the structural issues of deoxy HbS polymer, a brief introduction is provided here on the molecular pathology of sickle cell disease and our current appreciation of the quintic structure of the deoxy HbS polymer. The pathophysiology of sickle cell disease is a direct consequence of the single base mutation in the β-globin gene, an A $\rightarrow$ T mutation in the codon for the sixth residue of the β-chain. This single base mutation in gene results in the substitution of glutamic acid at the β^6 position of HbA by a Val residue. The resultant mutant Hb, sickle Hb (HbS), represents one of the most abundant naturally occurring mutant forms of Hb. This mutation leads to the polymerization of HbS when it is deoxygenated. The formation of the deoxy HbS polymer within the erythrocytes results in a considerable decrease in cell deformability and distortion of the cell shape; many of the erythrocytes take a sickle shape and hence the name sickle cell disease.[119,120] These deformed, rigid cells obstruct flow in the microcirculation.

The vaso-occlusive consequences of sickle cell disease are critically dependent on the intracellular concentration of deoxy HbS as well as the kinetics of polymerization. The understanding of the quintic structure of HbS polymer and the basic molecular alterations (solution active conformational differences) in deoxy Hb tetramer associated with the substitution of the acidic (charged) glutamic at the β_6 position by an uncharged Val residue with a hydrophobic side chain are the major objectives of current studies of HbS. It is generally believed that one or more solution active conformational differences of deoxy HbS, as compared with that of deoxy HbA, triggers the polymerization process. A clear understanding of these molecular details is a fundamental prerequisite either to design effective antisickling agents or to develop antisickling hemoglobins as candidates for gene therapy of sickle cell disease.

Electron microscopy coupled with the image reconstruction studies of the deoxy HbS polymer have established that the fibers present either in cells or generated *in vitro* consist of 14 strands, a closely packed inner core of four strands and an outer sheath of ten strands. The electron micrographic analysis, however, do not provide direct information per se on the orientation of individual HbS molecules in the fiber.[121-125] Definitive information on the intermolecular contacts in the sickle cell polymer has come from X-ray diffraction studies of deoxy HbS crystals,[126–129] molecular modeling of these structures into deoxy HbS polymers,[130,131] and the *in vitro* polymerization studies of binary mixtures of HbS and chemically modified HbS or HbA.[119, 120]

The X-ray crystallographic studies of deoxy HbS crystal have demonstrated that Wishner-Love (WL) double strand[126] (Figure 6.7) is the basic structural unit of the deoxy HbS fiber. The solubility studies with the binary mixtures of HbS and mutant HbA have established this fact.[69] Nonetheless, it should be noted that the double strands are linear in the crystal, whereas they are helical in the deoxy HbS polymers. The translation of the intra- and inter-double strand contacts, as reflected in the crystal structure to the helical format that exists in the polymer, has been a considerable challenge. A consequence of this difficulty is a discrepancy between the observations of the solution studies using mutant HbS (experimental observation) and the current molecular models of fiber (calculated or inferred). Only additional experimental observations using new mutant HbS can facilitate the resolution of this discrepancy.

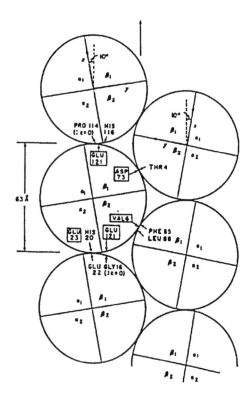

FIGURE 6.7 Wishner-Love double-strand model showing the arrangement of HbS tetramers in the fiber. Val-6(ß) of cis ß-chain (ß₂) of a tetramer from one strand makes the crucial lateral contact with the hydrophobic pocket (Phe[85] and Leu[88]) of the trans ß-chain (ß₁) of adjacent tetramer from the second strand. Some of the other lateral and axial contact residues involving α- and ß-chains are also shown.

The intermolecular contact regions within this basic structural unit (WL double strand) could be grouped into at least two prominent regions of the tetramer: the intra-double-strand lateral contacts and the intra-double-strand axial contacts. The intra-double-strand lateral contact region involves Val-6(β), the mutation that is responsible for the polymerization of deoxy HbS. The hydrophobic side chain of the Val-6(β) residue interacts with the acceptor pocket provided by the β–chain of another tetramer of the double strand, and this is predominantly an inter-tetrameric $\beta\beta$ interaction. Only one of the two Val-6(β) residues of HbS tetramer is involved in this interaction. The acceptor pocket is located at the EF corner. The Val-6(β) of each molecule in one filament makes contact with Ala-70(β), Thr-84(β), Phe-85(β), and Thr-87(β) of a molecule in a neighboring filament. Besides this interaction, the WL double strand is also stabilized by the inter-double-strand vertical contacts, known as intra-double-strand axial contacts, that involve inter-tetrameric interactions of the β-chain as well as the α-chains (Figure 6.8). The crystallographic analysis of the deoxy HbS crystal has assigned these two classes of interaction with reasonable amount of certainty. The fiber structure deduced from the crystallographic analysis is also consistent with the inhibition of polymerization seen with mutants of HbS.

The mutants of HbS for such studies are assembled *in vitro* as symmetric hybrids with a specific single mutation *per $\alpha\beta^S$dimer*, or generated as an asymmetric hybrid with a single site-specific mutation *per asymmetric tetramer* ($\alpha\beta^*\alpha\beta^S$), the β^* represents the chain with a single site-specific mutation. The asymmetric tetramers have facilitated the mapping of the cis and trans configuration of the intermolecular contact sites. The α- or the β-chain in the $\alpha\beta$ dimer of the tetramer that provides the Val-6(β) residue to form the polymer is referred to as the *cis dimer,* and the other dimer, the β-chain of which provides the acceptor pocket, is referred to as the *trans dimer.*

Besides these two regions of the tetramer, a third region of the tetramer also contributes to the stability of the deoxy HbS fiber by generating inter-double–strand interactions. This region is a unique domain specific for HbS fiber and is not found in the crystal. These interactions are predominantly from the α-chains of the tetramer. The inter-double-strand interactions are the least defined or understood in terms of the structure of the deoxy HbS fiber but are expected to be distinct between the various double strands. The inter-double-strand interactions need to be further refined.

Given the multiplicity of the intermolecular contact regions in the deoxy HbS polymer, the next level of structural analysis is the delineation of the linkage of the intermolecular contact site interactions. This can be achieved by the perturbation of two or more contact sites (regions) simultaneously and establishing the additivity and/or synergy of the polymerization inhibitory potential of these perturbations. The interaction linkages that are to be mapped could involve the amino acid residues within one contact domain (axial contact domain, or lateral contact domain) or two different domains or all the intermolecular contact domains. The first step for delineating the interaction linkage maps is the generation of HbS mutants with two or more sequence differences. The approaches of Benesch and Benesch for generating HbS with a mutation in the α-chain (i.e., combining naturally occurring mutant α-chains of Hb with the β^S-chain) could be extended for the generation of HbS with two mutations. HbS with two (or more) mutations at the implicated contact sites of the α-chain should help us to delineate the interaction linkage map of the contact sites in the polymer. The α-globin semisynthetic reaction is ideal to construct α-chains with two or multiple sequence differences needed for these studies. HbC Harlem,[132] HbS Antilles,[133] and HbS Osman[134] represent another set of naturally occurring mutant HbS with a second mutation in their β-chain. β^S-chains from these could be assembled with one or more mutant α-chains, and another set of chimeric HbS with additional sequence difference in the β^S-chain and multiple sequence difference in the α-chain could also be assembled.

The application of the α-globin semisynthetic reaction for the generation of an α-globin with two (or multiple) sequence differences is exemplified by exchange of complementary segments of the α-chains of two naturally occurring Hb variants with single α-chain mutations: HbI[135] [$_{16}$(Lys → Glu)] and Hb Sealy[136] [$_{47}$(Asp → His)]. Both Lys-16(α) and Asp-47(α) have been identified as intermolecular contact sites. Accordingly, generation of an α-chain with both of these mutations[137] will provide an opportunity to establish the additive/synergistic aspect of antipolymerization potential of these two sequence differences at α_{16} and α_{47}. The assembly of various chimeric α-chains by exchanging the complementary segments either between the two animal α-chains or between the human and animal α-chains is an extension of this approach to generate α-globin with multiple sequence differences.

6.8   ASSEMBLY, STRUCTURE, AND POLYMERIZATION OF CHIMERIC HEMOGLOBIN S

The tetramers assembled from the chimeric α-chain, or an animal α-chain with multiple sequence differences and β^A or β^S-chains, represent a class of molecular species of Hb with multiple sequence differences in the α-chain. These molecular species are referred to as *inter-species hybrids*. Since the tetrameric Hb is assembled by mixing the subunits of Hb from two different species, these are chimeric hemoglobins.

6.8.1   GENERATION OF CHIMERIC HEMOGLOBINS

Three classes of chimeric HbS have been generated as means of introducing a set of sequence differences into HbS: symmetric chimeric HbS are generated by hybridizing the β^S-chain with the animal α-chain, asymmetric HbS are generated *in situ* by mixing the chimeric HbS with either HbS or HbA, and chimeric HbS assembled from β^S-chains and chimeric α-chains. Asymmetric tetramers of these are also generated *in situ* in much the same way as with the symmetric chimeric HbS.

6.8.1.1   Symmetric Chimeric HbS

The chimeric HbS, i.e., hybrid Hb with multiple sequence differences, have been assembled *in vitro* previously as an attempt to establish the extent of the complementarity conserved between the α and β–chains of various vertebrate Hbs. Early attempts by Riggs and his colleagues[59] and the recent studies of Fronticelli and her colleagues[138] are examples of the preparation of such chimeric Hbs. These investigations of generation of chimeric Hbs represent an approach of introducing a set of sequence differences into the parent HbS *in vitro* by exchanging the subunits of proteins carrying out identical functions without the formation of the covalent bonds (Figure 6.9). These are, in principle, examples of noncovalent semisynthesis of Hb and are comparable to the formation of interspecies hybrid RNAse S discussed earlier.

There has been considerable interest in recent years in the development of transgenic animals expressing human Hb, particularly developing transgenic mice expressing HbS as an animal model for sickle cell disease.[139–143] Similarly, transgenic swine expressing HbA have also been developed as a source for the large-scale production of human pathogen-free HbA and could be used to generate Hb-based oxygen carriers.[144] The expression of human Hb in transgenic animals is expected to result in the production of two types of inter-species hybrids: $\alpha_2^A\beta_2^H$ and $\alpha_2^H\beta_2^A$ (A = animal; H = human), the chimeric Hbs along with the parent species of Hb. Thus, this represents an *in vivo* system for the generation of chimeric Hbs.

The chimeric Hbs could also be readily assembled *in vitro* by the *alloplex intermediate pathway*. We have now generated *in vitro* a number of chimeric HbS by assembling the vertebrate α-chains (or globin) with β^S-chains. The α-chains included in these studies are of mammalian origin, particularly from monkey, mouse, horse and swine.

6.8.1.2 Asymmetric Chimeric HbS

Another set of species of semisynthetic HbS, asymmetric hybrids (asymmetric chimeric HbS) have been generated *in situ* (Figure 6.9). These chimeric Hbs are

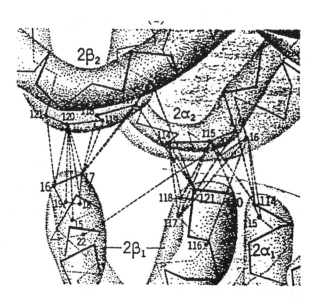

FIGURE 6.8 Intra-double-strand axial contact residues of HbS fiber. Most of these contacts involve the residues of α- and ß-chains coming from the AB and GH regions of the chains.

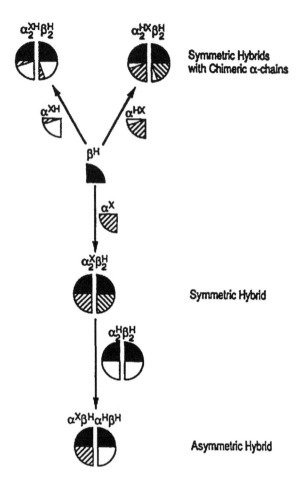

FIGURE 6.9 Schematic representation of the generation of various types of chimeric hemo-globins. α^H and $ß^H$ represent human α- and ß-chain, respectively. α^X represents an animal α-chain with multiple sequence differences, and α^{XH} and α^{HX} represent the chimeric α-chains with a set of sequence differences in either $\alpha(1–30)$ or $\alpha(31–141)$ region of the chain.

stable only under deoxy conditions, and in the oxy conformation these readily segregate into respective parent tetramers. When an equimolar mixture of two types of Hbs is deoxygenated, asymmetric hybrid containing an $\alpha\beta$ dimer from each of the two parent Hbs is formed with an equilibrium distribution of 25% each of parent Hbs and 50% of the asymmetric hybrid.[145,146] In this system of asymmetric hybrid, both α- and ß-chains of the two $\alpha\beta$ dimers of the tetramers are distinct. On the other hand, if an inter-species hybrid (chimeric Hb) is mixed with one of the parent Hbs, only one of the chains (either α- or β-chain) of the $\alpha\beta$ dimer will be distinct between the two $\alpha\beta$ dimers of the asymmetric hybrid. The approach of generating asymmetric hybrids has played a pivotal role in understanding the stereochemistry of the inter-molecular contacts of deoxy HbS polymer.

6.8.1.3 Semisynthesis of Chimeric α-Chain and the Generation of Symmetrical Chimeric HbS Containing Chimeric α-Chain

The α-globin semisynthetic reaction discussed above has been explored to exchange the complementary fragments of two α-chains to construct a chimeric α-globin containing a set of desired sequence differences (Figure 6.10). The exon/intron structure of the α-globin gene is conserved beyond the species boundary. The Glu-27, Glu-30, and Arg-31 are conserved in most of the vertebrate α-globin, certainly in the mammalian α-globin. Therefore, the structural principles involved in the *conformational trap* of the α-globin semisynthetic reaction can be extended to splice α_{1-30} and α_{31-141} segments of human α-globin with the respective complementary segments of the animal α-globin. In generating these chimeric α-chains, one has introduced a set of sequence differences of animal α-chain into the human α-chain. If the parent animal α-chain has enough complementarity with human β-chain to generate the tetramer, the chimeric α-chains can be hybridized with human β-chain (Figure 6.9). Chimeras of human and mouse, human and horse, as well as human and swine α-globins have been constructed. These chimeric α-globins have been reconstituted with β^S-chain, generating new species of HbS containing the chimeric α-chain.

Another set of chimeric α-chains could also be synthesized by exchanging the complementary fragments between two different animal α-chains. When these constructions are made, one of these chimeric α-chains may have considerably more number of sequence differences compared to both of the parent α-chains, depending on the distribution of sequence differences in α_{1-30} and α_{31-141} segments of the two parent chains. In these cases, it is also conceivable that the chimeric α-chains may end up with a lower degree of complementarity with human β-chain as compared to that in either of the parent animal α-chains. The sequence differences of chimeric α-chains are not chosen through evolutionary processes and, accordingly, the sequence of the human β–chain is unlikely to be ideally suited. To date, we have constructed one such chimeric α-chain generated from mouse α_{1-30} horse α_{31-141} chimeric α-chain. This has a total of 23 sequence differences, compared to the human α-chain, and 15 and 8 as compared to parent mouse and horse α-chains,[147] respectively. Interestingly, even this chimeric α-globin exhibited enough complementarity with the β^S-chain, and Hb assembled with these chains exhibited normal cooperative oxygen affinity.

6.8.2 Functional Properties of Chimeric HbS and HbS with Chimeric α-Chains

The various species of chimeric HbS (interspecies hybrid) and hybrids of β^S-chains with chimeric α-chains are listed in Table 6.1, along with their functional properties. All chimeric HbS with multiple sequence differences in their α-chains exhibited cooperative oxygen binding, the oxygen affinity comparable to that of human Hb. One exception is the chimeric HbS generated from the human-mouse chimeric α-chain.[148] These results support the fact that *native-like* quaternary structure of HbS

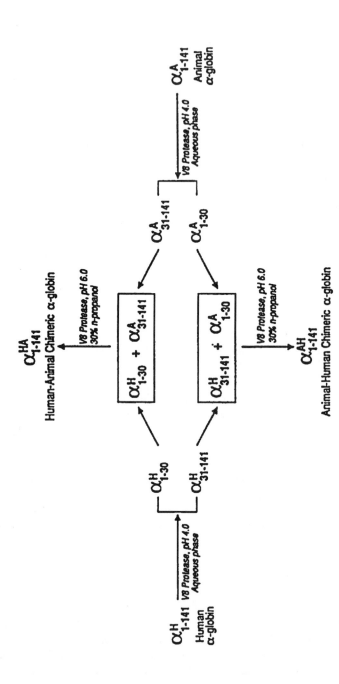

FIGURE 6.10 The generation of chimeric α-globin by exchanging α(1–30) and α(31–141) fragments of human and animal α-globin by V_8 protease catalyzed semisynthetic reaction.

TABLE 6.1
Oxygen Affinity of Hemoglobin Hybrids

Hemoglobin Hybrid	P_{50}	Hill Coefficient
$\alpha_2^H\beta_2^S$	4.5	2.5
$\alpha_2^{Mn}\beta_2^S$	4.5	2.3
$\alpha_2^M\beta_2^S$	4.5	2.4
$\alpha_2^M\beta_2^A$	3.4	2.5
$\alpha_2^{MH}\beta_2^S$	4.5	2.4
$\alpha_2^{HM}\beta_2^S$	3.0	2.4
$\alpha_2^{Hr}\beta_2^S$	4.5	2.3
$\alpha_2^{Hr}\beta_2^A$	4.7	2.5
$\alpha_2^{HrH}\beta_2^S$	4.7	2.3
$\alpha_2^{HHr}\beta_2^S$	4.3	2.3
$\alpha_2^{HrM}\beta_2^S$	3.4	2.3
$\alpha_2^{MHr}\beta_2^S$	3.7	2.3

The oxygen affinity of the hybrids has been determined at a hemoglobin concentration of 0.5 mM in 50 mM Bis-Tris acetate, pH 7.4, 37° C, using Heme-O-Scan.

is generated in these chimeric HbS, even though the α-chain of these tetramers, derived either from another mammalian source or the mammalian α-chain is partially humanized. The ability of these mammalian or chimeric α-chains to generate functional tetramer is consistent with the concept that the architectural principles for the generation of the basic Hb fold is evolutionarily conserved.

The generation of the basic Hb fold does not predict the intrinsic oxygen affinity of the chimeric Hb. A study of the modulation of the oxygen affinity of chimeric HbS by allosteric effectors has suggested that the alien α-chains of chimeric HbS transmit conformational changes into the DPG binding pocket. This suggests that a flexibility exists within the basic Hb fold to modulate the oxygen affinity either directly or through the modulation of the interaction of the tetramers with the allosteric effectors.[149] The high oxygen affinity of the chimeric HbS containing the chimeric α-chain generated from human α_{1-30} and mouse α_{31-141} reflects the *crosstalk* between the α_{1-30} and α_{31-141} segment of mouse α-chain, which is critical to modulate the conformational aspects of human β-chains and the DPG pocket to lower the oxygen affinity of the tetramer.

The significance of this type of architectural information for the future design of novel Hb-based oxygen carriers is self-apparent and represents a new area in which to embark. The current interest in this dimension is represented by recent studies of Nagai and his colleagues[150] to delineate and transplant the sequence differences responsible for the bicarbonate effect from crocodile Hb into human Hb, and that of Fronticelli and her colleagues[138] to identify and transplant the sequence differences responsible for the unusual chloride effect of bovine Hb into human Hb to generate a low oxygen affinity species of Hb.

6.8.3 INHIBITION OF β^S-CHAIN DEPENDENT POLYMERIZATION BY MOUSE α-CHAIN

As a prelude to the development of the transgenic mouse model for sickle cell disease, Beuzard and his colleagues assembled interspecies hybrids[151] of mouse α-chain and β^S-chain as well as mouse α-chain and $\beta^{S\ Antilles}$. The interspecies hybrids did not polymerize in 1.8 M phosphate buffer,[152,153] demonstrating the high polymerization inhibitory potential of mouse α-chain.[151] Since the mouse α-chain exhibits a very high polymerization inhibitory potential against the β^S-chain dependent polymerization, the identification of the sequence differences that endowed this inhibitory potential to the mouse α-chain will increase our appreciation for the quinary structural aspects of the HbS polymer on one hand, and the design of the antisickling Hb on the other.

There are 19 sequence differences in the mouse α-chain as compared to the human α-chain. Out of 19 sequence differences, only 3 are located at the implicated contact sites of the fiber, namely, α-48, α-78, and α-116 (see Figure 6.11 and Table 6.2). Of these three, α-48 and α-116 have been implicated as intra-double-strand contact sites,[130,131] and α-78 as an inter-double-strand contact site in the fiber. However, the amino acid residues present at these sites in the mouse α-chain are not the ones that have been shown to be inhibitory when present in the human α-chain. The polymerization inhibition by mouse α-chain could be due to one or more sequence differences located at these three contact sites, or it could be an additive/synergistic

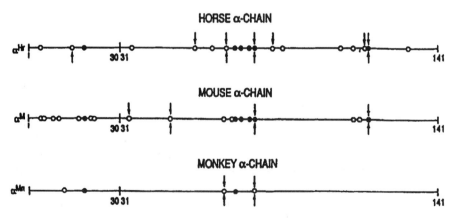

FIGURE 6.11 Comparison of the positions of the amino acid sequence differences of monkey, mouse and horse α-chains. Open and closed circles represent the positions of sequence differences in the animal α-chains as compared to human α-chain, closed circles represent the common sequence differences between the animal α-chains. The upward arrows indicate the human α-chain contact sites of deoxy HbS fiber implicated by solution studies, and downward arrows indicate the contact sites implicated by molecular modeling studies.

TABLE 6.2
Human, Monkey, Mouse, and House α–Sequence Differences

Sequence (Helix)	Human	Monkey	Mouse	Horse	Contact Site from Solution Studies	Contact Site from Molecular Modeling
4 (A2)	Pro		Gly	Ala	—	—
5 (A3)	Ala		Glu	—	—	—
8 (A6)	Thr	Ser	Ser	—	—	—
10 (A8)	Val	—	Ileu	—	—	—
15 (A13)	Gly	—	—	Ser	*	—
17 (A15)	Val	—	Ileu	—	—	—
19 (AB1)	Ala	Gly	Gly	Gly	—	—
21 (B2)	Ala	—	Gly	—	—	—
22 (B3)	Gly	—	Ala	—	—	—
34 (B15)	Leu	—	Ala	—	—	*
35 (B16)	Ser	—	—	Gly	—	—
48 (CD6)	Leu	—	Val	—	*	*
57 (E6)	Gly	—	—	Ala	—	*
63 (E12)	Ala	—	—	Gly	—	—
67 (E16)	Thr	—	Ala	—	—	—
68 (E17)	Asn	Leu	—	Leu	*	*
70 (E19)	Val	—	Ala	—	—	—
71 (E20)	Ala	Gly	Gly	Gly	—	—
73 (EF2)	Val	—	Leu	Leu	—	—
76 (EF5)	Met	—	Leu	Leu	—	—
78 (EF7)	Asn	Gln	Gly	Gly	*	*
82 (F3)	Ala	—	—	Asp	—	*
85 (F6)	Asp	—	—	Asn	—	—
107 (G14)	Val	—	—	Ser	—	—
111 (G18)	Ala	—	Ser	Val	—	—
113(GH1)	Leu	—	His	—	—	—
115(GH3)	Ala	—	—	Asn	—	*
116(GH4)	Glu	—	Asp	Asp	*	*
130 (H13)	Ala	—	—	Ser	—	—

*Indicates the sites in human α-chain implicated as contact sites of HbS fiber either by solution studies or by molecular modeling studies.

influence of the perturbations at all three sites. Possible contribution of one or more of other sequence differences, located at sites that have not been implicated so far as contact sites of HbS fiber, cannot be ignored.

However, identification of the contact site sequence differences of the mouse α-chain and the determination of the molecular basis of inhibition by these sequence differences is not an easy task. A direct approach is to introduce these mutations into the human α-chain, one residue at a time, and establish the influence on polymerization. As a second phase of these investigations, one has to establish the interaction linkage map of the intermolecular contact sites by simultaneous perturbation of one or more of the implicated intermolecular contact regions.

An alternative, simpler approach will be to dissect out the contribution of a set of sequence differences of mouse α-chain by transplanting these sequence differences into the human α-chain. Thus, one could establish whether a given segment of the mouse α-chain, when contiguous with the complementary region of the human α-chain, i.e., partial murinization of human α-chain, exhibits any polymerization inhibition potential.

The location of mouse α-chain sequence differences in its three-dimensional structure reflects some interesting structural aspects. Seven out of the 19 sequence differences of the mouse α-chain are located in the AB/GH corner (Table 6.2 and Figure 6.11). α_{116} and α_{113} are also located in this domain. Interestingly, this domain of Hb tetramer is the resident of a number of implicated intra-double-strand axial contact sites. Another five sequence differences are located at or around the EF corner. α-78 is located in this domain. This region is the residence of inter-double-strand contact sites of the tetramer. Thus, 12 of the 19 sequence differences are concentrated in two domains of Hb that are involved in the intermolecular interactions of the deoxy fiber. Therefore, it is conceivable that these regions of the chimeric HbS have a conformational dynamics distinct compared to that in the parent HbS, in spite of the fact that the chimeric HbS retains nearly the same overall basic Hb fold. Such a change in the conformational dynamics of this region could independently inhibit the assembly of polymers from deoxy HbS tetramers, or these should be coupled to the presence of other sequence difference(s) elsewhere in the α-chain to achieve the inhibition of the polymerization. An approach of transplanting a segment of mouse α-chain (namely a set of sequence differences) into human α-chain and establish the polymerization inhibition potential will resolve the issues of these molecular aspects. The information that can be gathered from these sets of experiments will be distinct from the information to be gained by introducing all the sequence differences of mouse α-chain individually into human α-chain by site directed mutagenesis to generate single site mutant α-chains.

A simple approach to introduce a set of sequence differences of mouse α-chain into human α-chain, i.e., partial murinization of human α-chain, is to screen the available vertebrate α-chains and select the ones that contain nearly the identical sequence differences in a given segment of interest. Such vertebrate α-chains could then be isolated and hybridized with the β^S-chain to generate a chimeric HbS with a limited set of sequence differences. Using these chimeric HbS, the contribution of the common sequence differences present between the two α-chains toward the inhibition of polymerization could be established.

6.8.3.1 Polymerization Behavior of Chimeric HbS Generated from Monkey α-Chain and β^S-Chain

The initial experiments on the partial murinization were focused on the EF corner sequence differences of mouse α-chain, since this region involves the inter-double-strand contacts in HbS fiber. The monkey α-chain has five sequence differences as compared to human, and three of these are located around the EF corner (Figure 9.11 and Table 6.2). Two other sequence differences are located near the N-terminal region, in the A Helix. Of the three sequence differences located near the EF region of the monkey α-chain, two are at the implicated contact sites, α-68 and α-78. The human α-chain has Asn residue at both of these sites, whereas monkey α-chain has Leu at α-68 and Gln at α-78. Three sequence differences of monkey α-chain, at α-8, α-19, and α-71 are identical to that present in mouse α-chain. The other common site of sequence difference between mouse and monkey Hb is α-78. Mouse has Gly at this site, whereas monkey has Gln. Given the fact that two of the sequence differences are at the implicated inter-double-strand contact region, one should anticipate the inhibition of the β^S-chain dependent polymerization by the monkey α-chain. However, the chimeric HbS with monkey α-chain exhibited a solubility that is comparable to that of HbS.[154] The results therefore demonstrate that the five sequence differences of monkey α-chain by themselves are not inhibitory to the β^S-chain dependent polymerization. The substitution of Leu for Asn at α-68 and Gln for Asn at α-78 appear to be consistent with the stereochemical requirements of deoxy HbS polymer formation. On the other hand, the presence of Lys either at α-68 or at α-78 has an inhibitory effect. The replacement of Asn at α-68 by Leu, or at α-78 by Gln, does not perturb the overall net charge of the EF corner. Therefore, it may be argued that the inhibition seen in the earlier studies on having the Lys residue at either of these two sites is reflective of a very specific polymerization inhibitory propensity of the positive charge of Lys at these sites.

A new mutant form of monkey α-chain containing His instead of Gln at α-78 has been identified recently (Roy et al., unpublished results) during the screening of the monkey α-chains.[155] This mutant monkey α-chain has also been hybridized with the β^S-chain to explore the role of a positively charged residue at this contact site. This is a nonconservative substitution and resulted in a lowered solubility for the chimeric HbS compared to HbS,[154] demonstrating the polymerization potentiating influence of His at α-78. As noted earlier, the nonconservative mutation of Asn-78 to Lys inhibits the polymerization. Both Lys and His residues are basic amino acid residues. However, the influence of the side chain of the positively charged residues Lys and His at α_{78} appears to be distinct. The contribution of the positive charge by a His residue at neutral pH will be dependent on the pK_a of this residue at α_{78}.

In view of the demonstration that replacement of Asn by a His residue is an isosteric substitution, and positive charge of the side chain of Lys at this site is inhibitory to polymerization, the polymerization potentiation behavior of His at this site apparently reflects a better packing of the side chain of His at the inter-double-strand contact domain with additional noncovalent interactions. Since an uncharged polar residue, Gln, is accommodated at α-78 without significant influence on the polymerization, it is unlikely that a Gly residue at this site seen in the mouse α-chain by itself has contributed to the high polymerization inhibitory potential of mouse α-chain. Therefore, it is reasonable to conclude that the sequence differences at α-8, α-19, α-71, and α-78 of the mouse α-chain are not the direct contributors for the polymerization inhibitory potential of mouse α-chain.

6.8.3.2 Polymerization of Chimeric HbS Generated from Horse α-Chain and β^S-Chain

Horse α-chain has 18 sequence differences as compared to human α-chain (Figure 6.11, Table 6.2). Eight of the sequence differences of the horse α-chain are located at the same sites as in the mouse α-chain. The residues at six of these sites are identical to those in the mouse. Four of these identical sequence differences are located in the EF corner, namely the sites α-71, α-73, α-76, and α-78. As noted earlier, this is the region of inter-double-strand interactions. The other two of these sequence differences are located at α-19 and α-116. Thus, the comparison of polymerization inhibitory potential of mouse and horse α-chains could establish the relative contribution of the common sequence differences of the two chains. Although the horse α-chain inhibited the β^S-chain dependent polymerization, its inhibitory potential is lower than that of mouse α-chain (Figure 6.12). Thus, the six common sequence differences between mouse and horse α-chains by themselves do not appear to be the primary determinants of the high polymerization inhibitory potential of mouse α-chain.

Mouse and horse α-chains have a sequence difference at α-4. In human α-chain α-4 is proline, in mouse it is Gly, and in horse it is an Ala residue. In the Edelstein and Cretegny double-strand model, α-4 has been postulated as a contact site.[156] However, it is not a contact site in the Watowitch model,[131] and polymerization studies are not available at this stage to establish the identity of this site as a contact site. The other common site of sequence difference between the mouse and horse α-chain is at α-111. Ala is the residue at this site in the human α-chain, whereas it is Ser in the mouse α-chain and Val in the horse α-chain. This is not an implicated contact site of the HbS fiber.

Of the remaining ten sequence differences of the horse α-chain, four are at implicated contact sites. Three of these, α-57, α-82, and α-115, are contact sites in the Watowich model.[131] However, solution studies confirming their implications are not yet available. Another site, α-15, has been suggested to be a contact site based on solution studies. The sequence difference at α-15 is a conservative substitution (Gly → Ser). This substitution leads to a slight increase in the volume of the side chain at this site. Nonetheless, it should be noted here that α-15 is not a contact site according to the Watowich model.[131] It is conceivable that some inhibitory potential of the horse α-chain may be coming from one or more of these sites. However, since the overall inhibitory potential of horse α-chain is smaller compared to the mouse α-chain, the contribution from these sites appears to be limited.

Since the contribution of the 8 common sequence differences appears to be small, the higher polymerization inhibitory potential of mouse α-chain should be contributed by the 11 sequence differences that are not common between mouse and horse α-chains. Of these 11 sequence differences, only α-48 (Leu → Val) is located at an implicated contact site. However, this is a conservative substitution, and the contribution of the side chain of Val at this site to the inhibition process is not clear.

Therefore, one may speculate that the higher polymerization inhibitory potential of the mouse α-chain is either a consequence of one or more of the remaining ten sequence differences. Their inhibitory effect could be because their location is at

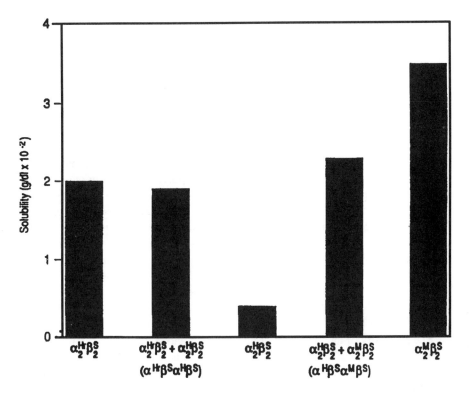

FIGURE 6.12 Comparison of polymerization inhibitory potential of mouse and horse α-chains. The solubility of the hemoglobin hybrids or hybrid mixtures (1:1) has been determined as follows: the hemoglobin hybrids or hybrid mixtures were taken (a total initial concentration of 1.5 mg/mL) in 2.0 M phosphate buffer, pH 7.4 at 0° C and deoxygenated in the presence of dithionite. The polymerization was induced by giving a temperature jump from 0 to 30^0 C. After the polymerization is over, as reflected by the optical density of the sample at 700 nm, the samples were spun at 9000 rpm for 20 min. The solubility of the hybrids was determined from the absorption of the supernatant at 555 nm using millimolar extinction coefficient of 12.5. The hybrids represented in the parentheses are the asymmetric hybrids generated in the copolymerization mixtures. α^{Hr} = horse α-chain; α^M = mouse α-chain, α^H = human α-chain.

the intermolecular contact region (sites) of the fiber. These sites are neither the assigned nor the postulated contact sites. Alternatively, the inhibition by the mouse α-chain could be a result of the synergy/additivity of the polymerization inhibitory potential of two or more of sequence differences at the implicated contact sites.

One type of synergy/additivity of the polymerization inhibitory potential is when, for example, α-chain contact site perturbations in the cis dimer act in concert with the perturbations of the α-chain contacts in the trans dimer. Similarly, the perturbations of two or more contact sites in the same α-chain, cis or trans, could also act in concert to inhibit the β^S-chain dependent polymerization.

The presence of such synergy/additivity of the inhibitory potential of the perturbations of contact sites could be resolved by segregating the sequence differences into the cis and trans dimers and determining the contribution of the segregated sequence differences in both location in the tetramer. The copolymerization approach of the type developed by Nagel and Bookchin[145,146] has been adopted to segregate the inhibitory potential of the mouse and horse α-chains in the cis and trans $\alpha\beta$ dimers.

6.8.3.3   Polymerization Inhibitory Potential of Horse α-Chain in cis and trans $\alpha\beta$ Dimers

As noted earlier, all the implicated contact sites of the tetramer do not make equivalent contributions to the structural stability of the polymer. For example, although there are two copies of Val-6(β) and two acceptor pockets per tetramer, the architectural design of the HbS polymer is such that it uses only one copy per tetramer. The β^S-chain of HbS that provides the Val-6(β) for the intermolecular interaction is referred as *cis chain,* and the dimer containing this β^S-chain as *cis $\alpha\beta$ dimer.* On the other hand, the β^S-chain that accommodates the hydrophobic side chain of Val-6(β) at its acceptor pocket is the *trans β chain,* and the $\alpha\beta$ dimer containing this trans β-chain is referred to as *trans $\alpha\beta$ dimer.* The intermolecular contact residues mapped into the α-chain located in the cis $\alpha\beta$ dimer are referred as *cis contact sites.* Similarly, the contact sites in the α-chain of the trans dimer as the *trans contact sites.*

The solubility of an equimolar mixture of HbS and $\alpha_2^{Hr}\beta_2^S$ is nearly the same[154] as that of the chimeric HbS, $\alpha_2^{Hr}\beta_2^S$ (Figure 6.12). This indicates that the asymmetric hybrid containing one horse α-chain and one human α-chain also has a solubility similar to $\alpha^{Hr}_2\beta^S_2$. This indicates that only one copy of horse α-chain in the chimeric tetramer is needed to inhibit the polymerization. The copolymerization studies of the chimeric HbS containing horse α-chain ($\alpha_2^{Hr}\beta_2^S$) with HbA and also with chimeric HbA [(containing horse α-chain ($\alpha_2^{Hr}\beta_2^A$)] have been carried out to determine whether the horse α-chain should be present in the cis or the trans dimer to exert its maximum inhibitory influence. The copolymerization studies have established that the polymerization inhibitory propensity of horse α-chain is much higher when it is present in the cis $\alpha\beta$ dimer[154] (Figure 6.13). However, all the polymerization inhibitory potential of horse α-chain does not come from the perturbation of the cis contact sites; the horse α-chain placed in the trans $\alpha\beta$ dimer also exhibits some inhibition of the polymerization.

6.8.3.4   Polymerization Inhibitory Potential of Mouse α-Chain in cis and trans $\alpha\beta$ Dimers

The solubility of an equimolar mixture of chimeric HbS containing mouse α-chain ($\alpha_2^M\beta_2^S$) and HbS is lower than that of the chimeric HbS ($\alpha_2^M\beta_2^S$) itself[154] (Figure 6.12). These results are distinct from those seen with the($\alpha_2^{Hr}\beta_2^S$), where the equimolar mixture HbS with the chimeric HbS exhibited a solubility comparable to that of the parent tetramer ($\alpha_2^{Hr}\beta_2^S$). The mouse α-chain appears to exhibit strong polymerization inhibition when present in either one of the dimers, and replacing one of the mouse α-chains in the tetramer with human α-chain reduces considerably the overall inhibitory influence.

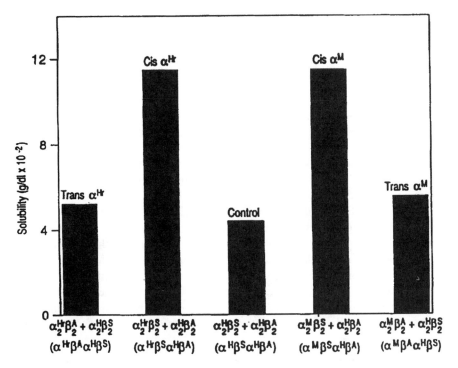

FIGURE 6.13 Comparison of the polymerization inhibitory potential of mouse and horse α-chains in cis and trans αβ dimers. The mixture of $\alpha_2^{Hr}\beta_2^A$ and $\alpha_2^H\beta_2^S$, on deoxygenation generates the asymmetric hybrid, $\alpha^{Hr}\beta^A\alpha^H\beta^S$, that has horse α-chain (α^{Hr}) associated with β^A-chain, i. e., trans αβ dimer with respect to β^S-chain. On the other hand, the mixture of $\alpha_2^{Hr}\beta_2^S$ and $\alpha_2^H\beta_2^A$, under deoxy conditions generates the asymmetric hybrid, $\alpha^{Hr}\beta^S\alpha^H\beta^A$, where horse α-chain is associated with β^S-chain, i. e., cis αβ dimer. Similar explanation is applicable to mouse α-chain hybrids. The experimental conditions and methodology employed are similar to the ones explained for Figure 6.12.

A comparison of the polymerization inhibitory propensity of the mouse α-chain with that of the horse α-chain in the cis and trans dimer suggests that the inhibitory potential of the chains in the cis and trans positions does not account for the total inhibitory potential of the chains when it is present in both of the dimers (Figure 6.12). This implies some degree of synergy in the cis and the trans inhibitory potential of the mouse or horse α-chain.

6.8.3.5 Segregation of the Inhibitory Potential of Mouse α–Chain into α(1–30) and α(31–141) Segments

The copolymerization experiments have demonstrated that the inter-chain complementation, namely the synergy/additivity of the perturbations of the intermolecular contact sites of the cis and trans dimer by the mouse α-chains placed in the respective αβ dimers of the tetramers, is responsible for the high polymerization inhibitory potential of mouse α-chain. Additional additivity and/or synergy of the inhibitory potential of sequence differences of the mouse α-chain come from within the α-chain, namely, between the sequence differences of segments within the contiguous chain (either in the cis or in the trans position), i.e., as an intra-chain complementation of the polymerization inhibitory potential of sequence differences. The latter structural aspect can be dissected out by transplanting a set of sequence differences of mouse α-chain into the human α-chain to generate chimeric α-chains.

Out of 19 sequence differences of the mouse α-chain, 8 are present in the α_{1-30} segment, and 11 are located in α_{31-141} segment of the chain. The α_{1-30} segment of human and mouse has been exchanged to generate human-mouse (α^{HM}) and mouse-human (α^{MH}) chimeric α-globins.[148] These chimeric α-globin chains have been assembled with the β^S-chains to generate the tetramer. These tetramers of β^S-chain containing chimeric α-chains are distinct from the parent chimeric HbS in that the α-chains are partially humanized. The human-mouse chimeric α-chain (α^{HM}) exhibited an inhibitory potential demonstrating the presence of polymerization inhibitory sequence differences in the mouse α_{31-141} segment. However, the inhibitory potential is considerably lower than that of mouse α-chain[148] (Figure 6.14). The other chimeric α-chain, mouse-human (α^{MH}), also retained some degree of inhibitory potential, even though none of the sequence differences of mouse α_{1-30} segment is at the implicated contact sites (Figure 6.11). Interestingly, these is a loss of considerable amount of inhibitory influence when the mouse α-chain segments are incorporated

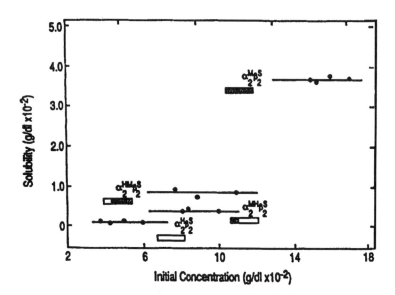

FIGURE 6.14 Segregation of the polymerization inhibitory potential of mouse α-chain into α(1–30) and α(31–141) segments. The open and striped bars indicate the human (α^H) and mouse α-chains (α^M), respectively, and the open-striped and striped-open bars indicate human α(1–30)-mouse α(31–141) chimeric α-chain (α^{HM}) and mouse α(1–30)-human (31–141) chimeric α-chain (α^{MH}), respectively. The solubility of the hybrids has been determined at different initial concentrations of the hybrids, employing the experimental protocols as described for Figure 6.12.

as pieces into the human α-chain; the two chimeric α-chains do not account for the total inhibitory potential of the mouse α-chain.

6.8.3.6   Polymerization Inhibitory Potential of Horse α(1–30) and α(31–141) Segments

Chimeric α-globins containing horse $α_{1-30}$ and human $α_{31-141}$ ($α^{HrH}$), as well as human $α_{1-30}$ and horse $α_{31-141}$ ($α^{HHr}$), have been generated and assembled with the $β^S$-chain to determine the contribution of the sequence differences of horse $α_{1-30}$ and $α_{31-141}$ segments in the polymerization inhibition. Chimeric HbS containing $α^{HrH}$ exhibits 3 sequence differences, and the one with $α^{HHr}$ exhibits 15 sequence differences, respectively, compared to human α-chain. These can be considered as partially equinized HbS. HbS equinized with the segment $α_{1-30}$, ($α_2^{HrH}β_2^S$) exhibited a polymerization behavior comparable to HbS (Figure 6.15). The contribution of the three sequence differences of horse $α_{1-30}$ is minimal in the inhibition of polymerization.[154] The site α-15 has been implicated as an intermolecular contact residue. The present results suggests that the Gly → Ser mutation at $α_{15}$ does not make any significant contribution to the polymerization inhibitory propensity of the horse α-chain. Solubility of HbS equinized with $α_{31-141}$, $α_2^{HHr}β_2^S$, is about 60% of the solubility of the chimeric Hbs with the full-length horse α-chain ($α_2^{Hr}β_2^S$). Thus, all the polymerization inhibitory sequence differences of horse α-chain appear to be located in the

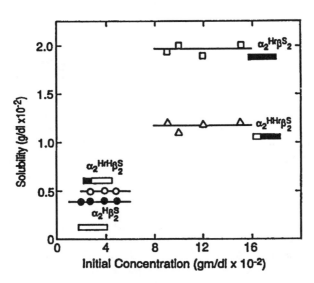

FIGURE 6.15   Segregation of the polymerization inhibitory potential of horse α-chain into α(1–30) and α(31–141) segments. The open and filled bars indicate the human ($α^H$) and horse α-chains ($α^{Hr}$), respectively and the open-filled and filled-open bars indicate human α(1–30)-horse α(31–141) chimeric α-chain ($α^{HHr}$) and horse α(1–30)-human α(31–141) chimeric α-chain ($α^{HrH}$), respectively. The solubility of the hybrids has been determined at different initial concentrations of the hybrids, employing the experimental protocols as described for Figure 6.12.

α_{31-141} segment, but the solubility of HbS equinized with horse α_{31-141} is not the same as the chimeric HbS with horse α-chain, even though the horse α_{1-30} segment exhibits very little polymerization inhibitory potential. This is indeed reminiscent of the results obtained with the mouse α-chain. Thus, it is clear that the sequence differences of α_{1-30} segment of mouse as well as of horse interact with the sequence differences of respective α_{31-141} to generate additional polymerization inhibitory influence that can not be accounted for by the inhibitory influence of the segregated segments.

6.8.3.7 Inter-dimeric (Intra-tetrameric) Complementation of the Polymerization Inhibitory Potential of Human α-Chain Murinized with α_{1-30} and Human α-Chain Murinized with α_{31-141}

The loss seen in the polymerization inhibitory potential of the mouse and horse α-chains on segregation of the sequence differences of the chains into α_{1-30} and α_{31-141} could be a consequence of the need for the sequence differences to be contiguous in the chain. Alternatively, if the sequence differences are segregated into two different α-chains of the tetramer, the polymerization inhibitory potential of the segments may be able to complement with each other and integrate their inhibitory potential to increase overall inhibitory potential of tetramer. The latter situation represents the a case when the polymerization inhibition occurs by linked multisite perturbation. One way of achieving such an integration is to establish the polymerization behavior of an equimolar mixture of two species of HbS containing the chimeric α-chains. Copolymerization studies of an equimolar mixture of HbS and $\alpha_2^M\beta_2^S$ have established that both mouse α-chains of chimeric HbS contribute to the inhibitory process.[154] Therefore, the loss of inhibitory potential of mouse α-chain on the segregation of its sequence differences into chimeric α-chains containing α_{1-30} and α_{31-141} could be a consequence of the need for the presence of one set of sequence differences in one α-chain and another set of sequence differences on the second α-chain of the tetramer. Under these circumstances, each α-chain could provide a different set of contact sites in the fiber. To resolve this issue, again the copolymerization approach of generating asymmetric hybrids *in situ* has been employed. Equimolar mixtures of HbS and hybrids containing chimeric α-chains $(\alpha^{MH}_2\beta^S_2)$ or $(\alpha^{HM}_2\beta^S_2)$ exhibited a considerably lower inhibitory potential[157] (Figure 6.16) than that of an 1:1 mixture of HbS and $(\alpha_2^M\beta_2^S)$. On the other hand, equimolar mixtures of two hybrids containing chimeric α-chains exhibited a higher inhibitory potential than that of 1:1 mixture of HbS and $\alpha_2^M\beta_2^S$. The copolymerization experiments permitted the presentation of the inhibitory potential of the full-length mouse α-chain as two murinized human α-chains. The asymmetric chimeric HbS generated *in situ* contains one copy of α_{1-30} murinized human α-chain and another copy of α_{31-141} murinized human α-chain. Apparently, the solubility of such an asymmetric chimeric HbS is higher than the one containing one copy of human α-chain and one copy of mouse α-chain.

These results suggest that the polymerization inhibitory potential of the α_{1-30} murinized human α-chain is able to complement intramolecularly with the polymer-

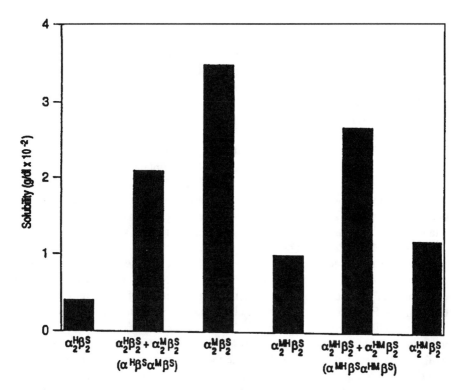

FIGURE 6.16 Complementation of the polymerization inhibitory potential of mouse $\alpha(1–30)$ and horse $\alpha(31–141)$. The asymmetric hybrid, $\alpha^H\beta^S\alpha^M\beta^S$, generated in the copolymerization mixture of $\alpha_2^H\beta_2^S$ and $\alpha_2^M\beta_2^S$ has all the 19 sequence differences of mouse α-chain on a single α-chain (α^M) whereas the asymmetric hybrid, $\alpha^{MH}\beta^S\alpha^{HM}\beta^S$, generated in the copolymerization mixture of $\alpha_2^{MH}\beta_2^S$ and $\alpha_2^{HM}\beta_2^S$ has 8 sequence differences of mouse $\alpha(1–30)$ on one α-chain and 11 sequence differences of mouse $\alpha(31–141)$ on another α-chain. The experimental conditions and methodology employed are similar to the ones explained for Figure 6.12.

ization inhibitory potential of $\alpha_{31–141}$ murinized human α-chain and generate an inhibitory potential that is higher than that expected for one copy of the murine α-chain. Accordingly, we speculate that the loss of polymerization inhibitory potential as a consequence on the generation of mouse-human and human-mouse chimeric α-chains reflects the loss of intramolecular complementation in the inhibitory influence of the mouse α-chains placed in the cis and the trans dimers.

6.8.3.8 Intra-chain Complementation of the Polymerization Inhibitory Potential of Equine and Murine α-Chains

The inter-dimeric complementation of the polymerization inhibitory potential of the mouse-human and human-mouse chimeric α-chains results in a significant recovery of the inhibitory potential of mouse α-chain that was lost on segregation of the sequence differences of the mouse α-chain into two chimeric α-chains. The inhibitory potential of an equimolar mixture of chimeric HbS containing α_{1-30} murinized human α-chain ($\alpha_2{}^{MH}\beta_2{}^S$) and chimeric HbS containing α_{31-141} murinized α-chain ($\alpha_2{}^{HM}\beta_2{}^S$) is still considerably lower than that of the Chimeric HbS containing murine α-chain ($\alpha_2{}^M\beta_2{}^S$). Thus, complementation of the polymerization inhibitory potential of the two pieces of a single copy of the murine α-chain is unable to exhibit the complete inhibitory potential of the intact murine α-chain. The presence of one or more sequence differences of the two pieces of mouse α-chain, when contiguous with those of complementary segment (either in the cis or trans position, or both), generates additional polymerization inhibitory potential.

As noted above, although the overall inhibitory potential of horse α-chain is considerably lower than that of mouse α-chain, the polymerization inhibitory potential of the mouse and horse α-chains are nearly equivalent when their inhibitory potential in cis and trans $\alpha\beta$ dimers are compared. The observation that horse α_{31-141} retains only 60% of the inhibitory potential of intact horse α-chain, even though the segment horse α_{1-30} does not posses any intrinsic polymerization inhibitory influence, suggests that the three sequence differences of horse α_{1-30} can activate additional polymerization inhibitory influence when they are contiguous with the sequence differences of horse α_{31-141}. The co-polymerization studies of an equimolar mixture of chimeric HbS containing horse α-chain and HbS containing α_{1-30} murinized human α-chain establish that the mouse-human chimeric α-chain and horse α-chain integrate inhibitory influence of their sequence differences to generate a very high inhibitory potential. This is in spite of the fact that the sequence differences are in two different α-chains of the tetramer. Accordingly, it was of interest to establish whether a contiguity in the sequence differences of the mouse α_{1-30} with those of the horse α_{31-141} will develop additional inhibition. The exchange of the segments of murine α-chain with the complementary segments of equine α-chain has been achieved to generate mouse-horse (α^{MHr}) and horse-mouse (α^{HrM}) chimeric α-chains. The antipolymerization potential of the horse-mouse chimeric α-chain (α^{HrM}) is lower than that of horse as well as mouse α-chain (Figure 6.17); however, it is higher than that of human-mouse chimeric α-chain.[158] Therefore, the sequence differences of horse α_{1-30} are unable to complement with that of mouse α_{31-141} as much as they do with the sequence differences of horse α_{31-141}. But the three sequence differences of the horse α_{1-30} do exhibit a complementation in terms of polymerization inhibition, as the polymerization inhibitory potential of horse-mouse chimeric α-chain is higher than that of human-mouse chimeric α-chain.

Contrary to the behavior of horse-mouse chimeric α-chain ($\alpha^{HrM}{}_2\beta^S{}_2$), the mouse-horse chimeric α-chain (α^{MHr}) exhibited a very high inhibitory potential; it is even higher than that of parent mouse α-chain itself (Figure 6.17). The hybrid containing mouse-horse chimeric α-chain ($\alpha^{MHr}{}_2\beta^S{}_2$) did not polymerize at all under the experimental conditions where mouse α-chain hybrid polymerizes (Figure 6.17). While the combination of the sequence differences of horse α_{1-30} and mouse α_{31-141} generates a chimeric α-chain with an inhibitory potential lower than that of the parent α-chains, the combination of the sequence differences of mouse α_{1-30} and horse α_{31-141} generates a chimeric α-chain that has much higher inhibitory potential, as compared to parent α-chains, indicating a better complementation of the sequence differences of mouse α_{1-30} and horse α_{31-141} in the inhibition of polymerization. The

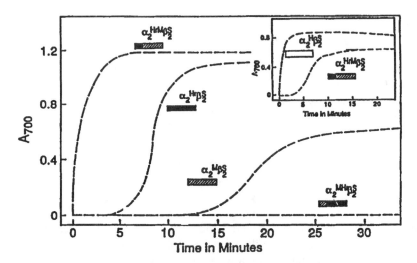

FIGURE 6.17   Comparison of the polymerization inhibitory potential of mouse α-chain and mouse-horse chimeric α-chain. The description for open, filled, and striped bars is as described for Figures 6.14 and 6.15. Filled-striped and striped-filled bars represent the horse α(1–30)-mouse α(31–141) chimeric α-chain (α^{HrM}) and mouse α(1–30)-horse α(31–141) chimeric α-chain (α^{MHr}), respectively. The polymerization of the hybrids has been carried out as described for Figure 6.12. The kinetics of polymerization was studied by recording the change in the turbidity continuously at 700 nm with Shimadzu UV 265 spectrophotometer. The initial concentration used for these hybrids is 1.5 mg/mL, whereas the results presented in the inset are at an initial concentration of 0.9 mg/mL.

combination of the sequence differences of these two segments has resulted in the generation of a super inhibitory α-chain.

6.8.3.9   Molecular Basis of Super Inhibitory Potential of Mouse-Horse Chimeric α-chain Molecular Modeling Studies

As discussed earlier, the super inhibitory potential of mouse-horse α-chain comes from the intra-dimeric complementation of the inhibitory potential of the sequence differences of mouse α_{1-30} and of horse α_{31-141}, *a linkage of cis and trans inhibitory potential (multisite perturbation of the intermolecular interactions of deoxy HbS polymer).* An additional inhibitory potential is generated as a consequence of establishing the chain contiguity, *an intra-chain linkage of the inhibitory influence of two segments.* The delineation of the molecular aspects of this synergy in the murine-equine chimera that elicits a super inhibitory potential could help with the design of super inhibitory α-chains. The super inhibitory α-chains, antisickling Hbs in general, are expected to have a profound influence on the future and success of gene therapy of sickle cell disease. A hybrid of super inhibitory α-chain with β^A-chains containing polymerization inhibitory sequence differences at the acceptor pocket should exhibit very high antisickling influence when expressed in cells generating

HbS. These antisickling Hbs will inhibit the polymerization of HbS by destabilizing the intermolecular contact sites of both cis and trans dimers. A gene construct of such a super inhibitory Hb will be an ideal candidate for the gene therapy of sickle cell disease. Engineering additional sequence difference into the β^A-chains at the $\alpha_1\beta_1$ interface that increases the stability of this interface could serve as the molecular trap for the endogenous normal α-chains. Since the affinity of β^A for the normal α-chain will be higher relative to that of β^S-chains for the normal α-chains, in cells expressing both these forms of β–chains, more of the β^S-chains would hybridize with the super inhibitory α-chains as a consequence of the higher stability of the $\alpha_1\beta_1$ of the dimer containing normal α-chain and the new engineered β-chain. This approach suggests further flexibility that exists in the gene therapy of sickle cell disease to modulate the pathophysiology of sickle cell disease.

As discussed earlier, two regions of common sequence differences between mouse α-chain and mouse-horse chimeric α-chain is the EF corner and AB/GH corner. There is a homology in the sequence differences of the EF region between the two parent chains, namely mouse and horse α-chains. The second region of a common sequence difference is the AB/GH corner. This region of the α-chain of HbS is the residence of many implicated contact sites of the HbS polymer and is involved extensively in the intra-double-strand axial contacts.

A close examination of the structure of HbS, and correlating it with the poly-merization inhibition achieved by generating a chimeric α-chain, indicates that one of the consequences of the formation of the chimeric α-chain and the concomitant assembly of the tetramers to generate the Hb fold is that AB and GH corners that are distal in the linear sequence become proximal in the three-dimensional structure of Hb. Accordingly, a proximity of new sequence differences is established that is distinct from that of two parent chains from which the segments have been obtained.

Some combination(s) of one or more common sequence differences of the region of mouse horse chimeric α-chains, α-17, α-19, α-21, α-22, α-107, α-111, α-115, and α-116, apparently generates a new strong polymerization inhibitory domain, besides the intrinsic inhibitory potential that existed in the horse α_{31-141} region of the horse α-chain.

The potential noncovalent interactions of the chimeric α-chain at the AB-GH corner of the HbS molecule, and the anticipated interactions as seen by molecular modeling on replacement of the amino acid side chains in the mouse-horse chimeric α-chain, are presented in Figure 6.18. These molecular modeling studies have shed new insight into the molecular aspects of the structure that might have resulted in a high polymerization inhibitory potential for the chimeric α-chain. The sequence differences of the segment α_{107} to α_{118} of horse α-chain and those in the segment α_{15} to α_{24} of mouse α-chain are considered for this molecular modeling studies. The side chains of the four sequence differences of the mouse α-chain (those located at α_{17}, α_{19}, α_{21}, and α_{22}) by themselves do not appear to result in any significant perturbations of the interactions of the side chains of AB GH corner residues. This analysis is consistent with the observation that the mouse-human chimeric α-chain exhibits very little inhibition of polymerization.

Out of the four sequence differences at the GH corner of the mouse-horse chimeric α-chain, the two sequence differences at α_{107} and α_{111} may be expected to

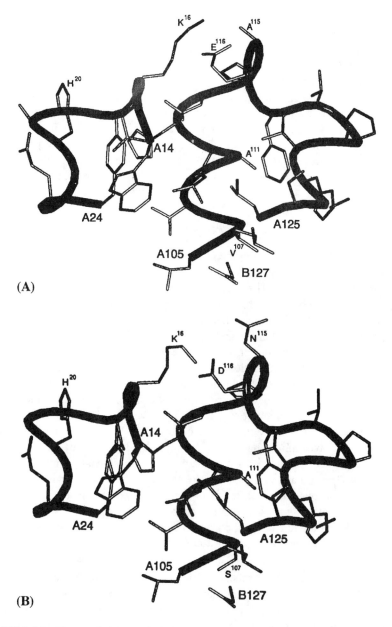

FIGURE 6.18 Noncovalent intra-chain interactions at the AB-GH corner of α-chain of HbS (A) and that of the mouse-horse chimeric α-chain of chimeric HbS (B). These molecular models were generated on an Indigo II from Silicon Graphics using the software package Insight II from Molecular Simulations. For generating the molecular models of the chimeric α-chain, amino acid replacements were done using the Biopolymer module in Insight II. Side-chain torsions were adjusted using Insight II followed by its default bump check. The torsions were explored for possible new hydrogen and/or ionic interactions. For presentation, segments A14 to A24 and A 105 to A125 of the α-chain have been selected.

perturb the $\alpha_1\beta_1$ interface interactions. The crystal structure of horse met hemoglobin indeed reflects this. These two residues are at the carboxyl terminal end of the G helix and in a position to interact along the axis of the helix. In HbS, Val-107(α) interacts with Gln-127(β), Ala-115(β), and Cys-112(β). In generating the chimeric HbS containing either horse α-chain or mouse horse chimeric α-chain, Cys-112(β) that was absent in the $\alpha_1\beta_1$ interface of horse hemoglobin has been engineered into the $\alpha_1\beta_1$ interface of chimeric HbS. Thus, the interactions at the $\alpha_1\beta_1$ interface of the chimeric HbS containing the mouse-horse chimeric α-chain is expected to contain elements of the interactions of both HbS, and horse Hb. The presence of Ser-107(α) (in place of Val in the human α-chain) in the chimeric HbS containing the mouse-horse chimeric α-chain, an increased interaction between the carboxyl end of the G-helix of the chimeric α-chain and the amino terminal end of the H-helix of the β^S-chain is anticipated. A repositioning of the side chain of Ser-107(α) can result in a favorable hydrogen bonding (Figure 6.19B) between this residue and Gln-127(β). Since all the chimeric HbS have the Cys residue at β_{112}, some perturbations in the interaction between the G_{14} residues of α and the β^S-chains of the $\alpha\beta$ dimers of the chimeric HbS relative to those of the parent molecules (horse Hb and HbS) are also anticipated. The presence of Cys-112(β) in the chimeric HbS in conjunction with the backbone residues of the $\alpha_1\beta_1$ interface coming from the horse α-chain is expected to result in an altered interaction between the carboxyl terminal region of G helix of the β^S-chain and the amino terminal region of the H-helix of the chimeric α-chain. The presence of Val at α_{111} in the chimeric α-chain is also expected to perturb the interactions involving the residues Ala-115(β) and Cys-112(β). Although Val-111(α) could increase the interaction with Ala-115(β), the presence of Cys-112(β) in the chimeric HbS could restore some of the elements of interaction with the side chain at α_{111} close to that seen in HbS. Thus, although the overall Hb globin fold is expected to be conserved in these chimeric HbS, considerable degree of dynamic changes could be anticipated in the $\alpha_1\beta_1$ interface region of the molecules containing either horse α-chain or mouse-horse chimeric α-chain. However, these perturbations at the $\alpha_1\beta_1$ interface are also anticipated in the chimeric HbS containing mouse-horse chimeric α-chain and expected to be present in the chimeric HbS containing horse α-chain. Accordingly, these perturbations at the $\alpha_1\beta_1$ interface of the chimeric HbS by themselves do not appear to be responsible for the super inhibitory influence of the mouse-horse chimeric α-chain.

While the sequence differences at α_{107} and α_{111} that are expected to perturb the intradimeric interactions, the sequence differences at α_{115} and α_{116} are expected to influence the intra-chain interactions, namely those between the AB and GH corners of the chimeric α-chain. Based on the crystal structure of horse Hb, the presence of Asn at α_{115} could generate additional interactions between the side chains at α_{115} and β_{123} and facilitate the anchoring of the amino terminus of the H-helix of the β-chain close to the GH corner. The molecular modeling studies suggest that the replacement of Glu-116(α) with Asp facilitates a better interactions between the amino group of Lys-16(α) with the β–carboxyl group of Asp-116(α). As seen in Figure 6.18B, the net effect is decreased accessibility of the ϵ-amino group Lys-16(α) (the side chains relatively more buried inside the crevice of the AB GH corner as compared to that in the human α-chain or mouse-human chimeric α-chain). It is of interest

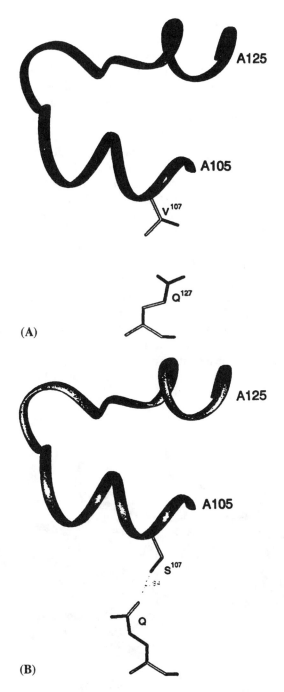

FIGURE 6.19 Formation of new hydrogen bond in the $\alpha_1\beta_1$ interface of chimeric Hbs between side chain of the Ser-107(α) of chimeric α-chain and the Gln-127(β). The orientation of Val-107(α) and Gln-127(β) in HbS is also shown for comparison to demonstrate the absence of such a noncovalent interaction in HbS. (A) = HbS, (B) = chimeric HbS.

to note that Lys-16(α) has been implicated as an intermolecular contact residue in the HbS fiber. Thus, although the perturbations of the interactions at the $\alpha_1\beta_1$ interface are relatively not so crucial as reflected by the horse and human-horse α-chain containing chimeric HbS, it is conceivable that the altered interactions of the $\alpha_1\beta_1$ interface of the chimeric HbS may be facilitating the interaction of Lys-16(α) with Asp-116(α) in such a way to decrease the accessibility of Lys-16(α) for inter-tetrameric interactions during the polymerization of chimeric HbS.

The molecular modeling of the AB-GH corner of mouse α-chain has also been carried out to determine the possible orientation of the side chain of Lys-16(α) of the chain. These studies suggest that the side chain Lys-16(α) of mouse α-chain will have nearly the same orientation as that in the mouse-horse chimeric α-chain (Figure 6.20). Accordingly, we speculate that the altered orientation of Lys-16(α) of the chimeric α-chain, coupled with the perturbation of the $\alpha_1\beta_1$ interface of the HbS containing this chimeric α-chain, endows the molecule an inhibitory potential that is higher than that of mouse α-chain, a super inhibitory influence.

Further detailed biophysical investigations of HbS containing super inhibitory α-chain using x-ray crystallography and molecular fingerprinting by NMR spectroscopy should be able to focus on the changes in local conformational dynamics of one or more domains of these molecule which, in turn, could shed further insights into the molecular basis of the generation of the super inhibitory α-chains. At any rate, identification of the exact sequence differences that perturb the interactions that synergistically modulate the inhibitory potential of α-chain will help us in designing potent antisickling α-chains and hence super inhibitory Hbs for gene therapy of sickle cell disease.

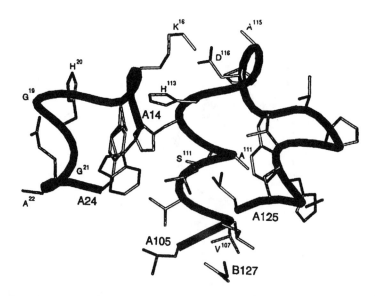

FIGURE 6.20 Noncovalent intra-chain interactions at the AB-GH corner of α-chain of chimeric HbS containing the mouse α-chain.

6.9 CONCLUDING REMARKS

The studies presented here discuss the development of a protease mediated semi-synthetic reaction and its application to understand the molecular aspects of the inhibition of the β^S-chain dependent polymerization by the mouse α-chain, and the semisynthesis of a new super inhibitory chimeric α-chain. The latter represents highly targeted research, focused on the design of antisickling Hbs for the gene therapy of sickle cell disease.

Protease catalyzed splicing of the peptide bonds to generate novel semisynthetic proteins has been generally considered as a difficult approach, nor has it been easy to develop generalized principles for the design of new semisynthetic strategies. Although these approaches have been appreciated by protein chemists, they have failed to receive wider appreciation from biochemists and molecular biologists. Accordingly, the field of protein semisynthesis has been growing only slowly since the demonstration by Laskowski and his colleagues of the semisynthesis of RNAse A from RNase S. New principles are being unraveled, and new manipulative approaches are also being devised, as is apparent from many chapters of this volume. In all the protease catalyzed splicing reactions a peptide bond is formed. The recognition that synthetic protein need not solely contain amide bonds within a polypeptide backbone has also resulted in the development of new innovative chemical ligation approaches. Chemical ligation strategies rely on highly selective reactions between the synthetic peptides, each bearing a unique and mutually reactive group. The chemoselectivity of these ligation reactions permits the coupling of the unprotected peptides in much the same way as the protease mediated splicing reactions.

The α-globin semisynthetic reaction discussed here is novel and distinct from other protease catalyzed protein semisynthetic reactions developed earlier. The facile splicing reaction is a consequence of the propensity of the contiguous segment generated *in situ* to assume an α-helical conformation that endows a degree of stability to the spliced peptide bond and serves as the *trap* of the splicing reaction. Thus, the present study has opened up an opportunity to use the cosolvent induced secondary structure of the ligated segment as the *conformational trap* of the splicing reaction. The generality of this concept has to be established with additional proteins and peptides. Two additional sites of globin has been identified that appear to follow the same principle to facilitate the splicing reaction. One is the trypsin mediated splicing of the tryptic fragments of the β-chain, βT_3 (residues 18 to 30 of β-chain) with βT_4 (residues 31 to 40 of β-chain). This reaction needed nearly 50% n-propanol to obtain about 25 to 30% yield in the synthetic reaction.[159] The other is the V8 protease catalyzed splicing of the complementary segments of αT_{12} (residues 100 to 127 of α-chain). The segments $\alpha_{100-116}$ could be spliced with $\alpha_{117-127}$ with about 15 to 20% yield under the conditions used for the α-globin semisynthetic reaction (Roy and Acharya, unpublished results).

Fontana and his colleagues have used V8 protease to splice the segments of thermolysin.[160] They have carried out the splicing both in the presence of n-propanol as well as glycerol. The peptide bond spliced by V8 protease is located in a helical segment in the parent protein; however, the contribution of the cosolvent induced secondary structure of the contiguous chain generated *in situ* has not been addressed.

The α-globin semisynthetic reaction is unique in that for the V8 protease catalyzed splicing at $Glu^{30} Arg^{31}$, a significant portion of the α-globin in not necessary.
The complementary segments of the B-helix of the α-chain are spliced by the enzyme
in equilibrium yields comparable to that obtained with the complementary fragments
of the α-globin chain. A decrease in the conformational entropy will result as a
consequence of the induction of α-helical conformation to contiguous segment in
the presence of n-propanol and the concomitant generation of an (i, i + 4) ionic
interaction in the helical conformation of the peptide. Accordingly, the B-helix of
the α-globin can be considered as an independently stabilized unit of α-globin, at
least in terms of the *conformational trap* of the semisynthetic reaction. The only
other example of a protease catalyzed splicing of the peptide bond in an equimolar
mixture of complementary peptides is the disulfide loop peptide of plasminogen.
The hydrolysis of the Arg-Val peptide bond of this peptide is readily reversible in
the presence of high concentrations of 1,4 butane diol as long as the disulfide bond
of the peptide is intact. The reduction of the disulfide bond renders tryptic hydrolysis
irreversible. Just as with the α-globin semisynthetic reaction, the rate and the efficiency of the condensation has been high. In the disulfide loop peptide, the splicing
reaction is an intramolecular reaction, in much the same way as with the splicing
of the peptide bonds in the fragment complementing systems. In the latter, the
noncovalent interactions of the complementary segments provide the proximity for
the α-amino and the α-carboxyl groups to be spliced. In the disulfide loop peptide,
this proximity is apparently afforded by the disulfide bond. At any rate, the disulfide
loop peptide and the B-helix of the α-globin represent the two simplest models of
protease catalyzed protein semisynthetic reactions, but each has distinct molecular
mechanisms (conformational traps) to facilitate the protease catalyzed splicing reaction. A structure-activity correlation study in this family of peptides should provide
better appreciation of the factors that control the kinetics and thermodynamics of
protease catalyzed fragment coupling reactions. Such information will be valuable
in designing broadly applicable methods of semisynthesis based on the enzymatic
coupling of peptide fragments.

The protease catalyzed formation of peptide bonds is not restricted to the in vitro
systems developed and discussed here. The recent demonstration of the *in vivo* protein
splicing reactions as a post translational molecular process generating mature functional proteins suggest the possible biological implications and significance of protein
splicing reactions.[161,162] The protein splicing reactions involve the precise excision
of an internal protein segment, from the precursor protein followed by the concomitant ligation of the flanking N- and C-terminal regions, exteins yielding the mature
protein. The *in vivo* protein splicing reactions[163] should be considered as equivalent
to RNA splicing reactions to remove the introns during the production of the mature
RNA. Site-directed mutagenesis of the parent proteins has established that the amino
acid sequence of the spacer domain dictates the *in vivo* splicing reaction. Interestingly,
the spacer domain inserted into an unrelated protein also facilitates the splicing of
that domain. The molecular aspects of the post-translational protein splicing reactions
are distinct compared to protease mediated splicing reactions. The splicing reaction
proceeds through a branched intermediate with two N-termini. This *in vivo* splicing
reaction has some elements of chemical ligation reactions.

Most of the protease catalyzed protein semisynthetic reactions are primarily the rejoining of the α-amino and the α-carboxyl groups of a *permissible discontinuity site* that has been introduced into the protein by limited proteolysis. Thus, it represents the bridging of the carboxyl and amino groups of a protein with two subunits that are proximal to one another. The fragment complementing system of a single-chain protein could be considered as the dimeric form of the parent protein, and the two segments of the fragment complementing system represent the subunits of the dimeric form of the protein. In this form, the amino and the carboxyl group of the permissible discontinuity site are proximal. In some multimeric proteins like Hb, the α-amino group of one α-unit is proximal to the α-carboxyl groups of the other subunit (α-chains); one could consider this as the permissible discontinuity site. If so, it raises an intriguing possibility that a *continuity* may be permissible in some multimeric proteins wherein a situation like this exists. It is of interest to note here that Somatogen has recently converted the normal tetrameric Hb into a trimeric protein by genetically fusing the two α-chains *head to tail* using a glycine as the bridge residue.[164,165] The molecular modeling studies have suggested that this is permissible site for bridging the carboxyl end of one α-chain with the α-amino group of the other α-chain of the Hb tetramer using Gly or Gly-Gly as the linkers. Consistent with this, Hb containing a bridged α-chains is functionally active.

The α-amino group and the α-carboxyl group of bovine pancreatic trypsin inhibitor are proximal to one another in the three-dimensional structure of the protein. A carbodiimide activated coupling of the α-carboxyl group with the α-amino group of this trypsin inhibitor has also been achieved to generate a circularized protein. Furthermore, this circularized trypsin inhibitor (the protein chain without a free α-amino and a free α-carboxyl group) has been opened at another site to generate a functional protein.[166] These results clearly reflect that a degree of plasticity is present in protein structures. These results, along with the *in vivo* splicing reaction, raise an intriguing question as to whether the proximity of an amino and a carboxyl group in a multimeric protein represents a *permissible site* of an ancestral protein either with the same function as the mature protein or a different function.

6.10 ACKNOWLEDGMENTS

The research discussed here is supported by NIH Grant HL 38665 and a Grant in Aid from American Heart Association, National Chapter. This semisynthetic program was initiated while A. S. Acharya was at Rockefeller University, New York. The direction, encouragement, and support that he received from late Prof. Stanford Moore during the formative years of his career is very much appreciated, and this article is dedicated to his memory. A number of present and past colleagues of Acharya have contributed to the success of this program, and it is with great pleasure that he acknowledges their contributions. These include Andy Dean, Ramnath Seetharam, Shabbir Khan, Subramonia Iyer, Krishna Mallia, Lesli Sussman, Youngnan J. Cho, Girish Sahni, Augusto Parante, Rajendra Prasad Roy, Laxmi Kandke, Kiran Khandke, Belur Manjula, M. Janardhan Rao, Ramesh Kumar, Ashok Malavali, and Ronald L. Nagel. The molecular models of chimeric α-chains were generated by Dr. Steve White of University of Oklahoma, Stillwater. The facilities extended to Acharya by Dr. R. L. Nagel are very much appreciated.

REFERENCES

1. Blundell, T. L., and Johnson, L. N., *Protein Crystallography,* Academic Press, New York, 1976.
2. Wyckoff, H. W., Hirs, C. H. W., and Timasheff, S. N., Eds., Diffraction methods for biological macromolecules, *Methods in Enzymology,* Vols. 114-115, Academic Press, Orlando, 1985.
3. Markley, J. L., and Ulrich, E. L., Detailed analysis of protein structure and function by NMR spectroscopy: Survey of resonance assignments, *Annu. Rev. Biophys. Bioeng.,* 13, 493, 1984.
4. Jardetzky, O., and Roberts, G. C. K., *NMR in Molecular Biology,* Academic Press, New York, 1981.
5. Clore, G. M., and Gronenborn, A. M., *NMR of Proteins,* CRC Press, Boca Raton, 1993.
6. Cohen, L. A., Chemical modification as a probe of structure and function. *The Enzymes,* Vol.1, Boyer, P. D., Ed., Academic Press, New York, 1970, 148.
7. Shaw, E., Chemical modification by active-site directed reagents. *The Enzymes,* Vol.1, Boyer, P. D., Ed., Academic Press, New York, 1970, 91.
8. Glazer, A. N., The chemical modification of proteins by group-specific and site-specific reagents. *The Proteins,* Vol II, Neurath, H., Hill, R. L., Eds., Academic Press, New York, 1976, 1.
9. Means, G. E., and Feeney, R. E., *Chemical Modification of Proteins,* Holden-Day, San Francisco, 1971.
10. Gutte, B., and Merrifield, R. B., The total synthesis of an enzyme with ribonuclease A activity, *J. Am. Chem. Soc.,* 91, 501, 1969.
11. Gutte, B., and Merrifield, R. B., The synthesis of ribonuclease A, *J. Biol. Chem.,* 246, 1922, 1971.
12. Erickson, B. W., and Merrifield, R. B., Solid-phase peptide synthesis, *The Proteins,* Vol II, Neurath, H., Hill, R. L., Eds., Academic Press, New York, 1976, 257.
13. Hofmann, K., Kisser, J. P., and Finn, F. M., Studies on polypeptides. 43. Synthesis of S-peptide 1–20 by two routes, *J. Am. Chem. Soc.,* 91, 4883, 1969.
14. Finn, F. M., and Hofmann, K., The synthesis of peptides by solution methods with emphasis on peptide hormones. *The Proteins,* Vol II, Neurath, H., Hill, R. L., Eds., Academic Press, New York, 1976, 106.
15. Kent, S. B. H., Chemical synthesis of peptides and proteins, *Annu. Rev. Biochem.,* 57, 957, 1988.
16. Schneider, J., and Kent, S. B. H., Enzymatic activity of a synthetic 99 residue protein corresponding to the putative HIV-1 protease, *Cell,* 54, 363, 1988.
17. Schnolzer, M. and Kent, S. B. H., Constructing proteins by dovetailing unprotected synthetic peptides: back bone-engineered HIV protease, *Science,* 256, 221, 1992.
18. Milton, R. C., Milton, S.c., and Kent, S. B. H., Total chemical synthesis of a D-enzyme: the enantiomers of HIV-1 protease show reciprocal chiral substrate specificity, *Science,* 256, 1445,1992.
19. Baca, M., and Kent, S. B. H., Catalytic contribution of flap-substrate hydrogen bonds in HIV-1 protease explored by chemical synthesis, *Proc. Natl. Acad. Sci.,* U.S.A., 90, 11638, 1993.

20. Baca, M., Alewood, P. F., and Kent, S. B. H., Structural engineering of the HIV-1 protease molecule with a beta-turn mimic of fixed geometry, *Protein Sci.,* 2, 1085, 1993.

21. Rees, A. R., Sternberg, M. J. E., and Wetzel, R., Eds., *Protein Engineering: A practical approach,* Oxford University Press, New York, 1992.

22. El-Gewely, M. R., Ed., *Mutagenesis and Protein Engineering,* Elsevier Science, Amsterdam, Netherlands, 1991.

23. Oxender, D. L., and Fox, C. F., Eds., *Protein Engineering,* Alan R. Liss, New York, 1987.

24. Wells, J. A., Powers, D. B., Bott, R. R., Graycar, T. P., and Estell, D. A., Designing substrate specificity by protein engineering of electrostatic interactions, *Proc. Natl. Acad. Sci., U.S.A.,* 84, 1219, 1987.

25. Neurath, H., Walsh, K. A., and Winter, W. P., Evolution of structure and function of proteases, *Science,* 158, 1638, 1967.

26. Estell, D. A., Graycar, T. P., and Wells, J. A., Engineering of an enzyme by site-directed mutagenesis to be resistant to chemical oxidation, *J. Biol. Chem.,* 260, 6518, 1985.

27. Perry, L. J., and Wetzel, R., Disulphide bond engineered into T_4 lusozyme: Stabilization of the protein toward thermal inactivation, *Science,* 226, 555, 1984.

28. George, P. M., and Carrell, R. W., *Serpins: engineered re-targeting of plasma proteinase inhibitors, in Design of Enzyme Inhibitors as Drugs,* Sandler, M., and Smith, H. J., Eds., Oxford University, New York, 1989, 581.

29. Jallat, S. Carvallo, D., Tessier, L. H., Roeklin, D., Roitsch, C., Ogushi, F., Crystal, R. G., and Courtney, M., Altered specificity of genetically engineered α_1 antitrypsin variants, *Protein Eng.,* 1, 29, 1986.

30. Johnson, K. A., and Benkovic, S. J., Analysis of protein function by mutagenesis, in *The Enzymes,* Vol. XIX, Sigman, D. S., and Boyer, P. D., Eds., Academy Press, San Diego, 1990, 159.

31. Dunn, B. M., DiBello, C., Kirk, K. L., Cohen, L. A., and Chaiken, I. M., Synthesis, purification, and properties of a semisynthetic ribonuclease S incorporating 4-fluoro-L-histidine at position 12, *J. Biol. Chem.,* 249, 6295, 1974.

32. Hofmann, K., Andreatta, R., and Bohn, H., Studies on polypeptides. XL. Synthetic routes to peptides containing beta- (pyrazolyl-1) -and beta- (pyrazolyl-3)-alanine, *J. Am. Chem. Soc.,* 90, 6207, 1968.

33. Andreatta, R., and Hofmann, K., Studies on polypeptides. XLI. The synthesis of [5-valine, 6-beta- (pyrazolyl-3) -alanine] -angiotensin II, a potent hypertensive peptide, *J. Am. Chem. Soc.,* 90, 7334, 1968.

34. Sanders, D. J., and Offord, R. E., The use of semisynthetically introduced 13C probes for nuclear magnetic resonance studies on insulin, *FEBS Lett.,* 26, 286, 1972.

35. Chaiken, I. M., Freedman, M. H., Lyerla, J. R. Jr., and Cohen, J. S., Preparation and studies of 19 F- labeled and enriched 13C-labeled semisynthetic ribonuclease-S' analogues, *J. Biol. Chem.,* 248, 884, 1973.

36. Davies, J. G., and Offord, R. E., The preparation of tritiated insulin specifically labelled by semisynthesis at glycine-A1, *Biochem. J.,* 231, 389, 1985.

37. Davies, J. G., Rose, K., Bradshaw C. G., and Offord, R. E., Enzymatic semisynthesis of insulin specifically labelled with tritium at position B-30, *Protein Eng.,* 1, 407, 1987.

38. Chaiken, I. M., Semisynthetic proteins and peptides, CRC, *Crit. Rev. Biochem.,* 11, 255, 1981.

39. Offord, R. E., and DiBello, C., Eds., *Semisynthetic Peptides and Proteins,* Academic Press, London, 1978.

40. Offord, R. E., *Semisynthetic Proteins,* John Wiley & Sons Ltd., Chichester, 1980.

41. Offord, R. E., Protein engineering by chemical means?, *Protein Eng.,* 1, 151, 1987.

42. Offord, R. E., Chemical approaches to protein engineering, in *Protein Design and the Development of New Therapeutics and Vaccines,* Hook, J. B., and Poste, G., Eds., Plenum, New York, 1990, 253.

43. Wallace, C. J. A., Understanding cytochrome *c* function: engineering protein structure by semisynthesis, *FASEB J.,* 7, 505, 1993.

44. Richards, F. M., and Vithayathil, P. J., The preparation of subtilisin modified ribonuclease and the separation of the peptide and protein components, *J. Biol. Chem.,* 234, 1459, 1959.

45. Taniuchi, H., Anfinsen, C. B., and Sodja, A., Nuclease-T: an active derivative of Staphylococcal nuclease composed of two noncovalently bonded peptide fragments, *Proc. Natl. Acad. Sci.,* U.S.A., 58, 1235, 1967.

46. Harris, D. E., and Offord, R. E., A functioning complex between tryptic fragments of cytochrome C: a route to the production of semisynthetic analogues, *Biochem. J.,* 161, 21, 1977.

47. Hantgan, R. R., and Taniuchi, H., Formation of a biologically active, ordered complex from two overlapping fragments of cytochrome C, *J. Biol. Chem.,* 252, 1367, 1977.

48. Li., C. H., Blake, J., and Hayashida, T., Human somatotropin: semisynthesis of the hormone by noncovalent interaction of the NH2-terminal fragment with synthetic analogs of the COOH-terminal fragment, *Biochem. Biophys. Res. Commun.,* 82, 217, 1978.

49. Birk, Y., and Li, C. H., Two fragments from fibrinolysin digests of ovin prolactin: characterization and recombination to generate full immunoreactivity, *Proc. Natl. Acad. Sci., U.S.A.,* 75, 2155, 1978.

50. Vita, C., Dalzoppo, D., and Fontana, A., Limited proteolysis of thermolysin by subtilisin: isolation and characterization of a partially active enzyme derivative, *Biochemistry,* 24, 1798, 1985.

51. Liao, M. J., London, E., and Khorana, H. G., Regeneration of the native bacteriorhodopsin structure from two chymotryptic fragments, *J. Biol. Chem.,* 258, 9949, 1983.

52. Liao, M. J., Huang, K. S., and Khorana, H. G., Regeneration of native bacteriorhodopsin structure from fragments, *J. Biol. Chem.,* 259, 4200, 1984.

53. Scaffone, F., Marchiori, F., Rocchi, R., Vidal, G., Tamburro, A., Scatturin, A., and Marzotto, A., Synthesis of an enzymatically active orn10-S-peptide of ribonuclease-S, *Tetrahedron Lett.,* No. 9, 943, 1966.

54. Pandin, M., Padlan, E. A., DiBello, C., and Chaiken, I. M., Crystalline semisynthetic ribonuclease-S', *Proc. Natl. Acad. Sci., U.S.A.,* 73, 1844, 1976.

55. Komoriya, A., and Homandberg, Chaiken, I. M., Enzymatic fragment condensation using kinetic traps, *Peptides 1980,* Brunfeldt, K., Ed., Scriptor Press, Copenhagen, 1981, 378.

56. Welling, G. W., Lenstra, J. A., and Beintema, J. J., Activity and antigenity of ribonuclease hybrids, *FEBS Lett.,* 63, 89, 1976.

57. Beintema, J. J., and Lenstra, J. A., Nuclear magnetic resonance study of a hybrid of bovine and rat ribonuclease, *Int. J. Peptide & Protein Res.,* 15, 455, 1980.

58. Voskuyl-Holtkamp, I., and Schattenkerk, C., Studies on polypeptides XXVI. Synthesis of the N-terminal 1–23 peptide sequence of rat pancreatic ribonuclease; enzymatic activity of the hybrid complex with bovine S-protein, *Int. J. Peptide & Protein Res.,* 11, 218, 1978.

59. Riggs, A., and Herner, A. E., The hybridization of donkey and mouse hemoglobins, *Proc. Natl. Acad. Sci., U.S.A.,* 48, 1664, 1962.

60. Gavish, M., Zakut, R., Wilchek, M., and Givol, D., Preparation of a semisynthetic antibody, *Biochemistry,* 17, 1345, 1978.

61. Burton, J., Topper, R., and Ehrlich, P., Semisynthesis of an antibody fragment: quantitation and correction of deletion sequences, in *Peptides - Structure and Biological Function,* Gross, E., and Meienhofer, J. Eds., Pierce Chem. Co., Rockford, IL., 1979, 605.

62. Benesch, R. E., Yung, S., Benesch, R., Mack, J., and Schneider, R. G., -chain contacts in the polymerization of sickle hemoglobin, *Nature,* 260, 219, 1976.

63. Benesch, R. E., Kwong, S., Benesch, R., and Edalji, R., Location and bond type of intermolecular contacts in the polymerization of haemoglobin S, *Nature,* 269, 772,1977.

64. Benesch, R. E., Kwong, S., Edalji, R., and Benesch, R., -chain mutation with opposite effects on the gelation of hemoglobin S, *J. Biol. Chem.,* 254, 8169,1979.

65. Benesch, R. E., Kwong, S., and Benesch, R., The effect of -chain mutations cis and trans to the ß6 mutation on the polymerization of sickle cell haemoglobin, *Nature,* 299, 231, 1982.

66. Crepeau, R. H., Edelstein, S. J., Szalay, M., Benesch, R. E., Benesch, R., Kwong, S., and Edalji, R., Sickle cell haemoglobin fiber structure altered by -chain mutation, *Proc. Natl. Acad. Sci., U.S.A.,* 78, 1406, 1981.

67. Nagel, R. L., and Bookchin, R. M., Areas of interaction in the Hb S polymer, in *Biological and Clinical Aspects of Hemoglobin Abnormalities,* Caughey, W. S., Ed., Academic Press, New York, 195, 1978.

68. Nagel, R. L., Bookchin, R. M., Johnson, J., Labie, D., Wajcman, H., Isaac-Sodeye, W. A., Honig, G. R., Schiliro, G., Crookston, J. H., and Matsutomo, K., Structural bases of the inhibitory effect of hemoglobin F and hemoglobin A2 on the polymerization of hemoglobin S, *Proc. Natl. Acad. Sci., U.S.A.,* 76, 670, 1979.

69. Nagel, R. L., Johnson, J., Bookchin, R. M., Garel, M. C., Rosa, J., Schiliro, G., Wajcman, H., Labie, D., Moopenn, W., and Castro, O., ß-chain contact sites in the hemoglobin S polymer, *Nature,* 283, 832, 1980.

70. Bookchin, R. M., and Nagel, R. L., Ligand-induced conformational dependence of hemoglobin in sickling interactions, *J. Mol. Biol.,* 60, 263, 1971.

71. Laskowski, M. Jr., The use of proteolytic enzymes for the synthesis of specific peptide bonds in globular proteins, in *Semisynthetic Peptides and Proteins,* Offord, R. E., and DiBello, C., Eds., Academic Press, London, 1978, 256.

72. Finkenstadt, W. R., and Laskowski, M. Jr., Resynthesis by trypsin of the cleaved peptide bond in modified soyabean trypsin inhibitor, *J. Biol. Chem.,* 242, 771, 1967.

73. Homandberg, G. A., and Laskowski, M. Jr., Enzymatic resynthesis of the hydrolyzed peptide bond(s) in ribonuclease S, *Biochemistry,* 18, 586, 1979.

74. Ardelt, W., and Laskowski, M. Jr., Thermodynamics and kinetics of the hydrolysis of and resynthesis of the reactive site peptide bond in turkey ovomucoid third domain by aspergillopeptidase B, *Acta Biochim. Polonica,* 30, 115, 1983.

75. Fruton, J. S., Proteinases as catalysts of peptide bond synthesis, *Trans. New York Acad. Sci.,* 41, 49, 1983.

76. Fruton, J. S., Proteinase-catalyzed synthesis of peptide bonds, *Adv. Enzymol. Relat. Areas Mol. Biol.,* 53, 239, 1983.

77. Bozler, H., Wayne, S. I., and Fruton, J. S., Specificity of pepsin-catalyzed bond synthesis, *Int. J. Peptide & Protein Res.,* 20, 102, 1982.

78. Wayne, S. I., and Fruton, J. S., Thermolysin-catalyzed peptide bond synthesis, *Proc. Natl. Acad. Sci., U.S.A.,* 80, 3241, 1983.

79. Corradin, G., and Harbury, H. A., Reconstitution of horse heart cytochrome C: interaction of the components obtained upon cleavage of the peptide bond following methionine 65, *Proc. Natl. Acad. Sci., U.S.A.,* 68, 3036, 1971.

80. Wallace C. J. A., and Offord, R. E., The semisynthesis of fragments corresponding to residues 66-104 of horse heart cytochrome C, *Biochem. J.,* 179, 169, 1979.

81. Proudfoot, A. E., and Wallace, C. J. A., Semisynthesis of cytochrome C analogues. The effect of modifying the conserved residues 38 and 39, *Biochem, J.,* 248, 965, 1987.

82. Proudfoot, A. E., Rose, K., and Wallace, C. J. A., Conformation-directed recombination of enzyme-activated peptide fragments: a simple and efficient means to protein engineering. Its use in the creation of cytochrome C analogues for structure-function studies, *J. Biol. Chem.,* 264, 8764, 1989.

83. Wallace, C. J. A., and Corthesy, B. E., Protein engineering of cytochrome C by semisynthesis: substitutions at glutamic acid 66, *Protein Eng.,* 1, 23, 1986.

84. Homandberg, G. A., Mattis, J. A., and Laskowski, M. Jr., Synthesis of peptide bonds by proteinases: Addition of organic cosolvents shifts peptide bond equilibria toward synthesis, *Biochemistry,* 17, 5220, 1978.

85. Homandberg, G. A., and Chaiken, I. M., Tripsin-catalyzed conversion of staphylococcal nuclease-T fragment complexes to covalent forms, *J. Biol. Chem.,* 255, 4903, 1980.

86. Komoriya, A., Homandberg, G. A., and Chaiken, I. M., Enzyme-catalyzed formation of semisynthetic staphylococcal nuclease using a new synthetic fragment, [48-glycine] synthetic-(6-49), *Int. J. Peptide Protein Res.,* 16, 433, 1980.

87. Juillerat, M. and Homandberg, G. A., Clostripain-catalyzed reformation of a peptide bond in a cytochrome C fragment complex, *Int. J. Peptide & Protein Res.,* 18, 335, 1981.

88. Graf, L., and Li, C. H., Human somatotropin: Covalent reconstitution of two polypeptide contiguous fragments with thrombin, *Proc. Natl. Acad. Sci., U.S.A.,* 10, 6135, 1981.

89. Jones, R. M., and Offord, R. E., The proteinase-catalyzed synthesis of peptide hydrazides, *Biochem. J.,* 203, 125, 1982.

90. Rose, K., De Pury, H., and Offord, R. E., Rapid preparation of human insulin and insulin analogues in high yield by enzyme-assisted semisynthesis, *Biochem. J.,* 211, 671, 1983.

91. Bongers, J., Offord, R. E., Felix, A. M., Campbell, R. M., and Heimer, E. P., Semisynthesis of human growth hormone-releasing factor by trypsin catalyzed coupling of leucine amide to a C-terminal acid precursor, *Int. J. Peptide & Protein Res.,* 40, 268, 1992.

92. Seetharam, R., and Acharya, A. S., Synthetic potential of Staphylococcus aureus V8 protease: an approach toward semisynthesis of covalent analogues of -chain of hemoglobin S, *J. Cell. Biochem.,* 30, 87, 1986.

93. Drapeau, G. R., and Houmard, J., Stphylococcal protease: a proteolytic enzyme specific for glutamoyl bonds, *Proc. Natl. Acad. Sci., U.S.A.,* 69, 3506, 1972.

94. Seetharam, R., Dean, A., Iyer, K. S., and Acharya, A. S., Permissible discontinuity region of the -chain of hemoglobin: Noncovalent interaction of heme and the complementary fragments α_{1-30} and α_{31-141}, *Biochemistry,* 25, 5949, 1986.

95. Sahni, G., Cho, Y. J., Iyer, K. S., Khan, S. A., Seetharam, R., and Acharya, A. S., Synthetic hemoglobin A: Reconstitution of functional tetramer from semisynthetic -globin, Biochemistry, 28, 5456, 1989.

96. Roy, R. P., and Acharya, A. S., Semisynthesis of hemoglobin, *Methods Enzymol.,* 231, 194, 1994.
97. Iyer, K. S., and Acharya, A. S., Conformational studies of -globin in 1-propanol: propensity of the alcohol to limit the sites of proteolytic cleavage, *Proc. Natl. Acad. Sci., U.S.A.,* 84, 7014, 1987.
98. Kullmann, W., Proteases as catalysts for enzymic syntheses of opioid peptides, *J. Biol. Chem.,* 255, 8234, 1980.
99. Konopinska, D., and Muzalewski, F., Proteolytic enzymes in peptide synthesis, *Mol. Cell. Biochem.,* 51, 165, 1983.
100. Sakina, K., Ueno, Y., Oka, T., and Morihara, K., Enzymatic semisynthesis of [LeuB30] insulin, *Int. J. Peptide & Protein Res.,* 28, 411, 1986.
101. Riechmann, L., and Kasche, V., Peptide synthesis catalyzed by serine proteinases chymotrypsin and trypsin, *Biochim. Biophys. Acta,* 830, 164, 1985.
102. Nagai, K., Enoki, Y., Tomita, S., and Teshima, T., trypsin-catalyzed synthesis of peptide bond in human hemoglobin: oxygen binding characteristics of Gly-NH$_2$(142α)Hb, *J. Biol. Chem.,* 257, 1622, 1982.
103. Inouye, K., Watanabe, K., Morihara, K., Tochino, Y., Kanaya, T., Emura, J., and Sakakibara, S., Enzyme-assisted semisynthesis of human insulin, *J. Am. Chem. Soc.,* 101, 751, 1979.
104. Morihara, K., Oka, T., and Tsuzuki, H., Semisynthesis of human insulin by trypsin-catalyzed replacement of Ala-B30 by Thr in porcine insulin, *Nature,* 280, 412, 1979.
105. Roy, R. P., Khandke, K. M., Manjula, B. N., and Acharya, A. S., Helix formation in enzymically ligated peptides as a driving force for the synthetic reaction: example of -globin semisynthetic reaction, *Biochemistry,* 31, 7249, 1992.
106. Lotan, N., Berger, A., and Katchalski, E., Conformation and conformational transitions of poly--amino acids in solution, *Annu. Rev. Biochem.,* 41, 869, 1972.
107. Barteri, M., and Pispisa, B., Influence of isopropanol-water solvent mixtures on the conformation of poly- L-lysine, *Biopolymers,* 12, 2309, 1973.
108. Brack, A., and Spach, G., Multiconformational synthetic peptides, *J. Am. Chem. Soc.,* 103, 6319, 1981.
109. Fasman, G. D., Ed., *Poly--α–amino acids,* Dekker, New York, 1967, 499.
110. Scheraga, H. A., Theoretical and experimental studies of conformations of polypeptides, *Chem. Rev.,* 71, 195, 1971.
111. Acharya, A. S., Sahni, G., Roy, R. P., and Manjula, B. N., Quality of the helix induced enzymically ligated peptides dictates the splicing of non-interacting peptides: A comparative study of the V8 protease catalyzed splicing of complementary segments of RNase S-peptide and α-globin, *Indian J. Chem.,* 32B, 1, 1993.
112. Sahni, G., Khan, S. A., and Acharya, A. S., Chemistry of the molecular trap of protease catalyzed splicing reaction of complementary segments of α-subunit of hemoglobin A, *J. Prot. Chem.* 17, 669, 1998.
113. Bucci, E., and Fronticelli, C., A new method for the preparation of and ß subunits of human hemoglobin, *J. Biol. Chem.,* 240, PC551, 1965.
114. Bucci, E., Preparation of isolated chains of human hemoglobin, *Methods Enzymol.,* 76, 97, 1981.
115. Rossi Fanelli, A., Antonini, E., and Caputo, A., Studies on the structure of hemoglobin: I. Physicochemical properties of human globin, *Biochim. Biophys. Acta,* 30, 608, 1958.
116. Rosenmeyer, M. A., and Huehns, E. R., On the mechanism of the dissociation of hemoglobin, *J. Mol. Biol.,* 25, 252, 1967.

117. Clegg, J. B., Naughton, M. A., and Weatherall, D. J., Abnormal hemoglobins: Separation and characterization of the α and ß chains by chromatography, and the determination of two new variants, Hb Chesapeake and HbJ (Bangkok), *J. Mol. Biol.,* 19, 91, 1966.

118. Yip, Y. K., Waks, M., and Beychok, S., Reconstitution of native human hemoglobin from separated globin chains and alloplex intermediates, *Proc. Natl. Acad. Sci., U.S.A.,* 74, 64, 1977.

119. Bunn, H. F., and Forget, B. G., *Hemoglobin: Molecular, genetic and clinical aspects,* W. B. Sanders, Philadelphia, 1986, 502-550.

120. Embury, S. H., Hebbel, R. P., Mohandas, N., and Steinberg, M. H., Eds., *Sickle Cell Disease: Basic principles and clinical practice,* Raven Press, New York, 1994.

121. Josephs, R., Jarosch, H., and Edelstein, S. J., Polymorphism of sickle cell hemoglobin fibers, *J. Mol. Biol.,* 102, 409, 1976.

122. Dykes, G., Crepeau, R. H., and Edelstein, S. J., Three dimensional reconstruction of the 14-filament fibers of hemoglobin S, *J. Mol. Biol.,* 130, 451, 1979.

123. Potel, M. G., Wellems, T. E., Vasser, R. J., Deer, B., and Josephs, R., Macrofiber structure and the dynamics of sickle hemoglobin crystallization, *J. Mol. Biol.,* 177, 819, 1984.

124. Rodgers, D. W., Crepeau, R. H., and Edelstein, S. J., Pairings and polarities of the 14 strands in sickle cell hemoglobin fibers, *Proc. Natl. Acad. Sci., U.S.A.,* 84, 6157, 1987.

125. Carragher, B., Bluemke, D. A., Gabriel, B., Potel, M. J., and Josephs, R., Structural analysis of sickle cell hemoglobin polymers. I. Sickle hemoglobin fibers, *J. Mol. Biol.,* 199, 315, 1988.

126. Wishner, B. C., Ward, K. B., Lattman, E. E., and Love, W. E., Crystal structure of sickle cell deoxyhemoglobin at 5 A resolution, *J. Mol. Biol.,* 98, 179, 1975.

127. Magdoff-Fairchild, B., and Chiu, C. C., X-ray diffraction studies of fibers and crystals of deoxygenated sickle cell hemoglobin, *Proc. Natl. Acad. Sci., U.S.A.,* 76, 223, 1979.

128. Padlan, E. A., and Love, W. E., Refined crystal structure of deoxyhemoglobin S, *J. Biol. Chem.,* 260, 8272, 1985.

129. Padlan, E. A., and Love, W. E., Refined crystal structure of deoxyhemoglobin S, *J. Biol. Chem.,* 260, 8280, 1985.

130. Watowich, S., Gross, L. J., and Josephs, R., Intermolecular contacts within sickle hemoglobin fibers, *J. Mol. Biol.,* 209, 821, 1989.

131. Watowich, S., Gross, L. J., and Josephs, R., Analysis of intermolecular contacts within sickle hemoglobin fibers: Effect of site-specific substitutions, fiber pitch and double-strand disorder, *J. Stru. Biol.,* 111, 161, 1993.

132. Bookchin, R. M., Nagel, R. L., and Ranney, H. M., Structure and properties of hemoglobin C Harlem, a human hemoglobin variant with amino acid substitutions in 2 residues of the ß-polypeptide chain, *J. Biol. Chem.,* 242, 248, 1967.

133. Monplaisir, N., Merault, G., Poyrat, C., Rhoda, M. D., Craescu, C., Vidaud, M., Galacteros, F., Blouquit, Y., and Rosa, J., Hemoglobin S Antilles: a variant with lower solubility than hemoglobin S and producing sickle cell disease in heterozygotes, *Proc. Natl. Acad. Sci., U.S.A.,* 83, 9363, 1986.

134. Langdown, J. V., Williamson, D., Knight, C. B., Rubenstein, D., and Carrell, R. W., Case report- Anew doubly substituted sickling hemoglobin; HbS-Osman, *Br. J. Haematol.,* 71, 443, 1989.

135. Beale, D., and Lehmann, H., Abnormal hemoglobins and the genetic code, *Nature,* 207, 259, 1965.

136. Schneider, R. G., Ueda, S., Alperin, J. B., Brimhall, B., and Jones, R. T., Hemoglobin Sealy (α_2 47His β_2): A new variant in a Jewish family, *Am. J. Hum. Genet.*, 20, 151, 1968.

137. Sahni, G. Mallia, A. K., and Acharya, A. S., Proteosynthetic activity of immobilized Staphylococcus aureus V_8 protease: Application in the semisynthesis of molecular variants of α-globin, *Anal. Biochem.*, 193, 178, 1991.

138. Fronticelli, C., Recombinant hemoglobins for the elucidation of a new mechanism of oxygen affinity modulation by chloride ions, in *Techniques in protein chemistry*, Vol III, Angeletti, R. H., Ed., Academic Press, New York, 1992, 399.

139. Costantini, F., Chada, K., and Magram, J., Correction of murine ß-thlassemia by gene transfer into the germ line, *Science*, 233, 1192, 1986.

140. Trudel, M., Magram, J., Chada, K., Wilson, R., Costantini, F., Expression of normal, mutant and hybrid human globin genes in transgenic mice, *Prog. Clin. Biol. Res.*, 251, 305, 1987.

141. Greaves, D. R., Fraser, P., Vidal, M. A., Hedges, M. J., Ropers, D., Luzzatto, L., and Grosveld, F., A transgenic mouse model of sickle cell disorder, *Nature*, 343, 183, 1990.

142. Ryan, T. M., Townes, T. M., Reilly, M. P., Asakura, T., Palmiter, R. D., Brinster, R. L., and Behringer, R. R., Human sickle hemoglobin in transgenic mice, *Science*, 247, 566, 1990.

143. Fabry, M. E., Nagel, R. L., Pachnis, A., Suzuka, S. M., Costantini, F., High expression of human β^S and α-globin in transgenic mice: hemoglobin composition and hematological consequences, *Proc. Natl. Acad. Sci., U.S.A.*, 89, 12150, 1992.

144. O'Donnell, J. K., Martin, M. J., Logan, J. S., and Kumar, R., Production of human hemoglobin in transgenic swine: an approach to a blood substitute, *Cancer Detect. Prev.*, 17, 307, 1993.

145. Bookchin, R. M., and Nagel, R. L., Conformational requirements for the polymerization of hemoglobin S: studies of mixed liganded hybrids, *J. Mol. Biol.*, 76, 233, 1973.

146. Bookchin, R. M., Balazs, T., Nagel, R. L., and Tellez, I., Polymerization of hemoglobin SA hybrid tetramers, *Nature*, 269, 526, 1977.

147. Nacharaju, P., and Acharya, A. S., unpublished data, 1994.

148. Roy, R. P., Nagel, R. L., and Acharya, A. S., Molecular basis of the inhibition of β^S-chain-dependent polymerization by mouse α-chain: Semisynthesis of chimeras of human and mouse α-chains, *J. Biol. Chem.*, 268, 16406, 1993.

149. Roy, R. P., Nacharaju, P., Nagel, R. L., and Acharya, A. S., Symmetric interspecies hybrids of mouse and human hemoglobin: Molecular basis of their abnormal oxygen affinity, *J. Protein Chem.*, 14, 81, 1995.

150. Komiyama, N. H., Miyazaki, G., Tame, J., and Nagai, K., Transplanting a unique allosteric effect from crocodile into human hemoglobin, *Nature*, 373, 244, 1995.

151. Rhoda, M., Domenget, C., Vidaud, M., Bardakdjian-Michau, J., Rpuyer-Fessard, P., and Beuzard, Y., Mouse α-chain inhibit polymerization of hemoglobin induced by human β^S or β^S [Antilles] chains, *Biochim. Biophis. Acta*, 952, 208, 1988.

152. Adachi, K., and Asakura, T., Nucleation controlled aggregation of deoxyhemoglobin S. Possible differences in the size of nuclei in different phosphate concentrations, *J. Biol. Chem.*, 254, 7765, 1979.

153. Adachi, K., Ozguc, M., and Asakura, T., Nucleation controlled aggregation of deoxyhemoglobin S. Participation of hemoglobin A in the aggregation of deoxyhemoglobin S in concentrated phosphate buffer, *J. Biol. Chem.*, 255, 3092, 1980.

154. Nacharaju, P., Roy, R. P., Nagel, R. L., and Acharya, A. S., Inhibition of sickle β-chain (β^S)-dependent polymerization by nonhuman α-chains, *J. Biol. Chem.* 272, 27869, 1997.

155. Roy, R. P., and Acharya, A. S., unpublished data, 1999.

156. Cretegny, I., and Edelstein, S. J., Double strand packing in hemoglobin S fibers, *J. Mol. Biol.,* 230, 733, 1993.

157. Nacharaju, P., Roy, R. P., Nagel, R. L., and Acharya, A. S., Complementation of the polymerization inhibitory potential of the sequence differences of the segments 1–30 and 31–141 of mouse α-chain when segregated into the two α-chains of the tetramer, presented at the 19th Annual Meeting of the National Sickle Cell Program, New York, March 23-26, 1994.

158. Nacharaju, P., Nagel, R. L., and Acharya, A. S., Inhibition of ß^S-chain dependent polymerization by chimeric α-chains: Potential anti-sickling sequence for gene therapy, presented at the 20th Annual Meeting of the National Sickle Cell Program, Boston, March 18-21, 1995.

159. Parante, A., Roy, R. P., and Acharya, A. S., unpublished data, 1990.

160. De Filippis, V., and Fontana, A., Semisynthesis of carboxyl-terminal fragments of thermolysin, *Int. J. Peptide & Protein Res.,* 35, 219, 1990.

161. Bowles, D. J., Marcus, S. E., Pappin, D. J. C., Findlay, J. B. C., Eliopoulos, E., Maycox, P. R., and Burgess, J., Post-transnational processing of concanavalin A precursors in jackbean cotyledons, *J. Cell. Biol.,* 102, 1284, 1986.

162. Kane, P. M., Yamashiro, C. T., Wolczyk, D. F., Goebl, M., Neff, N., and Stevens T. H., Protein splicing converts the yeast TFP1 gene product to the 69-kD subunit of the vacuolar H+-adenosine triphosphatase, *Science,* 250, 651, 1990.

163. Cooper, A. A., Chen, Y., Lindorfer, M. A., and Stevens, T. H., Protein splicing of the yeast TFP1 intervening protein sequence: a model for self-excision, *EMBO J.,* 12, 2575, 1993.

164. Looker, D., Abbott-Brown, D., Cozart, P., Durfee, S., Hoffman, S., Mathews A. J., Miller-Roehrich, J., Shoemaker, S., Trimble, S., Fermi, G., Komiyama, N. H., Nagai, K., and Stetler, G. L., A recombinant haemoglobin designed for use as a blood substitute, *Nature,* 356, 258, 1992.

165. Shoemaker, S. A., Gerber, M. J., Evans, G. L., Archer-Paik, L. E., and Scoggin, C. H., Initial clinical experience with a rationally designed, genetically engineered recombinant human hemoglobin, *Art. Cells, Blood Subs., and Immob. Biotech.,* 22, 457, 1994.

166. Goldenberg, D. P., Dissecting the roles of individual interactions in protein stability: lessons from a circularized protein, *J. Cell. Biochem.,* 29, 321, 1985.

7 Protease Inhibitors

Herbert R. Wenzel and Harald Tschesche

7.1 INTRODUCTION

Proteinases are known to play a crucial role in the normal functioning of many biological systems. These enzymes are involved in important processes such as food digestion, blood coagulation, fibrinolysis, tissue remodeling, blood pressure regulation, fertilization, phagocytosis, and complement immune reactions. Certain specialised proteinases release peptide hormones and neuromodulators from inactive precursors or degrade peptides that transmit messages, thus initiating or terminating a variety of biological responses.

Regarding the vast number of potential peptide or protein substrates in an organism, it is clear that besides a more or less pronounced specificity of cleavage an efficient regulation of proteolysis by endogenous proteinase inhibitors is mandatory. The number of these proteinaceous inhibitors so far isolated and characterized is very impressive and steadily growing. The great majority are directed toward members of 1 of the 4 classes of endopeptidases, the serine proteinases, and can be grouped into at least 16 inhibitor families based on sequence homology, topological relationships of disulfide bridging, and location of the proteinase binding site.[1,2]

The continuous interest of numerous research groups in naturally occurring serine proteinase inhibitors has emerged mainly for two reasons:

1. These proteins are especially suited to the study of general aspects of protein conformation, folding, and interaction by an array of experimental techniques.
2. Detailed knowledge of their structure and reactivity is indispensable for a thorough understanding of the controlling functions they exercise in many physiological proteolytic processes.

Moreover, several pathological situations, such as pulmonary emphysema or rheumatoid arthritis, have been related to the deleterious action of proteinases, especially those released from polymorphonuclear leukocytes.[3]

The mechanistically oriented studies have amply benefited from homologous protein inhibitors available from various natural sources, by total synthesis, semisynthesis, or recombinant approaches. The same holds true for the more applied research aimed at developing specific exogenous natural or synthetic proteinase inhibitors to control proteolytic activities and thereby regulate certain disease states.

During the last decade, recombinant techniques have become the most often used for the preparation of tailor-made protein homologs and should be considered as the first option. Nevertheless, the semisynthetic techniques have more to offer than mere historical interest, which is especially true in the field of proteinase inhibitors, where they have been exploited since the 1960s.[4–13] They permit, for example, the splitting and closure of peptide bonds at specific sites and the introduction of noncoded amino acids and amino acid analogs in a broad sense.

Inhibitors belonging to 3 of the about 16 families mentioned above have been the target of semisynthetic engineering. These are the soybean trypsin inhibitor (Kunitz) family, where several fundamental strategies were developed; the bovine pancreatic trypsin inhibitor (Kunitz) family, where the most extensive experience concerning the exchange of single amino acids and short peptides have been gathered; and the pancreatic secretory trypsin inhibitor (Kazal) family, including many ovomucoid third domains, where fragment exchanges have been carried out.

The purpose of this chapter is to describe the structure of these proteins, to summarize the semisynthetic strategies applied, and to discuss the properties of the engineered analogs. As background information, the catalytic enzyme mechanism and the fundamental concepts for binding of substrates and proteinaceous inhibitors of proteinases are briefly reviewed first.

7.2 CATALYTIC MECHANISM, SUBSTRATE SPECIFICITY, AND MECHANISM OF INHIBITION

The serine proteinases are a class of proteolytic enzymes whose central catalytic machinery is composed of three invariant residues, a uniquely reactive serine, giving rise to their name, a histidine and an aspartic acid. As a result of more than 40 years of intensive study, their mode of operation is now well understood in terms of their spatial atomic structures.[14] Hydrolysis of a peptide bond occurs by a two-step displacement with an amine produced first, followed by a carboxylic acid. The catalysis starts with the nucleophilic attack of the hydroxyl group in the side chain of a serine residue (Ser195 in chymotrypsin) on the partially positive carbonyl carbon of the amide bond in the substrate, thus forming a tetrahedral intermediate. This step is favored by the close proximity of the imidazole moiety of a histidine residue (His57 in chymotrypsin), which activates the hydroxyl group. The third component of the catalytic triad, the carboxyl group of an aspartic acid residue (Asp102 in chymotrypsin) is apparently necessary to properly position the imidazole moiety and to aid in proton transfer. Two backbone amide protons of the proteinase forming the so-called oxyanion hole are available for hydrogen bonding to the developing oxygen anion from the carbonyl group of the peptide bond undergoing attack. The tetrahedral intermediate then collapses to form the amine derived from the carboxyl end of the substrate and an ester intermediate between the serine hydroxyl group and the acyl portion of the substrate. Hydrolysis of the ester, which proceeds through a second tetrahedral intermediate, liberates the second product, the carboxylic acid derived from the amino end of the substrate.

All steps of this catalytic process are reversible. Thus, under proper experimental conditions, the reverse reaction, peptide bond formation can be brought about and used for synthetic purposes.[15–17]

The system of notation, introduced by Schechter and Berger, has generally become accepted for describing the interaction of proteinases and their substrates or inhibitors.[18] The peptide segment of the substrate or inhibitor, which is in contact with the active site of the enzyme, is designated, from the amino to the carboxyl terminus, as $\ldots$-P_3-P_2-P_1-P_1'-P_2'-$P_3'$$\ldots$, where the scissile bond is between P_1 and P_1'. The subsites of the proteinase, which accommodate the single residues, are termed $\ldots$-S_3-S_2-S_1-S_1'-S_2'-S_3'-$\ldots$ by analogy. For most serine proteinases, the residue P_1 interacting with the subsite S_1 has been found to be the primary determinant in binding and subsequent cleavage. The preference for certain P_1 side chains can be rationalized at the molecular level by the complementarity of the S_1 structure, which is often also called *specificity pocket.*

Some proteinases that have usually provided the screening system for the search for specific and potent inhibitors will illustrate the different specificities: Trypsin, as well as glandular and plasma kallikreins, preferentially cleave peptides with Lys or Arg in position P_1, which can be explained by a negatively charged residue at the bottom of their specificity pocket. Chymotrypsin, cathepsin G from azurophil granules of leukocytes, and the bacterial subtilisins are able to accommodate bulky hydrophobic side chains in their S_1 subsites. They typically cleave peptide bonds at the carboxyl group of Phe, Tyr, Trp, and Leu. The elastase from pancreas prefers Ala in position P_1, that from azurophil granules of leukocytes Val or Leu. Bacterial proteinases isolated from *Staphylococcus aureus* and *Streptomyces griseus* specifically cleave Glu-Xaa bonds.

A large fraction of small proteinaceous inhibitors of serine proteinases interact with their cognate enzymes according to the following common minimal standard or canonical mechanism:[19]

$$E + I \underset{k_{off}}{\overset{k_{on}}{\rightleftharpoons}} EI \underset{k^*_{on}}{\overset{k^*_{off}}{\rightleftharpoons}} E + I^*$$

E is the proteinase, I the native or virgin single-chain inhibitor, I* the so-called modified inhibitor specifically cleaved at its reactive site peptide bond P_1-P_1', EI the stable proteinase inhibitor complex, and k the different rate constants. When judging by typical numerical values for the ratio k^*_{off}/K_D with the dissociation constants K_D = k_{off}/k_{on}, the inhibitors paradoxically appear to be good substrates. Both k^*_{off} and K_D values are, however, many orders of magnitude lower than the values for typical substrates, so stable complexes are formed. The equilibrium constant K_{Hyd} = [I*]/[I] is typically close to unity at physiological pH values but increases sharply with increasing or decreasing pH.[20–22]

7.3   STI: THE SOYBEAN TRYPSIN INHIBITOR (KUNITZ)

Soybean contains several serine proteinase inhibitors, two of which have been well characterised: the highly cystine-crosslinked double-headed Bowman-Birk-inhibitor, consisting of 71 residues; and the Kunitz-inhibitor STI, consisting of 181 residues with two disulfide bridges.[23] In spite of its above-average size, the latter has been the subject of several fundamental semisynthetic transformations. In fact, the standard mechanism of inhibition was deduced from experiments with STI establishing Arg63-Ile64 as the reactive site of the inhibitor. STI was already crystallized in the complex with porcine trypsin more than 20 years ago,[24] but its complete three-dimensional structure could not be determined from the electron density maps obtained. Therefore, to get an idea of the overall protein topology, a nearly complete model of the homologous Kunitz trypsin inhibitor DE-3 from *Erythrina caffra* is shown in Figure 7.1.[25] The structure consists of 12 antiparallel β-strands joined by long loops. The reactive site, Arg63-Ser64 in this case, is located on an external loop that protrudes from the surface of the molecule. The β-sheets are obviously well conserved in STI, while some of the loops are different.

7.3.1   PIONEERING ENZYMATIC P_1 MUTATIONS IN STI

It was through the Kunitz soybean trypsin inhibitor that M. Laskowski's group developed the enzymatic mutation of single amino acid residues within a peptide chain. The classic trapping strategy, the implications of which for current work are not always remembered, is shortly summarized in Figure 7.2.

Incubation of native single-chain STI **1** with catalytic amounts of trypsin leads to formation of *seco*-STI **2**, in which the reactive site peptide bond is specifically hydrolysed. Compounds **1** and **2** are in pH-dependent equilibrium and both are active inhibitors that form the same complex with the target proteinase. Arg63 is easily removed from **2** by treatment with carboxypeptidase B, yielding inactive des-Arg63-*seco*-STI **3**. The equilibrium of this reaction lies far on the side of hydrolysis; its reversal requires both a relatively high concentration of arginine or lysine and coupling with the strong trypsin complex formation, the enzyme being present in molar excess. After kinetic control dissociation of the complexes with guanidinium chloride, either STI **1** or the homolog Lys63-STI **4** can be isolated in good yields.[26] The reaction sequence **1** → **2** → **3** → **1** was also successfully applied to prepare ^{13}C-Arg63-labeled STI for nuclear magnetic resonance studies.[27]

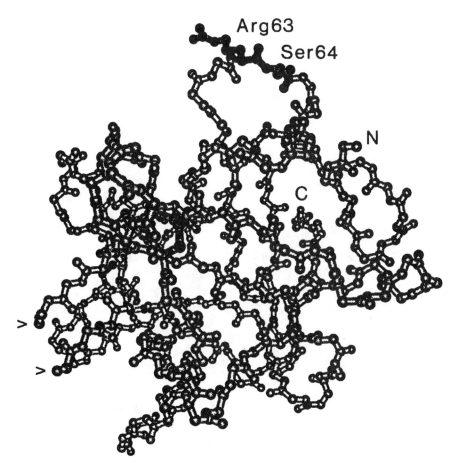

FIGURE 7.1   Spatial backbone of the Kunitz-type trypsin inhibitor from *Erythrina caffra,* consisting of 172 amino acid residues. The reactive site residues Arg63-Ser64, including their side chains, and the two disulfide bridges are drawn with solid bonds. N denotes the amino terminus and C the residue 170. Due to poor electron density, no atomic model could be built for the segment between residues 93 and 98 (>) and for the two C-terminal residues 171–172.

When using the combination carboxypeptidase A/chymotrypsin with their known preference for neutral, bulky side chains, aromatic residues can be incorporated to yield Xaa63-STI **5** (Xaa = Trp, Phe).[28,29] As predicted, both homologs are potent chymotrypsin inhibitors.

7.3.2   P$_1$' REPLACEMENTS AND REACTIVE SITE INSERTIONS IN STI DERIVATIVES

The pure enzymatic approach described above was combined with peptide-chemical operations to probe the role of the P'-residues in STI.[29] A flaw of the reaction sequences summarized in Figure 7.3 is that they have to start with irreversibly

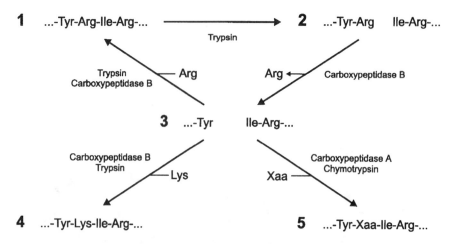

FIGURE 7.2 Enzymatic mutations of P_1 = Arg63 in the soybean trypsin inhibitor. Only residues Tyr62 through Arg65 of the reactive site region are shown.

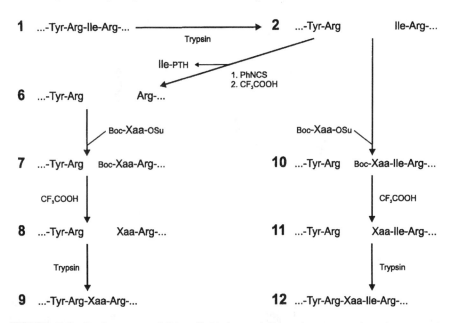

FIGURE 7.3 Replacement of P_1' = Ile64 in partially amino-protected soybean trypsin inhibitor and insertion of an amino acid in the reactive site. Nine out of ten lysine residues (one is presumably buried) are guanidinated, the amino terminus is carbamylated; only Tyr62 through Arg65 of the reactive site region are shown.

protected, i.e., α-carbamylated, ε-guanidinated STI **1** to limit reactions to a single amino group. Moreover, protein intermediates must repeatedly be renatured and purified because of the harsh experimental conditions. From *seco*-STI **2**, prepared by trypsin treatment, Ile64 is removed by a single Edman cycle, yielding des-Ile64-

seco-STI **6**. Chemical replacement (**6**→**7**→**8**→**9**) is now brought about by coupling with *tert*-butyloxycarbonyl-amino acid-N-hydroxysuccinimide esters (Xaa = Leu, Ala, Gly), Boc-removal in trifluoroacetic acid and kinetic control dissociation of the stoichiometric trypsin complexes. All three homologs prepared proved to be fully active trypsin inhibitors.[30] Thus, the P_1' side chain of STI appears to be of minor importance for the inhibitory activity.

Starting with *seco*-STI **2**, the last three reactions mentioned can be used to insert a single amino acid residue (Xaa = Ala, Ile) in the STI reactive site (**2** → **10** → **11** → **12**). The yield of the final enzymatic peptide bond synthesis, however, is very low, as no stable complex is formed.[31] Thus, the insertion of an amino acid between the P_1 and the P_1' residues of STI converts a trypsin inhibitor into a trypsin substrate.

7.4 BPTI: THE BOVINE PANCREATIC TRYPSIN INHIBITOR (KUNITZ)

The Kunitz trypsin inhibitor from bovine pancreas, BPTI, is the prototype of a wide-spread inhibitor family and probably the most extensively studied protein. Chemical reactivity, structure of the free inhibitor and its proteinase complexes determined by X-ray, neutron diffraction and NMR methods, thermodynamic stability, folding pathway, molecular dynamics, histochemical localization, biosynthesis, processing, evolution and therapeutic use have been the main subjects of research.[13,32]

The single polypeptide chain of BPTI consists of only 58 amino acid residues cross-linked by three disulfide bridges (Figure 7.4). It is folded in such a way that a pear-shaped molecular structure results. The main elements are an extended amino-

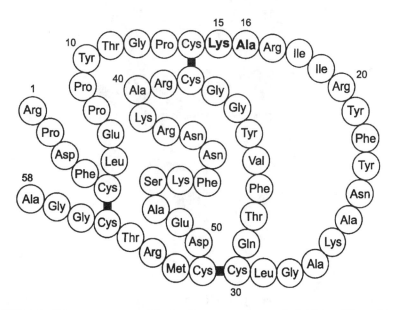

FIGURE 7.4 Primary structure of the bovine pancreatic trypsin inhibitor. The residues P_1 = Lys15 and P_1' = Ala16 appear in boldface.

terminal segment, a reactive site loop, a three-stranded ß-pleated sheet, and a short carboxyl-terminal α-helix (Figure 7.5).[33] The concentration of the negatively charged groups of Asp3, Glu7, Glu49, Asp50, and Ala58 at the bottom of the pear explains the highly dipolar character of the basic BPTI. The conformation of the rather rigid reactive site loop at the top adjacent to the scissile P_1-P_1' peptide bond, Lys15-Ala16, is complementary to proteinase binding sites, thus allowing an unconstrained, tight interaction with several enzymes.[34]

BPTI is an excellent inhibitor of bovine trypsin (K_D = 60 fM), mainly due to the positively charged side chain of Lys15, which is accommodated in the specificity pocket of the proteinase. For the same reason, relatively strong complexes are formed with porcine glandular kallikrein (K_D = 500 pM), human plasma kallikrein (K_D = 45 nM), and, surprisingly, when taking its different specificity into account, also with bovine chymotrypsin (K_D = 20 nM). Inhibition of human leukocyte elastase

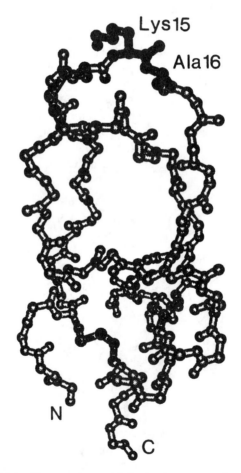

FIGURE 7.5 Spatial backbone structure of the bovine pancreatic trypsin inhibitor. The reactive site residues Lys15-Ala16, including their side chains, and the three disulfide bridges are drawn with solid bonds. N denotes the amino and C the carboxyl terminus.

$(K_D = 3 \ \mu M)$ and cathepsin G $(K_D = 4 \ \mu M)$ is very weak; porcine pancreatic elastase is not inhibited at all.

7.4.1 GENERAL PATHWAYS TO BPTI HOMOLOGS MUTATED AT P_1

After it had become evident that P_1 dominates the specificity of serine proteinase inhibitors, it was of course tempting to change this amino acid residue also in BPTI with a view to enlarging its field of application as a drug. The enzymatic mutation outlined above for STI could successfully be adopted with some modifications (Figure 7.6).[35]

Direct reactive site hydrolysis of BPTI **13** to give *seco*-BPTI **14** is catalysed only very slowly by bovine or porcine trypsin, the equilibrium at neutral pH, in which about half of the inhibitor molecules are nicked, is reached after several months.[21,36] This difficulty can be overcome by selective borohydride reduction of the Cys14/Cys38 disulfide bridge, followed by fast tryptic cleavage of the peptide bond Lys15-Ala16 and air reoxidation of the half cystine residues.[37] After cation-exchange chromatography, pure *seco*-BPTI **14** is obtained in about 70% yield even on a gram scale. Alternatively, a trypsin-like proteinase from the starfish species *Dermasterias imbricata* can be used to establish the equilibrium within a few days without disulfide reduction.[38]

Des-Lys15-*seco*-BPTI **15** is easily prepared by carboxypeptidase B treatment. As was shown with STI, aromatic amino acids can now be incorporated by the combined action of carboxypeptidase A and chymotrypsin to yield the single chain homologs Xaa15-BPTI **17** (Xaa = Phe, Trp). When Arg15-BPTI **16** is to be prepared semisynthetically, it is essential that kallikrein be used together with carboxypeptidase B in the trapping reaction, because trypsin unintentionally led to a product

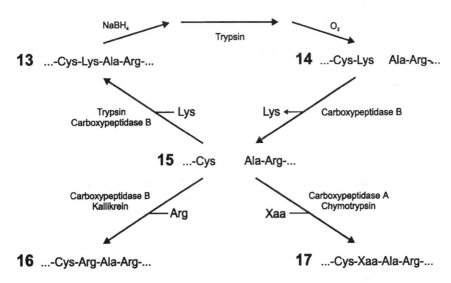

FIGURE 7.6 Enzymatic mutations of the $P_1 = $ Lys15 residue in BPTI. Only residues Cys14 through Arg17 of the reactive site region are shown.

lacking Arg39.[39] This was also the case, when ^{13}C-labeled lysine was introduced ($15 \rightarrow 13$) for nuclear magnetic resonance experiments.[40, 41]

Compared with BPTI, the Arg15-homolog is a substantially better inhibitor for several proteinases. For example, the K_D values decrease by a factor of about 20 for human trypsin and almost 80 for human plasma kallikrein.[42] Most notably, human coagulation factor XIa, although not inhibited by natural BPTI, has a K_D of 34 nM for the Arg15-compound.[43] Both Phe15- and Trp15-BPTI exhibit pronounced anti-chymotryptic activity, the K_D-values for the complexes with bovine chymotrypsin being decreased by one order of magnitude.[35]

Several attempts to extend the elegant enzymatic mutation method to the introduction of aliphatic amino acids with alkyl side chains have so far remained unsuccessful, mainly because a proper enzyme system could not be found. We have therefore developed more general strategies with more chemically based routes involving coupling reagents and reversible protection, which allow for replacement of Lys15 by virtually any other amino acid.[44–48] Briefly, the main reaction sequences consist of the following several principal steps (Figure 7.7).

- Reactive site cleavage of BPTI **13** as in the enzymatic mutation.
- Complete esterification of *seco*-BPTI **14** at its six carboxyl groups with acidified methanol to yield the hexamethyl ester **18**.
- Specific saponification of the Lys15 methyl ester with endoproteinase Lys-C in the presence of dioxane as organic cosolvent to yield the pentamethyl ester **19**.

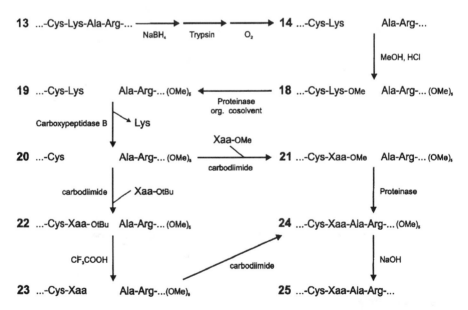

FIGURE 7.7   General chemical methods for the exchange of the P_1 = Lys15 residue in BPTI. Residues Cys14 through Arg17 are shown. The methyl ester formation, indicated by (OMe)$_5$, takes place at Asp3, Glu7, Glu49, Asp50, and Ala58 (cf. Figure 7.4).

- Removal of Lys15 by carboxypeptidase B.
- At this key intermediate **20**, with its single free carboxyl group, the reaction pathway branches out. In the more general route (**20** → **22** → **23** → **24** → **25**), the new P_1 amino acid Xaa as *tert*-butyl ester is coupled to Cys14 with a water soluble carbodiimide. Specific removal of the *tert*-butyl ester group with trifluoroacetic acid is followed by a second carbodiimide condensation joining Xaa15 and Ala16. Alkaline saponification of all methyl ester groups yields the Xaa15-BPTI homologs.
- In the shorter route (**20** → **21** → **24** → **25**), Xaa as methyl ester is coupled to Cys14. Resynthesis of the reactive site peptide bond, Xaa15-Ala16, depends on the availability of a suitable proteinase and can be achieved, for example, with trypsin (Xaa = Arg) or chymotrypsin (Xaa = Met, norleucine, Glu-γ-methyl ester). Saponification of the ester groups again yields the completely deprotected Xaa15-BPTI homologs.

About 20 P_1 substitutions have been performed using these enzymatic/peptide-chemical mutations. Among the Xaa15-BPTI homologs obtained is a series of 11 proteins that are also homologous in a chemical sense with regard to the structure of their varying alkyl side chain: Xaa = glycine, alanine, ß-aminobutyric acid, norvaline, valine, norleucine, leucine, isoleucine, *allo*-isoleucine, *tert*-leucine, *tert*-butylalanine. The K_D values for their complexes with two elastases certainly reflect the spatial structures of the two specificity pockets. BPTI homologs with P_1 residues having a short linear side chain are especially strong inhibitors of porcine pancreas elastase (norvaline15-BPTI: $K_D = 0.4$ nM, β-aminobutyric acid 15-BPTI: $K_D = 1.1$ nM), whereas homologs with branched residues have a significantly lower affinity (Val15-BPTI: $K_D = 57$ nM). Human leukocyte elastase, on the other hand, is able to accommodate bulky branched side chains in its S_1 site (Val15-BPTI: $K_D = 0.1$ nM, norvaline15-BPTI: $K_D = 0.2$ nM, β-aminobutyric acid 15-BPTI: $K_D = 0.3$ nM, Ile15-BPTI: $K_D = 0.4$ nM).

Glu15-BPTI was tested for inhibition of the two glutamate-specific proteinases from *Staphylococcus aureus* and *Streptomyces griseus*. No decrease in enzymatic activity was detected in either case.

7.4.2   GENERAL PATHWAYS TO BPTI HOMOLOGS MUTATED AT P'
POSITIONS

While the inhibitory specificity of a BPTI homolog is dominated by its P_1 amino acid residue, it can certainly also be influenced by the adjacent P' residues Ala16 and Arg17, which are both situated within the contact region. Thus, additional fine-tuning of the interaction between proteinase and inhibitor can be expected from variations of these residues. Several semisynthetic strategies for the required replacements have been exploited, which are often parallel to the reactions described in the previous sections. A simple enzymatic mutation method is outlined in Figure 7.8.[49,50]

Following the obligatory cleavage of BPTI **13** at its reactive site, Ala16 and Arg17 are successively removed from *seco*-BPTI **14** by the action of aminopeptidase K to yield the nicked derivatives **26** and **27**, respectively. From an incubation mixture

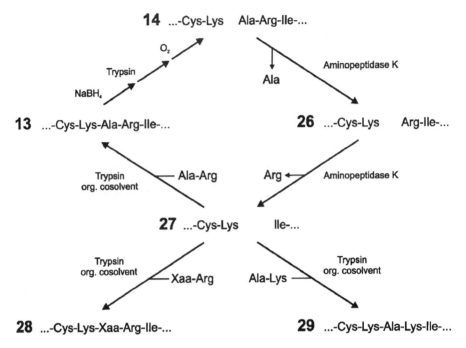

FIGURE 7.8 Enzymatic mutations of the $P_1' = $ Ala16 and $P_2' = $ Arg17 residues in BPTI. Only residues Cys14 through Ile18 of the reactive site region are shown.

of des-(Ala16-Arg17)-*seco*-BPTI **27** with a high excess of the dipeptide Ala-Arg and equimolar amounts of trypsin in the presence of 80% 1,4-butanediol, single-chain BPTI **13** can be isolated after some hours. Thus, the proteinase obviously closes two peptide bonds in the order Arg17-Ile18 preceding Lys15-Ala16 and traps the final product by complex formation. When using Ala-Lys as the dipeptide, Lys17-BPTI **29** is obtained. The dipeptides Xaa-Arg (Xaa = Gly, Leu) open up the way to Xaa16-BPTI homologs **28**. Because of the pronounced specificity of trypsin, this enzymatic fragment substitution is not generally applicable; it requires a carboxyl-terminal Arg or Lys in the dipeptides to be built in.

Other dipeptides or single amino acids can be incorporated in a general fashion, however, by peptide chemical coupling procedures and fully reversible protection with ester groups, as outlined in Figure 7.9.[51]

The key intermediate in the reaction sequence is des-(Ala16-Arg17)-*seco*-BPTI-pentamethyl ester **31**. It can be prepared from BPTI (**13** → **27** → **30** → **31**) by reactive site hydrolysis, aminopeptidase K treatment, complete esterification and specific enzymatic deblocking of Lys15. The last reaction is conveniently carried out using the esterolytic activity of trypsin which, compared with endoproteinase Lys-C, has the advantage of shorter reaction times. The organic cosolvent 1,4-butanediol is added to diminish the amidolytic activity of trypsin and to thus prevent the cleavage of the susceptible Arg39-Ala40 peptide bond. The gap between Lys15 and Ile18 in the BPTI sequence can be filled either directly with a dipeptide derivative

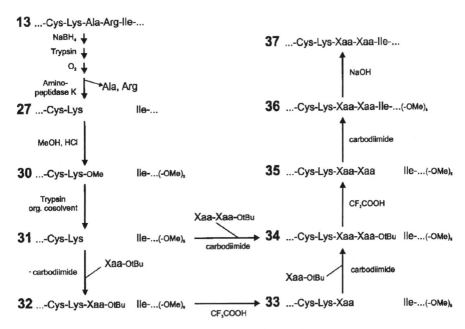

FIGURE 7.9 General reaction sequences for the replacement of residues P_1' = Ala16 and P_2' = Arg 17 in BPTI by other amino acids. Residues Cys14 to Ile18 are shown. The five methyl ester groups, (-OMe)$_5$, temporarily protect the four acidic side-chain functions of Asp3, Glu49, Asp50, and the C-terminus of Ala58 (cf. Figure 7.4).

($\mathbf{31} \rightarrow \mathbf{34} \rightarrow \mathbf{35} \rightarrow \mathbf{36} \rightarrow \mathbf{37}$) or successively with two amino acid derivatives ($\mathbf{31} \rightarrow \mathbf{32} \rightarrow \mathbf{33} \rightarrow \mathbf{34} \rightarrow \mathbf{35} \rightarrow \mathbf{36} \rightarrow \mathbf{37}$). Dipeptides and amino acids are C-protected as *tert*-butyl esters that can be split off with trifluoroacetic acid. All peptide bonds are formed with N-ethyl-N'-(3-dimethylaminopropyl) carbodiimide/1-hydroxy-benzotriazole, the final saponification being carried out at pH 10.5. Ala17-BPTI **37** (Xaa = Ala) was prepared as an example following either route.

The P_2' substitutions leading to Lys17-BPTI (enzymatically mutated) and Ala17-BPTI (chemically mutated) are only of minor to moderate significance for the inhibition of bovine trypsin, bovine chymotrypsin, and porcine kallikrein. The same holds true for the P_1' substitution in Gly16-BPTI (enzymatically mutated). Leu16-BPTI (enzymatically mutated), however, has considerably increased K_D values when compared with the parent inhibitor: trypsin, about 1 million-fold; chymotrypsin, 120-fold; kallikrein, 60-fold.

7.4.3 SEMISYNTHESIS OF BPTI BACKBONE VARIANTS

The rational design of inhibitors as potential drugs, be these peptides, proteins, or peptidomimetics, requires detailed knowledge of all aspects of their interaction with the target proteinases. This includes not only the arrangement of amino acid side chains in the contact region dealt with above, but also the underlying peptide backbone structure. We have therefore embarked on systematic variations by semi-synthesis of the elements making up the BPTI backbone around the reactive site.[52–55]

Figure 7.10 outlines two reaction sequences ending up with BPTI homologs that are lacking the side chains at P_1' and P_2' and thus are formally derived from Gly16, Gly17-BPTI. Their special feature is a non-peptidic bond linking P_1' and P_2'. Des-(Ala16-Arg17)-*seco*-BPTI **27**, prepared from BPTI **13** as described above, is the starting point for both pathways. In the shorter one (**27** → **38** → **39** → **44**), the Gly-Gly analogous ω-amino carboxylic acids are coupled, as Boc-protected N-hydroxy-succinimide esters, to the Ile18 amino group of **27**. Although this inhibitor fragment possesses a total of six potential sites for acylation, the reaction can be carried out with a high degree of selectivity at pH 4.75, which means that about 25% of the product is monoacylated at Ile18. After cleavage of the N-protecting Boc-groups, the second amide bond, that to Lys15, is closed by using trypsin. In the longer pathway (**27** → **40** → **41** → **42** → **43** → **44**), the order of peptide bond formations is reversed, and reversible carboxyl protection is applied. Trypsin is first used to couple the ω-amino carboxylic acid *tert*-butyl esters to the Ile18 amino group of **27**; the presence of 88% organic cosolvent (1,4-butanediol/dimethylsulfoxide 2:1) is essential in this case. The following reactions exploit an approved scheme: complete esterification with acidified methanol, removal of the *tert*-butyl ester group with trifluoroacetic acid, condensation with water-soluble carbodiimide, and saponification of all methyl ester groups.

Five Gly16,Gly17-BPTI homologs **44** have been prepared and characterized. The ω-amino acid residues substituting for Gly-Gly, their structure, and the disso-

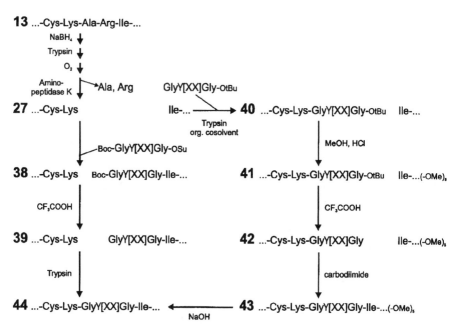

FIGURE 7.10   Scheme of the reaction sequences employed in the exchange of $P_1'–P_2' =$ Ala16-Arg17 in BPTI by peptide analogs. [XX] denotes a peptide bond surrogate (e.g., CH_2-CH_2, CH_2-NH, CO-CH_2, CH=CH), $(-OMe)_5$ the methyl ester formation at Asp3, Glu7, Glu49, Asp50, and Ala58 (cf. Figure 7.4).

ciation constants of the trypsin complexes compare with the data of the reference compound as follows:

Diglycine	-NH-CH$_2$-**CO-NH**-CH$_2$-CO-	K_D = 2 pM
5-Aminolevulinic acid	-NH-CH$_2$-**CO-CH$_2$**-CH$_2$-CO-	K_D = 50 pM
5-Aminoethyglycine	-NH-CH$_2$-**CH$_2$-NH**-CH$_2$-CO-	K_D = 7 nM
5-Aminovaleric acid	-NH-CH$_2$-**CH$_2$-CH$_2$**-CH$_2$-CO-	K_D = 2 nM
5-Amino-3-pentenoic acid *(trans)*	-NH-CH$_2$-**CH=CH**-CH$_2$-CO-	K_D = 1 nM
3-Aminomethylbenzoic acid	-NH-CH$_2$-**C$_6$H$_4$**-CO-	no inhibition

Despite the loss of two side chains in its contact region, Gly16, Gly17-BPTI remains an excellent permanent trypsin inhibitor. This statement also holds true for the COCH$_2$-derivative, although the K_D value increases by one order of magnitude. As the keto function is involved in neither intra- nor intermolecular hydrogen bonding, the activity is probably due to the retained backbone rigidity in this region. Both the CH$_2$NH- and the CH$_2$CH$_2$- derivative prove to be much weaker and, furthermore, temporary trypsin inhibitors, which are slowly degraded and thereby inactivated by the proteinase. A reason for this switch to a substrate may lie in the enhanced flexibility of the binding loops. This interpretation is corroborated by the properties of the CHCH-derivative with its rigid *trans* double bond, it is again a permanent trypsin inhibitor with a slightly decreased K_D-value. No inhibition is displayed by the C$_6$H$_4$-derivative with a *meta*-disubstituted benzene ring in which the α-carbon of the P$_2$' residue is also integrated. The fitting of the contact sites is obviously severely disturbed by the bulky aromatic ring system.

It is evident that the reaction sequences shown in Figure 7.10 are also applicable to other dipeptides and derivatives thereof. Moreover, the extension to shorter and longer fragments should be possible with some modifications. For example, current studies are aimed at introducing ester and hydroxyethylene moieties at different positions of the BPTI backbone, partly with retention of the side chain functions.

7.5   OM3: OVOMUCOID THIRD DOMAINS (KAZAL)

Small single-domain Kazal inhibitors have been found in all vertebrates examined. One of them, the human pancreatic trypsin inhibitor, has been considered a promising starting point for the engineering of useful therapeutic proteins with a minimum likelihood of an immunogenic reaction. A large natural set of Kazal inhibitors with great variability in the contact region is easily available from ovomucoids of avian eggs. Their polypeptide chains consist of three homologous Kazal domains that can, in favorable cases, be cleaved from one another without intradomain nicking. The production of ovomucoid third domains, OM3, by limited enzymatic hydrolysis proved to be most efficacious.[56]

Three-dimensional structures were determined for two of these OM3 inhibitors; [57,58] that from silver pheasant is shown in Figure 7.11 as an example. It consists of 56 amino acid residues cross-linked by three disulfide bridges. A three-stranded β-sheet and an α-helix approximately parallel to the β-strands form the nucleus of the

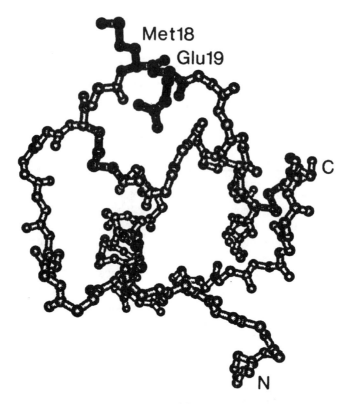

FIGURE 7.11   Spatial backbone structure of the silver pheasant ovomucoid third domain. The reactive site residues Met18-Glu19, including their side chains, and the three disulfide bridges, are drawn with solid bonds. N denotes the amino and C the carboxyl terminus.

molecule, to which the amino-terminal segment, the reactive site loop, and a segment connecting the α-helix with the carboxyl-terminal β-strand are attached.

The large set of 92 unique OM3 sequences mentioned above have originated from the massive effort of Laskowski's group, which, during the last two decades, isolated and sequenced OM3 inhibitors from more than 150 species of birds.[59–61] It turned out that the structurally important residues are strongly conserved, whereas the residues in contact with the cognate proteinases are by far the most variable ones in the molecule. K_D values for the complexes of virtually all of these OM3 inhibitors with several serine proteinases were measured with the following generalized results.[62,63] P_1 is predominant in the interaction with almost all enzymes. Changes in other residues may exert large effects on K_D provided that these residues make contact with the enzyme. Moreover, they often exert considerable differential effects on the different enzymes, which means that the same change can render the inhibitor stronger for one enzyme and weaker for another. Changes in surface residues, which do not contact the enzyme, are virtually without effect.

The ultimate goal, which is being approached, is to state a sequence to reactivity algorithm, which is a set of rules allowing prediction of the dissociation constants for various enzyme complexes from the amino acid sequence of the inhibitor alone.[64] The array of OM3 inhibitors available for comparison can be systematically enlarged by using semisynthetic strategies.

7.5.1 EXCHANGE OF LARGER PEPTIDE FRAGMENTS IN THE OM3 FAMILY

The synthetic methods outlined for BPTI in Chapter 4 comprise the exchange of one or two amino acid residues out of 58, so that the term *semisynthesis* in a literal sense appears somewhat exaggerated. However, it also proved feasible to substitute large fragments of proteinase inhibitors—for example, the complete P strand in members of the Kazal family. The method is as follows. OM3 inhibitors with 56 residues from two species of birds are enzymatically cleaved at their reactive sites between residues 18 and 19 (see Figure 7.11). After reduction of the three disulfide bonds, two peptides can be isolated in each case, the amino-terminal fragment 1–18 and the carboxyl-terminal fragment 19–56. The chains are then converted to mixed disulfides with glutathione. Upon incubation of the short fragment of one species with the long fragment of a second species, interchain disulfide bridges are formed. Enzymatic closure of the reactive site peptide bond finally yields a hybrid third OM3 with a new primary structure.[65] This strategy can be extended to the recombination of a natural long fragment 19–56 with a synthetic short fragment 1–18 or, more conveniently, 6–18, as the five amino-terminal residues have essentially no influence on the inhibitory activity.[66] Thus, the way is open for many interesting substitutions within the P strand, independent of their occurrence in natural ovomucoids and also including noncoded amino acids.

For obvious reasons, the main interest is again concentrated on the predominant P_1 residue. Therefore, a set of 13 homologs of turkey OM3 has recently been prepared, including those with the following P_1 residues, which form a homologous series in a chemical sense: alanine, α-aminobutyric acid, norvaline, norleucine, α-aminoheptanoic acid.[67] K_D values for the complexes of these variants with a selection of six serine proteinases were determined. Discussing the details of this comprehensive study would be beyond the scope of this review. It suffices here to mention that, for the two elastases, a reasonable agreement exists between the conclusions drawn in this study[67] and those drawn in the study with the corresponding BPTI homologs.[47] A striking difference between the two inhibitor families becomes apparent, however, when assaying the Glu-specific proteinases. Glu18-turkey OM3 is an excellent inhibitor of the enzyme from *Streptomyces griseus* with K_D = 18 pM,[68] while Glu15-BPTI is at least five orders of magnitude weaker. Neither homolog inhibits the enzyme from *Staphylococcus aureus*.

7.6 SUMMARY, CONCLUSIONS, AND OUTLOOK

Semisynthetic engineering has now been exploited with proteinase inhibitors for about 30 years. The numerous results obtained not only decisively contributed to the establishment of the standard mechanism of inhibition, they also provided several clues for the understanding of protein-protein interactions in general. In fact, the enzymatic mutations were probably the first examples of the rational creation of novel biological activities based on prior knowledge of structure-function relationships in proteins. Recent advances include the subtle stepwise enlargement of a crucial side chain within an inhibitor to probe the active site of cognate proteinases or the systematic variation of a single peptide bond of an inhibitor. A special feature of many underlying semisynthetic strategies is that they critically depend on the special relationship between the proteinases and their substrate-like inhibitors. This relationship allows both preferential single cleavage and efficient resynthesis under appropriate experimental conditions. It is also responsible, however, for the fact that the methods cannot generally be applied to any site within an inhibitor molecule.

In principle, this restriction does not apply to two other techniques. On the one hand, the total chemical synthesis of small proteinase inhibitors with up to about 80 residues is clearly within the capabilities of current automated solid-phase methods.[69-71] On the other hand, efficient expression systems for the production of recombinant proteinase inhibitors, for example homologs of BPTI[72-75] and OM3[76-79] are available, complemented by powerful phage display techniques.[80,82] Thus, amino acid replacements, denied to semisynthetic reaction sequences, have become possible. The really interesting and very promising fact, however, is that in many cases the maximum profit can be expected from strategies combining recombinant, enzymatic, and chemical approaches.[11,83]

ACKNOWLEDGMENTS

The work in our laboratory has been supported by the Deutsche Forschungsgemeinschaft (SFB 223, SFB 549, TS 8/37). Thanks are due to Rainer Beckmann for technical assistance and Grainne Delany for linguistic advice. Refined X-ray crystal structures of STI and its complex with porcine trypsin were recently determined.[84]

REFERENCES

1. Laskowski, M., Jr., and Kato, I., Protein inhibitors of proteinases, *Annu. Rev. Biochem.,* 49, 593, 1980.
2. Bode, W., and Huber, R., Natural protein proteinase inhibitors and their interaction with proteinases, *Eur. J. Biochem.* 204, 433, 1992.
3. Taylor, J. C., and Mittman, C., Eds., *Pulmonary Emphysema and Proteolysis: 1986,* Academic Press, Orlando, 1987.
4. Fritz, H., Tschesche, H., Greene, L. J., and Truscheit, E., Eds., *Proteinase Inhibitors,* Springer-Verlag, Berlin, 1974.
5. Offord, R. E., and Di Bello, C., Eds., *Semisynthetic Peptides and Proteins,* Academic Press, London, 1978.
6. Sheppard, R. C., Partial synthesis of peptides and proteins, in *The Peptides: Analysis, Synthesis, Biology,* Vol. 2, Gross, E., and Meienhofer, J., Eds., Academic Press, New York, 1980, 441.
7. Offord, R. E., *Semisynthetic Proteins,* Wiley, Chichester, 1980.

8. Chaiken, I. M., Semisynthetic peptides and proteins, *CRC Crit. Rev. Biochem.*, 11, 255, 1981.

9. Medvedkin, V. N., Semisynthesis of proteins and peptides, *Bioorg. Khim.*, 15, 581, 1989.

10. Tesser, G. I., Analogues of some small proteins by semisynthesis (Part 2), *Kontakte (Darmstadt)*, 1989 (3), 29.

11. Offord, R. E., Chemical approaches to protein engineering, in *Protein Design and the Development of New Therapeutics and Vaccines*, Hook, J. B., and Poste, G., Eds., Plenum, New York, 1990, 253.

12. Wallace, C. J. A., Peptide ligation and semisynthesis, *Curr. Opin. Biotechnol.*, 6, 403, 1995.

13. Wenzel, H. R., and Tschesche, H., Reversible inhibitors of serine proteinases: Naturally occurring miniproteins, semisynthetic variants, recombinant homologs, and synthetic peptides, in *Peptides: Synthesis, Structures, and Applications*, Gutte, B., Ed., Academic Press, San Diego, 1995, 321.

14. Steitz, T. A., and Shulman, R. G., Crystallographic and NMR studies of the serine proteases, *Annu. Rev. Biophys. Bioeng.*, 11, 419, 1982.

15. Homandberg, G. A., Mattis, J. A., and Laskowski, M., Jr., Synthesis of peptide bonds by proteinases. Addition of organic cosolvents shifts peptide bond equilibria toward synthesis, *Biochemistry*, 17, 5220, 1978.

16. Kullmann, W., Proteases as catalytic agents in peptide synthetic chemistry: Shifting the extent of peptide bond synthesis from a "quantité négligeable" to a "quantité considérable", *J. Protein Chem.*, 4, 1, 1985.

17. Schellenberger, V., and Jakubke, H.-D., Protease-catalyzed kinetically controlled peptide-synthesis, *Angew. Chem., Int. Ed. Engl.*, 30, 1437, 1991.

18. Schechter, I., and Berger, A., On the size of the active site in proteases. I. Papain, *Biochem. Biophys. Res. Commun.*, 27, 157, 1967.

19. Laskowski, M., Jr., and Sealock, R. W., Protein proteinase inhibitors - Molecular aspects, in *The Enzymes*, Vol. 3, 3rd Ed., Boyer, P. D., Ed., Academic Press, New York, 1971, 375.

20. Niekamp, C. W., Hixson, H. F., Jr., and Laskowski, M., Jr., Peptide-bond hydrolysis equilibria in native proteins. Conversion of virgin into modified soybean trypsin inhibitor, *Biochemistry*, 8, 16, 1969.

21. Siekmann, J., Wenzel, H. R., Matuszak, E., von Goldammer, E., and Tschesche, H., The pH dependence of the equilibrium constant K_{Hyd} for the hydrolysis of the Lys15-Ala16 reactive-site peptide bond in bovine pancreatic trypsin inhibitor (aprotinin), *J. Protein Chem.*, 7, 633, 1988.

22. Ardelt, W., and Laskowski, M., Jr., Effect of single amino acid replacements on the thermodynamics of the reactive site peptide bond hydrolysis in ovomucoid third domain, *J. Mol. Biol.*, 220, 1041, 1991.

23. Ikenaka, T., Odani, S., and Koide, T., Chemical structure and inhibitory activities of soybean proteinase inhibitors, in *Proteinase Inhibitors*, Fritz, H., Tschesche, H., Greene, L. J., and Truscheit, E., Eds., Springer-Verlag, Berlin, 1974, 325.

24. Sweet, R. M., Wright, H. T., Janin, J., Chothia, C. H., and Blow, D. M., Crystal structure of the complex of porcine trypsin with soybean trypsin inhibitor (Kunitz) at 2.6-Å resolution, *Biochemistry*, 13, 4212, 1974.

25. Onesti, S., Brick, P., and Blow, D. M., Crystal structure of a Kunitz-type trypsin inhibitor from *Erythrina caffra* seeds, *J. Mol. Biol.*, 217, 153, 1991.

26. Sealock, R. W., and Laskowski, M., Jr., Enzymatic replacement of the arginyl by a lysyl residue in the reactive site of soybean trypsin inhibitor, *Biochemistry,* 8, 3703, 1969.

27. Baillargeon, M. W., Laskowski, M., Jr., Neves, D. E., Porubcan, M. A., Santini, R. E., and Markley, J. L., Soybean trypsin inhibitor (Kunitz) and its complex with trypsin. Carbon-13 nuclear magnetic resonance studies of the reactive site arginine, *Biochemistry,* 19, 5703, 1980.

28. Leary, T. R., and Laskowski, M., Jr., Enzymatic replacement of Arg63 by Trp63 in the reactive site of soybean trypsin inhibitor (Kunitz) - An intentional change from tryptic to chymotryptic specificity, *Fed. Proc.,* 32, 465, 1973.

29. Kowalski, D., Leary, T. R., McKee, R. E., Sealock, R. W., Wang, D., and Laskowski, M., Jr., Replacements, insertions, and modifications of amino acid residues in the reactive site of soybean trypsin inhibitor (Kunitz), in *Proteinase Inhibitors,* Fritz, H., Tschesche, H., Greene, L. J., and Truscheit, E., Eds., Springer-Verlag, Berlin, 1974, 311.

30. Kowalski, D., and Laskowski, M., Jr., Chemical-enzymatic replacement of Ile64 in the reactive site of soybean trypsin inhibitor (Kunitz), *Biochemistry,* 15, 1300, 1976.

31. Kowalski, D., and Laskowski, M., Jr., Chemical-enzymatic insertion of an amino acid residue in the reactive site of soybean trypsin inhibitor (Kunitz), *Biochemistry,* 15, 1309, 1976.

32. Fritz, H., and Wunderer, G., Biochemistry and applications of aprotinin, the kallikrein inhibitor from bovine organs, *Arzneim.-Forsch./Drug Res.,* 33, 479, 1983.

33. Wlodawer, A., Deisenhofer, J., and Huber, R., Comparison of two highly refined structures of bovine pancreatic trypsin inhibitor, *J. Mol. Biol.,* 193, 145, 1987.

34. Huber, R., Kukla, D., Bode, W., Schwager, P., Bartels, K., Deisenhofer, J., and Steigemann, W., Structure of the complex formed by bovine trypsin and bovine pancreatic trypsin inhibitor. II. Crystallographic refinement at 1.9 Å resolution, *J. Mol. Biol.,* 89, 73, 1974.

35. Jering, H., and Tschesche, H., Replacement of lysine by arginine, phenylalanine and tryptophan in the reactive site of the bovine trypsin-kallikrein inhibitor (Kunitz) and change of the inhibitory properties, *Eur. J. Biochem.,* 61, 453, 1976.

36. Tschesche, H., and Kupfer, S., Hydrolysis-resynthesis equilibrium of the lysine-15-alanine-16 peptide bond in bovine trypsin inhibitor (Kunitz), *Hoppe-Seyler's Z. Physiol. Chem.,* 357, 769, 1976.

37. Jering, H., and Tschesche, H., Preparation and characterization of the active derivative of bovine trypsin-kallikrein inhibitor (Kunitz) with the reactive site lysine-15-alanine-16 hydrolyzed, *Eur. J. Biochem.,* 61, 443, 1976.

38. Estell, D. A., Wilson, K. A., and Laskowski, M., Jr., Thermodynamics and kinetics of the hydrolysis of the reactive-site peptide bond in pancreas trypsin inhibitor (Kunitz) by *Dermasterias imbricata* trypsin 1, *Biochemistry,* 19, 131, 1980.

39. Bode, W., Walter, J., Huber, R., Wenzel, H. R., and Tschesche, H., The refined 2.2-Å (0.22-nm) X-ray crystal structure of the ternary complex formed by bovine trypsinogen, valine-valine and the Arg15 analogue of bovine pancreatic trypsin inhibitor, *Eur. J. Biochem.,* 144, 185, 1984.

40. Richarz, R., Tschesche, H., and Wüthrich, K., Structural characterization by nuclear magnetic resonance of a reactive-site [13]carbon-labelled basic pancreatic trypsin inhibitor with the peptide bond Arg-39-Ala-40 cleaved and Arg-39 removed, *Eur. J. Biochem.,* 102, 563, 1979.

41. Richarz, R., Tschesche, H., and Wüthrich, K. Carbon-13 nuclear magnetic resonance studies of the selectively isotope-labeled reactive site peptide bond of the basic pancreatic trypsin inhibitor in the complexes with trypsin, trypsinogen, and anhydrotrypsin, *Biochemistry*, 19, 5711, 1980.

42. Tschesche, H., Beckmann, J., Mehlich, A., Feldmann, A., Wenzel, H. R., Scott, C. F., and Colman, R. W., Semisynthetic arginine-15-aprotinin, an improved inhibitor for human plasma kallikrein, *Adv. Exp. Med. Biol.,* 247B, 15, 1989.

43. Scott, C. F., Wenzel, H. R., Tschesche, H., and Colman, R. W., Kinetics of inhibition of human plasma kallikrein by a site-specific modified inhibitor Arg15-aprotinin: Evaluation using a microplate system and comparison with other proteases, *Blood,* 69, 1431, 1987.

44. Wenzel, H. R., Beckmann, J., Mehlich, A., Schnabel, E., and Tschesche, H., Semisynthetic conversion of the bovine trypsin inhibitor (Kunitz) into an efficient leukocyte-elastase inhibitor by specific valine for lysine substitution in the reactive site, in *Chemistry of Peptides and Proteins,* Vol. 3, Voelter, W., Bayer, E., Ovchinnikov, Y. A., and Ivanov, V. T., Eds., de Gruyter, Berlin, 1986, 105.

45. Tschesche, H., Beckmann, J., Mehlich, A., Schnabel, E., Truscheit, E., and Wenzel, H. R., Semisynthetic engineering of proteinase inhibitor homologues, *Biochim. Biophys. Acta*, 913, 97, 1987.

46. Mehlich, A., Beckmann, J., Wenzel, H. R., and Tschesche, H., Aprotinin derivatives with chromophoric leaving groups can be used as highly selective active-site titrants for serine proteinases and permit the determination of kinetic constants of enzyme-inhibitor complexes, *Biochim. Biophys. Acta,* 957, 420, 1988.

47. Beckmann, J., Mehlich, A., Schröder, W., Wenzel, H. R., and Tschesche, H., Preparation of chemically "mutated" aprotinin homologues by semisynthesis. P_1 substitutions change inhibitory specificity, *Eur. J. Biochem.,* 176, 675, 1988.

48. Beckmann, J., Mehlich, A., Schröder, W., Wenzel, H. R., and Tschesche, H., Semisynthesis of Arg15, Glu15, Met15, and Nle15-aprotinin involving enzymatic peptide bond resynthesis, *J. Protein Chem.,* 8, 101, 1989.

49. Groeger, C., Wenzel, H. R., and Tschesche, H., Enzymatic semisynthesis of aprotinin homologues mutated in P' positions, *J. Protein Chem.,* 10, 245, 1991.

50. Tschesche, H., Groeger, C., and Wenzel, H. R., Enzymatic fragment substitution as a tool in protein design, *Biomed. Biochim. Acta,* 50, S175, 1991.

51. Groeger, C., Wenzel, H. R., and Tschesche, H., Chemical semisynthesis of aprotinin homologues and derivatives mutated in P' positions, *J. Protein Chem.,* 10, 527, 1991.

52. Groeger, C., Wenzel, H. R., and Tschesche, H., The importance of the rigidity of the peptide backbone for the inhibitory properties of BPTI demonstrated by semisynthetic structural variants, *Angew. Chem., Int. Ed. Engl.,* 32, 898, 1993.

53. Groeger, C., Wenzel, H. R., and Tschesche, H., BPTI backbone variants and implications for inhibitory activity, *Int. J. Peptide Protein Res.,* 44, 166, 1994.

54. Groeger, C., Structure-function-variants of the trypsin-kallikrein-inhibitor, *Doctoral thesis,* University of Bielefeld, 1994.

55. Deitermann, M., Preparation of backbone-variants of the trypsin-kallikrein-inhibitor, *Diploma thesis,* University of Bielefeld, 1994.

56. Kato, I., Schrode, J., Kohr, W. J., and Laskowski, M., Jr., Chicken ovomucoid: Determination of its amino acid sequence, determination of the trypsin reactive site, and preparation of all three of its domains. *Biochemistry,* 26, 193, 1987.

57. Weber, E., Papamokos, E., Bode, W., Huber, R., Kato, I., and Laskowski, M., Jr., Crystallization, crystal structure analysis and molecular model of the third domain of Japanese quail ovomucoid, a Kazal type inhibitor, *J. Mol. Biol.,* 149, 109, 1981.

58. Bode, W., Epp, O., Huber, R., Laskowski, M., Jr., and Ardelt, W., The crystal and molecular structure of the third domain of silver pheasant ovomucoid (OMSVP3), *Eur. J. Biochem.,* 147, 387, 1985.

59. Laskowski, M., Jr., Kato, I., Ardelt, W., Cook, J., Denton, A., Empie, M. W., Kohr, W. J., Park, S. J., Parks, K., Schatzley, B. L., Schoenberger, O. L., Tashiro, M., Vichot, G., Whatley, H. E., Wieczorek, A., and Wieczorek, M., Ovomucoid third domains from 100 avian species: Isolation, sequences, and hypervariability of enzyme-inhibitor contact residues, *Biochemistry,* 26, 202, 1987.

60. Laskowski, M., Jr., Apostol, I., Ardelt, W., Cook, J., Giletto, A., Kelly, C. A., Lu, W., Park, S. J., Qasim, M. A., Whatley, H. E., Wieczorek, A., and Wynn, R., Amino acid sequences of ovomucoid third domain from 25 additional species of birds, *J. Protein Chem.,* 9, 715, 1990.

61. Apostol, I., Giletto, A., Komiyama, T., Zhang, W., and Laskowski, M., Jr., Amino acid sequences of ovomucoid third domains from 27 additional species of birds, *J. Protein Chem.,* 12, 419, 1993.

62. Empie, M. W., and Laskowski, M., Jr., Thermodynamics and kinetics of single residue replacements in avian ovomucoid third domains: Effect on inhibitor interactions with serine proteinases, *Biochemistry,* 21, 2274, 1982.

63. Park, S. J., Effect of amino acid replacements in ovomucoid third domains upon their association with serine proteinases, *Ph.D. thesis,* Purdue University, West Lafayette, 1985.

64. Laskowski, M., Jr., An algorithmic approach to sequence → reactivity of proteins. Specificity of protein inhibitors of serine proteinases, *Biochem. Pharmacol.,* 29, 2089, 1980.

65. Wieczorek, M., and Laskowski, M., Jr., Formation of covalent hybrids from amino-terminal and carboxy-terminal fragments of two ovomucoid third domains, *Biochemistry,* 22, 2630, 1983.

66. Wieczorek, M., Park, S. J., and Laskowski, M., Jr., Covalent hybrids of ovomucoid third domains made from one synthetic and one natural peptide chain, *Biochem. Biophys. Res. Commun.,* 144, 499, 1987.

67. Bigler, T. L., Lu, W., Park, S. J., Tashiro, M., Wieczorek, M., Wynn, R., and Laskowski, M., Jr., Binding of amino acid side chains to preformed cavities: Interaction of serine proteinases with turkey ovomucoid third domains with coded and noncoded P$_1$ residues, *Protein Sci.,* 2, 786, 1993.

68. Komiyama, T., Bigler, T. L., Yoshida, N., Noda, K., and Laskowski, M., Jr., Replacement of P$_1$ Leu[18] by Glu[18] in the reactive site of turkey ovomucoid third domain converts it into a strong inhibitor of Glu-specific *Streptomyces griseus* proteinase (GluSGP), *J. Biol. Chem.,* 266, 10727, 1991.

69. Ferrer, M., Woodward, C., and Barany, G., Solid-phase synthesis of bovine pancreatic trypsin inhibitor (BPTI) and two analogues. A chemical approach for evaluating the role of disulfide bridges in protein folding and stability, *Int. J. Peptide Protein Res.,* 40, 194, 1992.

70. Lu, W., Qasim, M. A., and Kent, S. B. H., Comparative total syntheses of turkey ovomucoid third domain by both stepwise solid phase peptide synthesis and native chemical ligation, *J. Am. Chem. Soc.,* 118, 8518, 1996.

71. Lu, W., Starovasnik, M. A., and Kent, S. B. H., Total chemical synthesis of bovine pancreatic trypsin inhibitor by native chemical ligation, *FEBS Lett.,* 429, 31, 1998.

72. Marks, C. B., Vasser, M., Ng, P., Henzel, W., and Anderson, S., Production of native, correctly folded bovine pancreatic trypsin inhibitor by *Escherichia coli, J. Biol. Chem.,* 261, 7115, 1986.

73. Auerswald, E.-A., Schröder, W., and Kotick, M., Synthesis, cloning and expression of recombinant aprotinin, *Biol. Chem. Hoppe-Seyler,* 368, 1413, 1987.

74. Norris, K., Norris, F., Bjørn, S. E., Diers, I., and Petersen, L. C., Aprotinin and aprotinin analogues expressed in yeast, *Biol. Chem. Hoppe-Seyler,* 371, Suppl., 37, 1990.

75. Brinkmann, T., Schnierer, S., and Tschesche, H., Recombinant aprotinin homologue with new inhibitory specificity for cathepsin G, *Eur. J. Biochem.,* 202, 95, 1991.

76. Lu, W., Zhang, W., Molloy, S. S., Thomas, G., Ryan, K., Chiang, Y., Anderson, S., and Laskowski, M., Jr., Arg^{15}-Lys^{17}-Arg^{18} turkey ovomucoid third domain inhibits human furin, *J. Biol. Chem.,* 268, 14583, 1993.

77. Hinck, A. P., Walkenhorst, W. F., Westler, W. M., Choe, S., and Markley, J. L., Overexpression and purification of avian ovomucoid third domains in *Escherichia coli, Protein Eng.,* 6, 221, 1993.

78. Kojima, S., Fushimi, N., Ikeda, A., Kumagai, I., and Miura, K., Secretory production of chicken ovomucoid domain 3 by *Escherichia coli* and alteration of inhibitory specificity toward proteases by substitution of the P1 site residue, *Gene,* 143, 239, 1994.

79. Lu, W., Apostol, I., Qasim, M. A., Warne, N., Wynn, R., Zhang, W. L., Anderson, S., Chiang, Y. W., Ogin, E., Rothberg, I., Ryan, K., and Laskowski, M., Jr., Binding of amino acid side-chains to S_1 cavities of serine proteinases, *J. Mol. Biol.,* 266, 441, 1997.

80. Roberts, B. L., Markland, W., Ley, A. C., Kent, R. B., White, D. W., Guterman, S. K., and Ladner, R. C., Directed evolution of a protein: Selection of potent neutrophil elastase inhibitors displayed on M13 fusion phage, *Proc. Natl. Acad. Sci. USA,* 89, 2429, 1992.

81. Roberts, B. L., Markland, W., Siranosian, K., Saxena, M. J., Guterman, S. K., and Ladner, R. C., Protease inhibitor display M13 phage: Selection of high-affinity neutrophil elastase inhibitors, *Gene,* 121, 9, 1992.

82. Kiczak, L., Koscielska, K., Otlewski, J., Czerwinski, M., and Dadlez, M., Phage display selection of P_1 mutants of BPTI directed against five different serine proteinases, *Biol. Chem.,* 380, 101, 1999.

83. Wallace, C. J. A., Guillemette, J. G., Hibiya, Y., and Smith, M., Enhancing protein engineering capabilities by combining mutagenesis and semisynthesis, *J. Biol. Chem.,* 266, 21355, 1991.

84. Song, H. K., and Suh, S. W., Kunitz-type soybean trypsin inhibitor revisited: Refined structure of its complex with porcine trypsin reveals an insight into the interaction between a homologous inhibitor from *Erythrina caffra* and tissue-type plasminogen activator, *J. Biol. Chem.,* 275, 347, 1998.

8 Cytochrome c Semisynthesis

Carmichael J.A. Wallace

0-8493-4727-0/00/$0.00+$.50
© 2000 by CRC Press LLC

8.1   HISTORICAL

A rational protein engineering program requires knowledge of the three-dimensional structure of the protein and of the evolutionary conserved residues within it. For a number of reasons, cytochrome *c* was an early and popular choice for study by protein biochemists[1] and, by the early 1970s, sequence comparisons and high resolution X-ray crystallography were so well developed that the structure and function of cytochrome *c* appeared as a *Scientific American* centerfold,[2] featuring paintings by Irving Geis that are still used on the covers of textbooks[3] and journals.[4] This first structural determination, of the horse protein, was followed by others representing a wide evolutionary range, including plants and fungi, and served to illustrate that a strong conservation in primary structure led to an almost unvarying three-dimensional fold.[5] Figure 8.1 shows the horse protein, the starting point of most semisyntheses; the minor differences among the total set of structures are discussed by Murphy and Brayer.[5] The amino acid sequences of cytochromes from more than

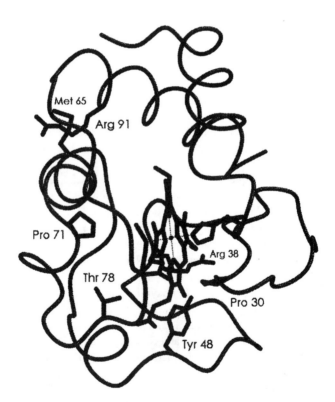

FIGURE 8.1   A molecular model of the three-dimensional structure of cytochrome *c*. The model was constructed using the co-ordinates of the horse protein as determined by Bushnell et al.,[5] and the modeling software Insight II (MSI). The view presented here is directed toward the exposed heme edge, through which electron input/output is believed to occur. Some, but not all, significant side chains are included on the α-carbon skeleton. Modeling undertaken by Christian Blouin.

100 eukaryotic species are known. Figure 8.2 shows a comparison that permits calculation of the evolutionary conservation at each position, which is presumed to be an indicator of its occupant's functional importance.[5] As soon as structures were available, they were followed by mechanistic proposals which, although often incorrect in many or all aspects, did serve to stimulate attempts to verify the hypotheses by the study of site-specific structural analogs, and protein engineering of cytochrome c was initiated.

Obviously, at that time, only protein chemical techniques were available for the generation of such derivatives, by synthesis or chemical modification. Specific chemical reactions were already much used in studies of cytochrome c but suffered from a number of limitations. Nonetheless, some extremely valuable observations have been made, especially with singly modified lysine derivatives.[6] However, the only synthetic methods that provide the possibility of replacing any residue with a wide range of alternatives are total chemical synthesis or semisynthesis. Total synthesis can be attempted in one of two ways. The stepwise approach uses an insoluble support (hence the term *solid-phase peptide synthesis*) upon which the peptide chain is built by the repetitive addition of amino acids. The fragment condensation method calls for the construction of short peptides by solution chemistry and their subsequent coupling in the correct order to yield the desired sequence.

Both strategies have been essayed for the preparation of cytochrome c analogs. A synthesis of the tuna sequence by the solid-phase method was reported by Sano.[7] Given the state of development of the technology, this was an extremely courageous attempt, and the results point out the difficulties he faced. These included the sensitivity to side reactions of some amino acids, especially Met and Trp, the lack of suitable side-chain protecting groups (for His), the harsh conditions for covalent heme reinsertion, and the physical breakdown of the solid support during an extended synthesis. The project was abandoned, but in the subsequent 20 years, the techniques of solid-phase peptide synthesis have been so much improved that it now has a major role (to be discussed below) in the production of specific protein analogs, including cytochrome c.[8]

A strategy for the synthesis of the entire sequence of yeast apocytochrome by fragment condensation was developed by Scoffone's group, and substantial development work was performed over a number of years,[9,10] but final reassembly was not achieved. Nevertheless, the strategy, and the intermediates produced, have also proved to be of ultimate use in the goal of producing informative analogs via the intermediacy of semisynthesis. The limitations and technical difficulty of the other chemical approaches to specific analog preparation have meant that the main source of informative cytochrome c derivatives has been protein semisynthesis.

The advent of site-directed mutagenesis of cloned genes resulted in a comparatively simple and highly efficient method for protein engineering. Cytochrome c was adopted by Michael Smith's group as a model for their pioneering work in the area,[11] and a great deal has already been learned from the many mutants of yeast and mammalian cytochromes c produced in this way. However, in the foreseeable future, semisynthesis will remain the only possible, or only economically viable, means of inserting noncoded amino acids, or labeled amino acids at specific sites in a sequence. Sometimes, for reasons yet unexplained, a cloned gene, mutant or

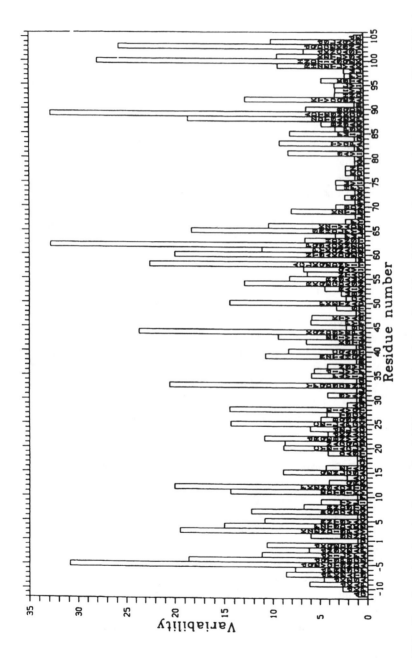

FIGURE 8.2 Bar chart showing the degree of variability encountered at each residue of the cytochrome *c* structure among 92 published sequences covering the full range of eukaryotic diversity. Variability is computed by dividing the number of different amino acids observed at a position by the frequency of occurrence of the most common amino acid. Taken from Ref. 5 with permission of University Science Books, Sausalito, CA.

not, is simply not expressed by a heterologous host. As well as remaining an alternative, semisynthesis can be a true partner of oligonucleotide-directed mutagenesis, since the latter technology can be used to generate large amounts of an otherwise scarce protein for the semisynthetic purposes described above, or, as described in Chapter 3 and below, to build into a protein specific cleavage and resynthesis sites to facilitate the semisynthetic process.

8.2   CYTOCHROME C FUNCTION

The protein contains covalently bound heme, the visible-light spectrum of which led to both its early detection in cells, and to recognition of its function. The realization that this heme, unlike that of the oxygen-carriers, cycled between Fe^{II} and Fe^{III} oxidation states, resulted in its assignment as an electron carrier in cellular respiration within the mitochondrion.[12]

Subsequent physiological studies and molecular dissection of the respiratory chain—the complex sequence of carriers that mediate electron transfer from NADH to O_2, and in the process do work in the form of chemiosmotic gradient generation—have revealed much about cytochrome c's role within that system. An up-to-date view of the electron transport system is shown in Figure 8.3. In addition to the basic role of shuttling electrons between the proton-pumping complexes III and IV, cytochrome c is physically positioned to act as source or sink for a number of secondary redox enzyme systems located either in the inter-membrane space or on the outer mitochondrial membrane. Because these enzymes are often both soluble and simpler than cytochrome c reductase or oxidase they have frequently been employed as models for respiratory protein-protein interactions, including a recent co-crystallization and structure determination of the complex between cytochrome c and cytochrome c peroxidase that has proved informative on both how the cytochrome binds and how it exchanges electrons with its partners.[13]

Further details of structure-function relationships, where illustrated by semisynthetic analogs, are detailed below. For a general and historical overview, the reader is directed to the comprehensive review, in two volumes, written by Moore and Pettigrew.[14,15]

Consideration of the implications of this role reveals a number of attributes the protein must possess to be effective:

1. *Specificity.* In a complex system of redox active proteins and small molecules, it is important that cytochrome c couple only with appropriate partners. While this regime is to some extent determined by topology (Figure 8.3), inspection of the surface of the protein suggests that it has evolved to achieve complementarity with them. A major feature of that surface is a *docking ring* of positive charge that surrounds the small proportion of the heme group that is exposed to solvent.
2. *Appropriate energetics.* The oxidation-reduction potential of a carrier in a chain that spans more than 1000 mV must harmonize with its neighbors to ensure smooth flow and efficient energy conversion. The redox potential of cytochrome c (+260 mV), and indeed all other cytochromes, is very

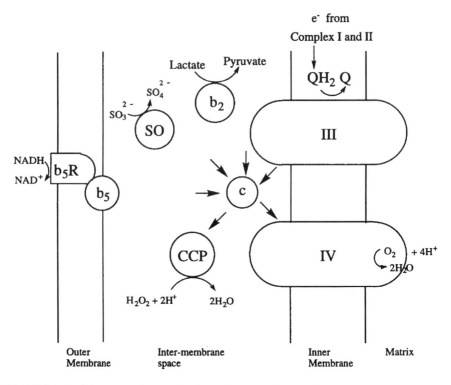

FIGURE 8.3 Schematic of the role of cytochrome c in mitochondrial electron transport. The protein is peripheral to the inner mitochondrial membrane, in equilibrium between association with the phospholipid surface and free in the intermembrane space. As such, it can act as an intermediary not only with the integral membrane cytochromes [complexes III and IV of the respiratory chain] but also with subsidiary redox enzymes. These include cytochrome c peroxidase (CCP), an NADH oxidase (cyt b_5 reductase), a sulphite oxidase (SO), and a lactate dehydrogenase (cyt b_2).

different from that of free heme (−230 mV), and thus the protein coat, via a wide range of influences, substantially modulates the potential to set it at an appropriate level. Among the more obvious ways that the polypeptide chain can modify the physicochemical properties of the heme are by providing specific axial ligands and a hydrophobic coat.

3. *Catalytic potential.* Long-range electron transfer through an organic chemical matrix is not easily achieved, hence it is assumed that the protein has evolved to assist, or at least minimize the energetic barriers to, the passage. The strong dipole generated by the asymmetric placement of lysine residues in the protein could orient it relative to the phospholipid bilayer and the redox partners embedded in it such that transfer distances are minimized and rates are diffusion limited, and also ensure association with the membrane so that diffusion between partners occurs in two, not three, dimensions.

Additionally, the evolution of catalytic function may have involved placement of residues to ensure a pathway of relatively high conductance between redox centers.

4. *Susceptibility to controlling influences.* Although it has long been recognized that the *in vitro* reaction rates of cytochrome *c* with its partners were quite sensitive to such factors as ionic strength and composition, it is not known for certain whether this is fortuitous or if these influences might be used to regulate *in vivo* activity. Certainly, cytochrome *c* is ideally positioned within the cell to do so, for the intermembrane space of the mitochondrion is continuous with the cytoplasm, at least for small molecules, so that the protein can *sense* the larger environment, and its loose association with the inner membrane can be readily modulated by these influences. The most potent of them are H^+ and ATP, and the end products of respiration and phosphorylation. Thus, these interactions could be used to provide a rheostat-like control of electron transfer in the circuit.

Many of these ideas on function, and others, are still open to vigorous debate and have raised many questions to which protein engineering in recent years has been able to provide answers.

8.3   LIMITED PROTEOLYSIS AND FRAGMENT COMPLEX FORMATION

As was discussed in earlier chapters, the propensity of fragments obtained by limited proteolysis to form non-covalent complexes has become a cornerstone of semisynthetic technique. The first semisynthetic analogs were themselves modified forms of such complexes, and this remains a useful approach to mutant generation. Just as importantly, complexation has become an essential catalytic feature of most recent semisyntheses. For cytochrome *c*, a great many complexes are known, and have been exploited in semisynthesis as well as being the subject of structure-function studies in their own right.

A non-covalent complex, as the name implies, is the association of fragments representing, more or less, the entire primary sequence of the native protein to form a structure approximately the conformation of the parent protein. *More or less* is used advisedly, since viable non-covalent complexes may be formed that lack some fraction of the sequence, or indeed they may be composed of overlapping fragments. In any case, the normal non-covalent forces that define and maintain tertiary structure operate, in the absence of a peptide or disulfide bond linking the pieces, to hold the fragments together, and these constructs generally exhibit normal or partial enzymatic activity. They are stable enough to survive chromatography by a variety of methods and so can be obtained in highly purified form.

8.3.1 Preparation

These complexes are formed when two or more protein fragments associate strongly. Typical values of K_D are 10^{-6} to 10^{-8} M. It is assumed that, if the complex functions like the parent protein, then its structure closely mimics the native conformation. The first example of a functional complex to be discovered was the well known ribonuclease S system,[16] which was later shown by crystallography to have a similar structure to ribonuclease A. This report was followed by others covering a wide range of proteins, including cytochrome c.[17] The techniques used for limiting cleavage to yield small numbers of fragments for semisynthesis are also likely to give rise to complexing sets of fragments, and this has proved particularly so for the cytochrome.[17] The question is raised of whether cytochrome c is particularly suited to complex formation, and it may be that the heme group does act as a nucleation center for the cooperative refolding of the fragments. However, the range of other protein types in which the phenomenon has been noted and the number of differing cleavage sites within that range suggest that the apparent propensity of cytochrome c may be simply a consequence of the intensive study given it (as in so many other areas) by a number of groups.

The first reported complex resulted from limited cyanogen bromide cleavage of the protein. Of the two methionine residues in the protein, that at position 65 is not essential, so that purified fragments (1–65)H and (66–104) were seen to complex strongly.[18] Furthermore, on standing this initial complex underwent a change that restored full biological activity. Only later was it realized that this change was in fact the spontaneous reformation of the missing peptide bond (65–66),[19] which has since been so extensively exploited (see Chapter 3 and below).

Shortly after this first example, it was shown that overlapping complexes, as well as contiguous ones, could be stable.[20] (1–65)H derived from the CNBr cleavage was combined with the apoprotein produced when heme is cleaved from the holo-protein. The spectrum of the product was like that of native cytochrome c and not the (1–65)H fragment.

A novel, stable contiguous complex was derived after lysine residues of horse cytochrome c were protected by acetimidylation.[21] The action of trypsin is thus limited to arginyl residues of which there are two in the protein. In fact, even after denaturation, only one, Arg38, proves susceptible.[22] The product complex, (1–38)H:(39–104), showed spectral properties identical to those of the parent protein but had only limited biological activity in cytochrome c-depleted mitochondria. Similar conclusions were reached by other workers[23] using a different ε-amino group protection. They further showed that, while activity with reductase is diminished, the complex is more efficient at electron transfer to cytochrome oxidase.

Stellwagen's group undertook a thorough investigation of the complementarity of both contiguous and overlapping fragments by examining absorbance and CD spectra, redox potentials, and biological activity. They concluded that, in the systems they examined, extensive overlap seemed necessary for functionality.[24] Although obviously unaware of the results with (1–38)H:(39–104),[22] they cited the work of Taniuchi's group showing that a new [(1–53)H:(54–104)] contiguous complex could have substantial biological activity.[25]

Taniuchi and co-workers had made the important observation that overlapping complexes could be trimmed by a strictly limited exposure to trypsin to remove all or much of the redundant portions and yield fully or near-contiguous complexes. Thus, the mixture of (1–65)H and apo (1–104) gave (1–53)H:(39–104),

(1–53)H:(54–104), and (1–53)H:(56–104). They rightly surmised that if the cleavage point were in the 38–57 region (the so-called bottom loop), then contiguous complexes would retain high functionality. This has received considerable support from later work.[17]

Shortly thereafter, Taniuchi's group showed that break points in the 20s loop were also allowable,[26] in a study where 1–38 and apo 1–104 were combined and trypsin treated. Two alternative products, (1–25)H:(23–104) and (1–38)H:(56–104), were recovered: the viability of the latter complex implied that the bottom loop was in fact irrelevant to at least some of the functions of the protein. They were also the first to report a three-fragment complex exhibiting some biological activity,[27] which was prepared by tryptic trimming of the redundant portions of an overlap complex, yielding (1–25)H:(28–38):(56–104). This complex was reduced by yeast lactate dehydrogenase at 16% of the rate for native cytochrome c. (1–25)H:(56–104) was not reduced. Wallace and Proudfoot[17] also used the trimming of overlap complexes to prepare (1–55)H:(56–104) from (1–65)H:(56–104), and showed that partial acid hydrolysis would yield yet another novel contiguous complex, (1–50)H:(51–104).

Confirmation that break points outside the bottom and 20s loops will be much less well tolerated than those within has been obtained. A new overlap complex (1–65)H:(60–104) was proteolytically trimmed with chymotrypsin, giving a complex, (1–59)H:(60–104), that exhibited no activity in depleted mitochondria.[28]

Our recently developed strategy of using site-directed mutagenesis to shuffle residues that are potentially sites of cleavage and religation for semisynthesis (see Chapter 3) also lends itself to the generation of novel complexes. Indeed, those that can be derived from a set of six such mutants (with cleavage at residues 25, 28, 35, 55, 68, or 75)[30] have been checked for ascorbate reducibility and, as expected, all but the latter two are active (C.J.A. Wallace, unpublished results). The known two-fragment contiguous or near-contiguous complexes of horse cytochrome c (*primary complexes*) are compiled in Table 8.1. These primary complexes have been the precursors of both secondary complexes in which the native sequence is maintained while the break point is shifted and structural analogs of both primary and secondary complexes: these semisynthetic non-covalent complexes are discussed in Section 8.3.2. They have also proved valuable in their own right for studies of the thermodynamics and kinetics of protein folding and of electron transfer.

8.3.2 FUNCTIONAL STUDIES

Taniuchi's group has used the (1–53)H:(54–104) complex to investigate the conformational dynamics of cytochrome c.[31] By labeling fragment (54–104), fragment exchange kinetics could be examined over a range of temperatures. Thus, thermodynamic parameters for the unfolding (and refolding) process were determined for both ferric and ferrous forms of the complex. The results implied that despite the apparent similarity of static conformations, the dynamic conformation of the protein differed greatly between oxidation states.

This system and others were used in their kinetic studies of complex formation,[32–34] to reveal second-order, biphasic kinetics and a very high degree of sensitivity of these rates to the sequence of the heme-containing fragment. Kinetic studies on

TABLE 8.1

Two-Fragment Non-covalent Complexes of Horse Cytochrome *c* Generated by Limited Proteolysis

	Composition	Derivation	Functionality	Ref.
1.	1–25H+ 23–104	Trypsin trimming of overlap complex 1–38H + Apo 1–104	60% in yeast LDH assay	26
2.	1–38H + 56–104	As 1, or trypsin action on 1–38H + 39–104	46% in yeast LDH assay	26, 17
3.	1–38H + 39–104	Trypsin on ε–NH$_2$-protected cytochrome *c*, then deprotection	25% in succinate oxidase assay, 200% in cytochrome oxidase assay (V$_{max}$)	22, 23
4.	1–50H + 51–104	Limited acidolysis	30% in succinate oxidase assay	17
5.	1–53H + 54–104	Trypsin trimming of 1–65H + Apo 1–104	25% in yeast LDH assay	25
6.	1–53H + 56–104	Minor product in 5 or 7	25% in yeast LDH assay	25
7.	1–55H + 56–104	Trypsin trimming of 1–65H and 56–104 derived from 2	40% in succinate oxidase assay	17
8.	1–56H + 57–104	Limited thermolysin digest in aqueous trifluoroethanol	Not determined	29
9.	1–59H + 60–104	Chymotrypsin trimming of 1–65H and 60–104 from BNPS-skatole cleavage	Inactive	28
10.	1–65 + 66–104	Limited CNBr cleavage	Inactive	24
11.	1–80H + 81–104	Limited CNBr cleavage	Inactive [residue 80 is homoserine]	24

the three-fragment complex revealed the nature of the folding pathway,[27] and a conformational dynamics study revealed a thermal transition at 30° in the mechanism of unfolding, and presumably folding, of the three-fragment complex.[35] Fisher and Taniuchi have recently examined a set of hybrid complexes composed of vertebrate and yeast components to define the *core domains* for the folding process, and to establish a folding order.[36] This, and related work on other protein complexes, and the preparation methods have been conveniently reviewed.[37]

Such studies provided insights into the folding and unfolding that occur in native proteins, while those of Wallace and co-workers also examined the relationship between the exact position of the missing peptide bond in the complex and biological activity.[17,38] Taking the (1–38)H:(39–104) complex as starting point, semisynthesis was used to create a secondary complex with shifted break point (1–37)H:(38–104). It had already been noted that, although differing complexes contained identical sequences, some variation in functionality could be observed.[25] We reasoned that comparative studies of related complexes could provide useful information on the functional role of the residues at the break points, and of the secondary structural elements that contained them. In fact, the seemingly minor change from (1–38)H:(39–104) to (1–37)H:(38–104) produced striking consequences. Biological activity increased from 25 to 75% of that of the parent protein, and the redox potential increased from 150 to 220 mV. It was also noted that, while the ferrous complex (1–38)H:(39–104) was completely converted to (1–38)H:(56–104) upon trypsin treatment, the nick-shifted product was to a large extent resistant to the enzyme.

The latter result indicated that the 40s and 50s loop of Figure 8.1 is much more tightly integrated into the overall protein structure in the derivative and that this stabilization must be due to the presence in the loop structure of the residue Arginine 38. In the crystal structure, this residue in seen to make extensive hydrogen bonds, and a salt bridge, with surrounding residues. The rare occurrence of a buried arginine residue performing such a role led us to suggest that this was the primary function of the residue—important enough to ensure its evolutionary invariance. This conclusion is confirmed by differences noted in the pKs for both acid and alkaline conformational transitions in site-directed mutants of position 38. For example, [Gly38] human cytochrome c has a pK_{695} of 8.05, more than 1 pH unit lower than that of the native protein.[39]

The concerted changes noted above in activity in the depleted mitochondria assay and in redox potential had been apparent with other cytochrome c analogs,[40,41] so a systematic study of this phenomenon was undertaken with a wide range of two-fragment complexes.[17] The results confirmed that there was indeed a relationship between these two parameters, and that the logarithm of activity varied directly with redox potential (Figure 8.4). In any rate equation for electron transfer, there should be a driving force term, expressed as the difference between redox potentials of donor and acceptor.[42] In the respiratory chain, for steps not coupled to proton translocation, such values are normally just slightly negative. A more positive value means a lower driving force, and the values of the difference between redox potentials of these complexes and reductase steadily increase with decreasing electron transfer rates. The implication is that the reductase-cytochrome c step is rate limiting in this system, a view that is supported by rate measurements of the isolated components.[43] If this reasoning is correct, then these analogs should have values of potential difference from oxidase more negative than normal and a greater driving force. As a result, electron transfer rates to oxidase should increase. Measurements of activities of the complex (1–38)H:(39–104) with isolated components show lower activity with reductase and higher activity with oxidase,[23] compared with the native protein, clearly supporting the above interpretation. The variable, and sometimes substantial, change in redox potential is a function of the position of the break point in a complex. This raises the questions, "Why should loss of a single peptide bond lead to a change at all?" and "Why is the effect position dependent?" Of the factors postulated to influence redox potential of a buried heme center (see Section 8.6.5), the most likely would seem to be solvent exposure. The break would allow *breathing*-partial unfolding and refolding—in the bottom loop, increasing solvent accessibility to the heme edge. Supportive evidence is provided by the increased proteolytic susceptibility of the loop structure in many of the complexes, and the much more rapid autoxidation of complexes than of native protein. The latter observation implies

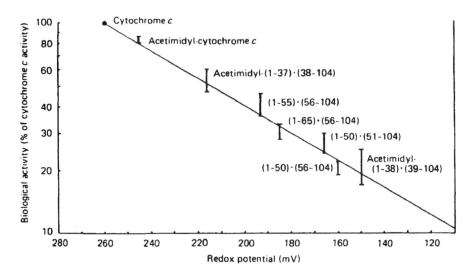

FIGURE 8.4 The relationship between the biological activities of a set of two-fragment non-covalent complexes of cytochrome c and their midpoint oxidation-reduction potentials. Biological activity is the initial rate of electron transfer in a cytochrome c-depleted mito-chondrial preparation upon repletion with the cytochrome derivative, relative to that seen with the intact protein. This technique assays activity with the reductase (complex III). Redox potential was determined by the method of mixtures, at pH 7. Reproduced with permission from C.J.A. Wallace and A.E.I. Proudfoot, 1987, *Biochemical Journal,* 245, 773–779, © the Biochemical Society..

that the protein structure is now much more penetrable by O_2, in the same way that we are postulating for H_2O.

To further probe this question, and that of the positional effect on redox potential variation, we have prepared and compared a wide range of complexes with bottom loop-located break points, including some with more extended gaps at the junctions. The data are shown in Table 8.2, and they reveal that, although the presence of a break causes substantial destabilization of the heme-crevice to the ligand exchange reaction (pK 695 nm), there is little variation with the position or size of the gap. However, redox potential varies quite widely, between extremes that are either close to the native value or much nearer to that of a mutant in which the entire loop is deleted and solvent accessibility is necessarily high.

Thus, it is clear from these data that some break points are much more desta-bilizing than others, and that gaps (or overlaps) lead to significantly larger effects on redox potential than clean breaks. We have shown (P.A. Adams and C.J.A. Wallace, unpublished results) that the 1–50:51–104 complex has a much greater peroxidase activity than the native protein, which we assume is due to the greater accessibility of H_2O_2 also to the heme center via the bottom loop. A complete correlation of peroxidative activity with redox potential for this entire set of com-plexes should prove the point. Non-covalent complexes have also been used in studies of the conformational factors underlying cytochrome c antigenicity,[44] and of the anion-binding properties of the protein.[45]

TABLE 8.2
A Comparison of Physical Properties of Two-Fragment Complexes of Related Structure

Group 1	pK695 nm	E'_m
1–50H:51–104	7.1	190
1–50H:56–104	7.2	175
1–53H:56–104	7.1	183
1–55H:56–104	7.2	216
1–65H:56–104	7.3	184
Cytochrome c	9.3	266
Des [40–55] Cyt c	7.4	130
Group 2		
Acim 1–37H:38–104	8.2	246
Acim 1–37H:39–104	8.5	184
Acim 1–38H:39–104	8.4	190
Acim 1–38H:40–104	8.4	184
Acim 1–38H:41–104	8.5	177
Acim 1–39H:40–104	8.3	183
Acim Cytochrome c	9.7	250
1–38H:39–104	7.7	195

8.3.3 COMPLEX FORMATION AND THE CATALYSIS OF FRAGMENT CONDENSATION

The realization that the (1–65)H:(66–104) complex of cytochrome c spontaneously undergoes conversion to the covalent form [Hse65] cytochrome c was a consequence of the discovery of a similar phenomenon in pancreatic trypsin inhibitor.[19,46] Peptide bond formation results from aminolysis of the internal ester, homoserine lactone, generated at the C-terminus of fragments by the CNBr cleavage reaction. In free solution, using model compounds, this reaction proceeds very slowly and is not competitive with hydrolysis.[47] Hence, the high efficiency of peptide bond synthesis (60 to 90%) must be a consequence of catalysis by the structure of the complex itself. Harbury suggested that the conformational requirements for this kind of resynthesis are stringent,[48] based on the observation that reaction only proceeds in the ferrous, and not the ferric, complex. This view has been amply confirmed since. A one residue gap or overlap at the break point,[24,41,48] or even a one-residue shift in the break point is enough to eliminate religation. While other complex systems have been discovered in which spontaneous peptide bond formation between fragments will occur,[49] others [e.g., lactalbumin and superoxide dismutase (C.J.A. Wallace, unpublished results)] show no evidence of recombination. Attempts to induce the phenomenon artificially have also proved largely unrewarding. R.C. Sheppard built homoserine lactone into the C-terminus of S peptide but observed only 7% reformation of ribonuclease A from the resulting ribonuclease S analog.[50] We have incorporated the lactone into the cytochrome c (1–38)H:(39–104) complex system,[51] but very little religation of the fragments resulted.

As discussed above, complex formation is a necessary condition for most examples of fragment condensation by reverse proteolysis, so it is assumed that an analogous autocatalysis operates in these circumstances. Indeed, the conformational requirements might be less stringent, given the wider range of examples of the phenomenon.[52] A protease-mediated resynthesis of the (1–38)H:(39–104) complex was reported using the Arg-specific clostripain.[53] Others were unable to show peptide bond formation between 38 and 39, although the two fragments were held together by acrolein (a contaminant of some glycerol) mediated cross-links.[54]

Despite these initial disappointments, the basic principle revealed by the early studies has since been the inspiration of general methods that use the autocatalytic properties of noncovalent complexes to promote fragment condensation. These are extensively discussed in Chapter 3.

8.4 SOME PRACTICAL CONSIDERATIONS IN THE SEMISYNTHESIS OF CYTOCHROME C

8.4.1 THE HEME GROUP

Cytochromes c are unique in containing covalently bound heme, and careful consideration must be given to this sometimes chemically sensitive group in planning a semisynthesis. In practice, this has meant that manipulations of the sequence of heme-containing fragments have not been attempted, and efforts to achieve classical chemical fragment condensation of heme fragments to the remainder of the molecule have been abandoned.[55]

An alternative strategy would be to reinsert heme into a semisynthetic apoprotein in the final step. This can be achieved chemically; in trial experiments with apoproteins derived from native bovine cytochrome c, protoporphyrinogen was reacted with the protein, and the product porphyrin apoprotein c was combined with ferrous acetate under anaerobic conditions.[56] The yield of crude product was 7 to 10%. The purified material exhibited 85% of native biological activity. Alternatively, there exists in yeast mitochondria an enzyme catalysing the specific insertion of heme into yeast apoprotein once it has crossed the outer membrane.[57] The enzyme, which can be solubilized,[58] is also capable of reincorporating heme into the horse apoprotein,[59] so the suggestion was made that the determinants of specificity were simple and located near the attachment points, Cysteines 14 and 17.[60] If so, then the enzyme would be capable of inserting heme into semisynthetic apoproteins or smaller apofragments destined for inclusion in non-covalent complexes. Both approaches have been shown to be feasible,[60,61] although at present the low activity of the mitochondrial cytochrome c synthetase preparations inevitably meant that product yields are low. Recently, however, the gene has been cloned and expressed in *E. Coli,*[151] making this route a potentially attractive option.

8.4.2 SIDE-CHAIN PROTECTION

Preliminary ε-amino protection of cytochrome *c* has proved useful in many semi-syntheses in three roles. Restriction of tryptic cleavage to arginine residues allows strictly limited fragmentation.[22] In the remodeling of fragments by sequential degradation and resynthesis, the Edman reaction has been frequently employed,[40,62] requiring that the ε-NH$_2$ groups be blocked to prevent permanent derivatization. Finally, in chemical fragment condensations (e.g., 62) α- and ε-amino groups must be differentiated. A number of blocking groups have been exploited for these objectives, the first of which requires retention of water solubility. These are the citraconyl group which because of acid lability is only suitable for the first role,[27] the methyl-sulphonylethyloxy-carbonyl (Msc) group,[63] the acetimidyl (Acim) group,[21] and the trifluoracetyl (TFA) group. These latter three are all in principle base-labile, but experience had shown incomplete removal of TFA groups.[64] The other two groups have been extensively used, Msc by Tesser and co-workers, and Acim by ourselves. Both have characteristics to recommend them. The Msc group is very rapidly deprotected in strong base, but since it gives an uncharged derivative, it tends to reduce the water-solubility of the protein and fragments. The Acim group is only slowly removed, at high pH but, if care is taken, cytochrome *c* is undamaged.[21] Because the Acim group preserves the positive charge of the amino group, solubility is maintained, and deprotection is not strictly necessary. This measure of convenience has been questioned[65] but, by very many criteria, the acetimidylated protein is so little different from the native protein[66] that it constitutes a valid point of comparison for structure-function studies of analogs.

One group has experimented with post-cleavage amino group protection[64] using the conventional blocking agents of total chemical synthesis for temporary use in those steps where water solubility is not a priority. They found conditions in which a significant differentiation of α- and ε-amino groups could be achieved. The same tactic has also been used where temporary side-chain carboxyl protection is required, as in the carboxyl component of a chemical fragment condensation. Where this fragment is prepared by solid phase peptide synthesis, specific ω-COOH protection can be built in.[67] If the peptide is from a natural source, then CNBr fragments provide an elegant solution to the problem,[68] for the C-terminal homoserine is in the lactone form, which may be simply hydrolysed to the free carboxyl form after protection of the side-chain groups. For this purpose, use has been made of the p-methoxyphenyl ester or the methyl ester.[62,69] Tesser's group's ingenious approach avoided the need for side-chain carboxyl protection in *classical* fragment condensation altogether.[70] In a solution synthesis of (66–79) they introduced the C-terminal amino acid as a protected hydrazide; at the completion of the synthesis, the peptide hydrazide could be converted directly into the peptide azide, an activated species that will combine with the α-NH$_2$ of fragment 80–104.

8.4.3 ACCESSIBILITY OF RESIDUES

Table 8.1 showed that the variety of limited cleavage techniques available to protein chemists could give, even considering two-fragment systems alone, break points well distributed throughout the sequence. Obviously, when we consider methods that may yield three or more fragments, accessibility to most regions of the sequence is further increased. The importance of this factor lies in the limitations of the sequential degradation and resynthesis techniques, which means that an amino acid intended for substitution should lie, at most, three or four residues from a break point in a naturally derived fragment. If no convenient cleavage site can be found, then a residue remains inaccessible. Fortunately, the current availability of reliable solid-phase peptide synthesis means that now a synthetic fragment would normally be deployed in these circumstances, and the realization that we can use site-directed mutagenesis, as described elsewhere, to create cleavage and religation sites at many desirable sequence positions should allow easy access to almost any residue.

8.4.4 Fragment Reassembly

The difficulty and inefficiency of the chemical coupling of large peptide fragments has already been touched upon, as has the major drawback of the cytochrome c system, the sensitivity of heme to the coupling conditions. The fortuitous discovery of the spontaneous resynthesis of CNBr fragments provided one way to avoid the problem,[19] so that, until 1987, all fragment condensation semisyntheses of cytochrome c depended on this phenomenon for the final coupling step.

An automatic limitation to residues in the 66–104 region ensues. Although it proved possible to prepare semisynthetic noncovalent complexes with substitutions outside this area, and useful insights were gained, there has always been a prejudice toward analogs of the full covalent sequence of a protein. As related above, the first attempts to artificially induce the conformationally assisted spontaneous religation at other sites in proteins met with little success. We considered that perhaps homoserine lactone was insufficiently activating of the carbonyl carbon for all but a very few special cases and that a more activating group might promote religation generally.

The problem of introducing the ideal activating agent was solved by Keith Rose and collaborators.[71,72] Using trypsin, a set of aminoacyl esters was added by reverse proteolysis to the C-terminus of fragment (1–38). The esters employed varied in activating power. Some, such as the widely-used N-hydroxysuccinimide ester, proved unstable under reverse proteolysis conditions. Of the others, the dichlorophenyl ester proved to combine sufficient stability with adequate activating power for the next step. In this step, the now (1–39) dichlorophenyl ester was mixed with equimolar (40–104) in dilute neutral aqueous solution. Raising the pH led to rapid aminolysis of the ester to give about 40% of product cytochrome c. Thus, a conformationally assisted religation had been achieved, demonstrating that where a looser complex structure meant homoserine lactone was an ineffective activating agent, the principle of conformational assistance could nonetheless be harnessed by use of stronger but still mild reagents. It could also be shown that the system would operate when noncontiguous complexes were prepared, as long as the termini were in reasonable proximity in the tertiary structure; thus (1–39)H and (56–104) were combined in 60% yield to give the bottom loop-deleted analog of cytochrome c.[73] We termed the method *autocatalytic fragment religation* (AFR).

These promising results were followed up by an extensive study aimed at optimizing the system.[74] The parameters examined for the activation step (protease-catalysed addition of amino acid esters) included pH, solvent, nucleophile concentration, enzyme:substrate ratio, and (most importantly) the nature of the added amino acid. A wide range of substrates were found to give almost uniformly high yields for the addition. In the coupling step, the influence of the C-terminal residue and the location of the break point were considered. In the great majority of the tested systems, coupling yields in the 30 to 60% range were noted. These latter two observations hinted that the approach that had been developed could prove to be truly universal—that as long as the fragments complexed, any break point in any protein could prove a suitable site for conformationally-catalysed peptide bond synthesis.

To sustain a claim of true generality for AFR as a tool for protein engineering, three conditions should be met.

1. It should be possible to introduce break points in any region of a sequence. Table 8.1 showed the range of primary complexes that could be prepared by the use of just a few limited cleavage methods.
2. Meeting the first condition means that any residue might form the C-terminus of a fragment, and it is therefore necessary to be able to activate many if not all such residues. To satisfy this condition, enzymes other than trypsin should be competent for reverse proteolysis. We have been examining this point and find that serine proteases are generally suitable.[28,75] Within this class, an extremely wide range of substrate specificities exists.
3. Activated fragments should always combine in high yield with contiguous partners of a non-covalent association. In almost every case so far examined, fragments activated by dichlorophenyl esters religate with high efficiency. The two exceptions that have been noted occur in complexes with break points at tight bends, so that it is possible that in such structures, the termini are held apart, rather than together. If this proves general, it will be necessary to avoid this class of complex for resynthesis.

The fact that only a few of the two-fragment systems in which homoserine lactone is found at the C-terminus will undergo AFR has been noted above as the prime motivation for the development of the dichlorophenyl ester-based system. However, it is also possible that functions other than the weak activation provided by homoserine lactone could be responsible for negligible religation. A case in point is that of the cytochrome c from Saccharomyces cerevisiae, where methionine is found at position 64 in the vertebrate numerotation, but not at 65. This protein can be cleaved to give a two fragment complex, but no religation occurs. We had, in the past, speculated on why that should be.[41]

We had realized, as discussed in Chapter 3, the potential synergy of site-directed mutagenesis and semisynthesis in that the former could be used to introduce AFR sites at any desired position in a sequence to facilitate a final religation. The yeast system seemed an ideal test case for this premise, and for the suggestion that the break point location relative to the overall 3D structure would be an important determinant of efficiency, and so Met64 was changed to Leu, and Ser65 to Met.[76,77] Subsequent cleavage and efficient religation have proved the point that the two technologies can operate synergistically and that, contrary to our earlier fears, spontaneous religation of CNBr fragments can be a versatile general method if care is taken in the choice of break point. In addition, the synergistic approach can also clearly be exploited to introduce sites of cleavage and activation by proteases so that the generalized AFR strategy that we have developed can be further enhanced.

8.5 STRATEGIES FOR THE GENERATION OF SEMISYNTHETIC CYTOCHROME C ANALOGS

8.5.1 INTRODUCTION

The primary goal of protein semisynthesis has always been the generation of defined analogs for structure-function studies, although both clinical studies with and industrial production of semisynthetic insulins have been reported.[78,79] Cytochrome c was chosen as a target for semisynthetic studies for its intrinsic interest as a model for both general questions concerning protein folding and structural stabilization and for the specific study of biological electron transfer reactions. The availability of high-resolution structures of both oxidation states was an essential prerequisite for these goals, but clearly the choice was also influenced by the relative abundance and rugged nature of the protein. In addition to providing information about known functions, such studies may also reveal the existence of novel ones. The creation of these analogs has involved the use of most of the strategies of fragment-condensation semisynthesis discussed in Chapters 1 through 3.

8.5.2 CHIMERIC SEQUENCES

The first true semisynthetic analogs of the protein followed rapidly upon the discovery of the spontaneous religation of CNBr fragments.[19] Taking homologous fragments from different species, the same workers were able to show that a high degree of recombination could also occur in heterologous pairs to give fully active cytochromes.[80,81] The disadvantage of the approach is that the substitutions achievable are only those that are well tolerated in the course of evolutionary history. An important point made by these chimerae and the original religated material is that the product [Hse65] cytochrome c is functionally equivalent to the native Met65 form (a point confirmed by many authors), and so conclusions can be confidently drawn from any analog prepared by use of this reaction.

8.5.3 FRAGMENT-SPECIFIC CHEMICAL MODIFICATION

A completely or partially specific modification of a protein sequence can be achieved by a combination of semisynthesis and traditional chemical modification techniques. By confining the reaction to a single fragment, only one or a few defined residues will be altered in the reassembled protein. With cytochrome c, this approach has been taken, not unexpectedly, with the (1–65)H + (66–104) pair, derivatizing both common and rare residues. Only two arginine residues are encountered in the horse sequence, at positions 38 and 91, so that this tactic, which was employed with three different guanidino group-specific reagents, allows independent derivatization of each.[82] Similarly, fragment (1–65)H contains only one of the four tyrosine residues of the horse sequence and has been modified by iodination or acetylation prior to recombination with 66–104.[83] Clearly, the success of this method also depends on the availability of highly specific reagents, but a broader specificity may be enhanced by selective protection. For example, acetylation of (1–65)H by acetylimidazole will modify both tyrosine and lysine residues. Prior protection by acetimidylation confines the modification to Tyr48. The acetimidylation reaction itself has been used to produce specific partial modification of the protein following an earlier study employing the guanidinyl group.[60,84]

8.5.4 TOTAL SYNTHESIS OF FRAGMENT 66–104

As noted above, the most successful strategy for cytochrome analog generation has been via the spontaneous religation of CNBr fragment (1–65)H and synthetic or modified (66–104). The most direct route to sequence modifications of this peptide was first attempted in 1977, when synthesis of a 39-residue peptide was still an ambitious undertaking with solid-phase techniques, so that only 1 to 2% of the crude product would couple with (1–65)H.[85] A later attempt by a different group, using alternative resins and reagents, was more successful, but side reactions due to the unavailability of suitable arginine protection led to an heterogeneous product.[86]

The attempt by the Scoffone group, described in Section 8.1, to synthesise the whole molecule by solution methods[9,10] was eventually to become a semisynthesis project. Borin adapted the fragment condensation strategy to yield the yeast homolog of (66–104), (71–108), and an analog of the yeast CNBr fragment (70–108) (see Figure 8.5).[87] The former was combined in good yield with horse (1–65)H to give a horse-yeast chimeric sequence,[88] but the (70–108) fragment of yeast would not recombine with the corresponding yeast heme fragment (1–69) even though the break point is shifted by just one residue.

Solution total synthesis was also adopted by Tesser's group for the preparation of (66–104) incorporating substitutions in the 81–104 segment (see also Section 8.5 2).[89–92] The scheme adopted was modeled on that which they had used earlier for semisynthesis of 66–104 described in the next section. Low yields of final product were recorded, but it should be remembered that the nature of the introduced replacement amino acid, as much as any difficulties with the synthesis, can have a profound effect on the spontaneous religation yield.

Most recently, interest has refocused on solid-phase methods for (66–104) synthesis using automated systems. In one study, the native sequence was prepared in pure form at 24% overall yield, which recombined with (1–65)H as efficiently as native (66–104).[93,94] Both this and the Met(SO$_2$)80 analog sequence were also prepared by manual methods in somewhat lower yields. The analog would not religate with (1–65)H.

We first used automated methods to make the native sequence and five analogs incorporating change at three conserved amino acids[67,78,83] in the 66–104 sequence[95,96] and later to prepare a set of nine different substitutions (shown in Figure

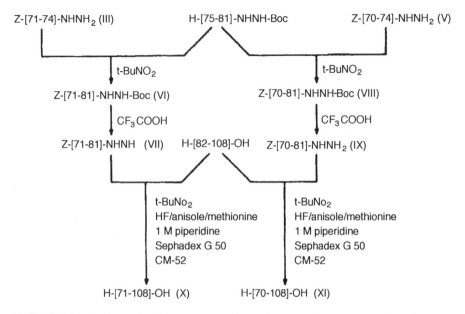

FIGURE 8.5 A scheme for the generation of cytochrome *c* fragments based on the yeast sequence by a solution synthesis-fragment condensation strategy. Both [70–108] and [71–108] analogs (yeast numbering) were created. The former corresponds to the naturally obtained non-heme cyanogen bromide fragment of the yeast protein, the latter to that of the horse protein, but with the yeast sequence, thus permitting production of a horse-yeast chimeric cytochrome. Reprinted from Borin et al., *Biopolymers*, 25, 2269, copyright 1986, by permission of John Wiley & Sons, Inc.

8.6) at the crucial Met80 position.[97] In these syntheses, the protocols developed by the Caltech group produce the target peptide as 80% of the crude product,[8] which permitted purification in good yield. Religation yields varied from very low to (in many cases) as high as the natural fragment. Replacement of Met80 by His, Cys and Leu, and of Tyr67 by Phe and p-F-Phe, have also been reported by others.[98–100] Recently, syntheses of peptides incorporating changes at residues 68,70,71,74,91, and 97 have been undertaken using the same methodologies.[101–103]

8.5.5   SEMISYNTHESIS OF FRAGMENT 66–104

Analogs of this fragment could also be made either by manipulating the sequence of natural (66–104) directly or by total synthesis of a part of the fragment and its recombination with a naturally derived peptide. The partial synthesis route has been favored by the groups of Tesser and Warme. For both, the strategy has been to prepare synthetic (66–79) by, respectively, solution or solid-phase methods, followed by coupling to (80–104) derived from the natural CNBr fragment (81–104). The first report of a successful synthesis of this type and reconstitution of the native sequence of cytochrome *c* came from Tesser's laboratory.[70,104,105] The strategy was then exploited to make analogs of conserved residues in the 66–79 region: Lysines 72, 73, and 79, Tyrosine 74,[106] and the invariant Tyrosine 67.[107] They have also

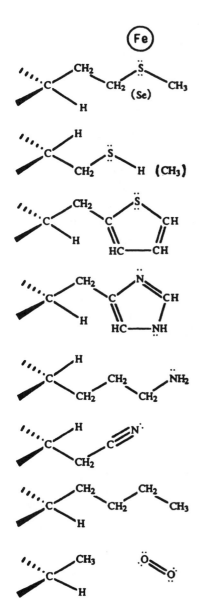

FIGURE 8.6 Structural formulae of the various side chains introduced as replacement for the heme iron-ligating methionine residue 80 in horse cytochrome *c*, using a fragment-condensation semisynthesis protocol.[97] All but one are noncoded amino acid residues. The exception, alanine, cannot be introduced directly by site-directed mutagenesis methods, since the mutant is nonfunctional. Taken from Ref. 97 with permission.

reported other analogs at position 67, a substitution at Thr78,[92] and the sequential degradation and resynthesis of natural (81–104) to prepare modifications to residue 81.[108] A flow diagram of this synthesis, which illustrates the Tesser group's general strategy for both semisynthetic and wholly synthetic fragments 66–104, is shown in Figure 8.7.

The alternative, partial solid-phase strategy was demonstrated first with the native sequence[56] and then exploited to swap phenylalanine and p-fluorophenylala- nine for tyrosine at 67 and leucine for tyrosine at 74.[67,109] Unfortunately, the very

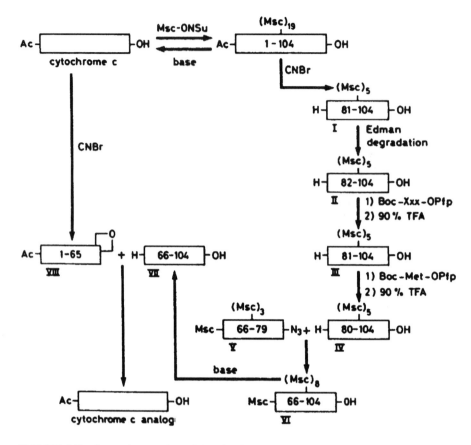

FIGURE 8.7 One scheme adopted by the Tesser group for the fragment condensation semisynthesis of horse cytochrome c.[108] Sequential degradation and resynthesis replaced the residue at position 81 in the C-terminal CNBr fragment. This was then linked to the synthetic 66–79 peptide via a "conventional" azide coupling, in protected form. After deprotection, a 66–104 fragment is available for the stereoselective spontaneous religation of CNBr fragments used by many groups. Taken from Ref. 108 with permission.

small yields of final products did not permit their complete purification for biological tests.

In contrast, we chose to attempt a semisynthesis of the protein entirely based on the three natural fragments formed by CNBr cleavage. Therefore, reconstruction of the native sequence required restoring the essential Met80, which was achieved by removal of the C-terminal homoserine of 66–80 by carboxypeptidase A, and coupling of methionine to the N-terminus of 81–104, prior to chemical coupling to give 66–104 of natural sequence.[69,110] This material religated with 1–65, but we found that an [o-fluoroPhe82] analog of (66–104) prepared at the same time would not,[69] but an improved route yielded this and an [ε-Cbz Lys81] substitution.[110] The same study reported the shifting of the essential methionine from position 80 to 81. A flow diagram summarizing the essential steps of all these semisynthesis appears in Figure

8.8. We have also used the sequential degradation and resynthesis techniques to replace [Glu66] in the natural fragment (66-104) by Gln, Lys, and norvaline.[40]

8.5.6   SOLID-PHASE SYNTHESIS OF FRAGMENT 1–65

Di Bello and co-workers have suggested an elegant way to avoid the problems posed to such a synthesis by the need to insert the heme group before spontaneous religation of (1–65)H and (66–104) will occur. They observed that in the presence of (1–25)H, the apofragments (1–65) or (23–65) would combine with natural (66–104) in up to 80% yield.[111,112] Thus, substitutions could be built into the 1–65 segment, and the apoprotein would then, hopefully, be a substrate for the enzyme of mitochondria, cytochrome c-heme lyase.[57-58] Alternatively, for substitutions in 23–65, the semisynthetic complex (1–25)H: (23–104) could be studied. In fact, incubation of the synthetic apoprotein with beef heart mitochondria and hemin gave an 8% yield of the holoprotein, which appeared to be functional.[61] Thus, a route has been pioneered for a total synthesis of the protein (Figure 8.9).

8.5.7   SEMISYNTHETIC MODIFICATIONS OF NONCOVALENT
COMPLEXES

The discovery of the first functional contiguous complex of cytochrome c, (1–38)H:(39–104), was followed by its use as the starting point for semisynthesis.[22,113] Residue 39 was removed and replaced, and residue 38 was removed, to make a set of complexes with substitutions at residue 39, or in which one or two of residues 38 and 39 were absent. Subsequent work led to the reincorporation of Arg38 at the N-terminus of (39–104), to give the first nick-shifted or secondary complex.[34,114] We later extended this work to incorporate substitute residues into the complex at position 38 and 39, as shown in Figure 8.10,[115] and made hybrid complexes of lysine-modified and unmodified fragments.[60]

It was noted that the stable truncated complex (1-38)H:(60-104),[116] unlike the previously reported (1–38)H:(56–104),[26] was quite inactive. The suggestion that tryptophan 59 was essential to a functional complex,[117] in ordering the correct iron ligation through its hydrophobic interaction with the heme, was supported by the quenching by heme of the tryptophan fluorescence seen in such complexes.[29] The proposition was tested and supported in the truncated complex system by the addition of N-terminal tryptophan to (60–104) giving the active complex (1–38)H:(59–104).[116]

A third system in which fragment structural manipulation has been used to generate analogs is in the three-fragment complex (1–25)H:(28–38):(39–104). Here, the incorporation of synthetic fragments (28–38) prepared by solid-phase methods has been used to test the roles of Leucines 32 and 35 and Threonine 28,[118] and of Proline 30 and Glycine 34.[119]

8.5.8   MODIFICATIONS TO FRAGMENT 39–104

The development of the conformationally assisted coupling of protease-activated fragments described in Chapter 3 has opened up new regions of the sequence for

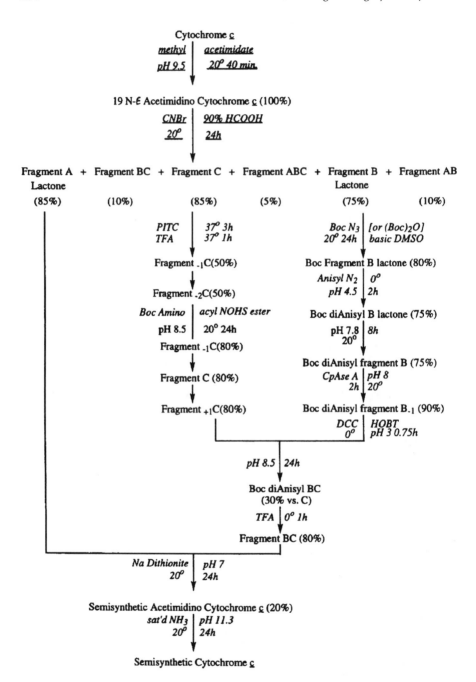

FIGURE 8.8 The fragment condensation semisynthesis strategy of Wallace and Offord.[62, 69] This scheme employs entirely naturally derived peptide fragments [A = 1–65, B = 66–80, C = 81–104], subject to multiple sequential degradation and resynthesis steps, to introduce a replacement residue at position 80, 81, or 82. The sheer complexity of the procedure was a major driving force for the development of the simpler approaches that superceded it.

I. Chemical synthesis of the fragments and activation to [Hse 65] (1-65) lactone

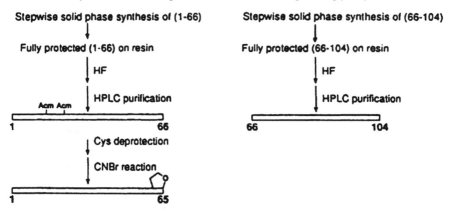

II. Complexation assisted joining between activated [Hse 65] (1-65) lactone and (66-104)

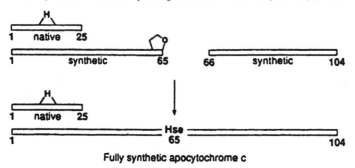

Fully synthetic apocytochrome c

III. Heme insertion by mitochondria

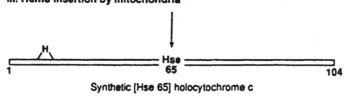

Synthetic [Hse 65] holocytochrome c

FIGURE 8.9 The protocol developed by Di Bello et al. to facilitate semisynthetic modifications within the heme-containing fragment of cytochrome c.[61] The sensitivity of the heme prevents the use of conventional solid-phase chemistry, so a synthetic apo fragment is prepared and ligated to the contiguous 66–104 fragment by the stereoselective method using a short natural heme fragment as a structural template. The religated apoprotein is then a substrate for a heme lyase-containing mitochondrial extract. Taken from Ref. 61 with permission.

substitution by covalent semisynthesis. Obvious targets for modification by this technique were residues 39 and 40 at the N-terminus of (39–104). Having been removed by the Edman degradation, these residues were reinstated by reverse proteolysis at the C-terminus of (1–38)H, prior to the religation reaction.[51]

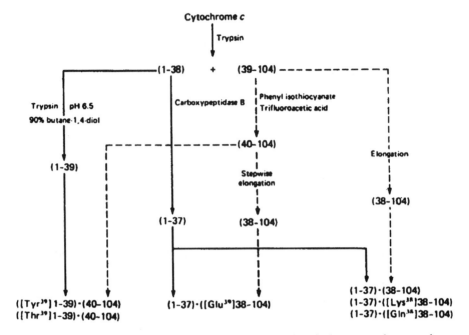

FIGURE 8.10 A simple strategy for the production of semisynthetic non-covalent complexes of the horse protein based on limited tryptic cleavage at residue 38.[115] Stepwise degradation and resynthesis is undertaken with either chemical or enzymatic procedures. Since the complexes produced have activities of up to 75% of that of holoprotein in bioassays, comparison of semisynthetic with unmodified complexes can give meaningful data on structure-function relationships.

8.6 FUNCTIONAL CHARACTERIZATION OF SEMISYNTHETICALLY ENGINEERED CYTOCHROME C ANALOGS

8.6.1 EXPERIMENTAL SYSTEMS

Semisynthesis has provided point mutations of 31 of the 104 residues in the horse sequence, including 14 of the 21 invariant residues (Table 8.3). In most cases, more than one replacement amino acid has been introduced. If one included semisynthetic non-covalent complexes, more than 100 analogs have been studied, and it comes as no surprise that a very wide range of experimental techniques have been applied to them.

Absorbance spectroscopy has, of course, been widely applied in the study of derivatives; characteristic bands in the visible region can reveal much about the coordination and spin states of the heme iron and hence the conformational or electronic changes consequent on modification. In particular, perturbations in spectra of analogs induced by changing environmental, especially chaotropic, conditions are revealing about the forces that stabilize protein structure and the residues that provide them.[96] In particular, titration of the 695 nm Fe-S charge transfer band is used as a measure

TABLE 8.3
A Compilation of Amino Acid Substitutions Introduced by Semisynthesis into Horse Cytochrome c

Residue	Evolutionary Conservatism	Mutation	Functional Effects	Suggested Cause	Reference
Thr 28	Variable	I→,deletion	No effect	Lower E'_m; Fe^{2+} destabilized more than Fe^{3+}	118
Pro 30	Invariant	P→G	Decreased bioactivity	Hydrophobic core destabilized	119
Leu 32	Invariant	L→V,F,I,NVal	Decreased Stability	Side-chain stereochemistry crucial	118
Gly 34	Invariant	G→A,S	Internal destabilization	Increased bulk causes more fluctuation	119
Leu 35	Highly conserved	L→I	No effect		118
		L→K,T	Some decrease in stability	Hydrophobic core destabilized	118
Gly 37	Highly conserved	G→A	Not tested	Stereochemistry not crucial	28
Arg 38	Invariant	R→Q,K,G,etc	Decreased E'_m	α-loop destabilized	38,82,115
Lys 39	Conserved	K→A,F,G,E,etc	Decreased E'_m	α-loop destabilized	51,113,115
Thr 40	Highly conserved	T→K,V,F	Decreased E'_m	H-bond to 157 eliminated	51
Tyr 48	Invariant	Y→I Tyr, O-Ac-Tyr	Increased E'_m	Increased hydrophobicity at core	83
Lys 53	Conserved	K→Acim Lys	Decreased E'_m	Conformational Change (by NMR)	60
Trp 59	Invariant	W→deletion	Inactive	Trp packs Haem	116
Lys 60	Variable	K→A	Reduced activity	Shifts dipole axis?	28
Met 65	Variable	M→Homoserine	No effect		69,104
Glu 66	Highly conserved	E→K,Q	Increased E'_m	General electrostatic effect	40
			ATP affinity reduced	Part of binding site	
		E→NVal	Inactive	60s helix disrupted	
Tyr 67	Highly conserved	Y→L,F,pFPhe	Decreased E'_m	Crevice H_2O accessible, polarity change	92,96,107,109
Leu 68	Invariant	L→SeM	pK_{695m} falls	SeM packs heme crevice less well than Leu	102
Asn 70	Highly conserved	N→Homoserine	Mild destabilization	Together, these two residues form a	101
Pro 71	Invariant	P→NVal	Non-functional	tight bend initiating 70s loop?	101
Lys 72	Invariant	K→Acetyl Lys	Increased Km with oxidase	+ve charge for oxidase binding	106
Lys 73	Invariant	K→Acetyl Lys	Increased Km with oxidase	+ve charge for oxidase binding	106
Tyr 74	Highly conserved	Y→L	Decreased thermostability	Loss of structural rigidity?	106,109
Ile 75	Highly conserved	I→K	Decreased E'_m, 695 band lost	Severe destabilization	109
Thr 78	Highly conserved	T→N, Aba, V	Decreased E'_m	Structural destabilization	92,96
Lys 79	Invariant	K→Acetyl Lys	Increased Km with oxidase	+ve charge for oxidase binding	106
Met 80	Invariant	M→Very many	Varied	Met sets E'_m, assists electron transfer	97,98
Ile 81	Highly conserved	I→CbzK	Low Activity	Increased polarity/bulk in electron port	62
		I→A,L,V	Little effect		109
Phe 82	Invariant	F→o-FPhe	Diminished acivity	Polarity change at electron port	62
		F→L	Increased Km with oxidase	Aromatic ring aids binding?	92
Ala 83	Variable	A→P	Normal E'_m, Low e-transfer	Loss of flexibility in electron port	96
Arg 91	Invariant	R→Dmp Orn etc	High affinity ATP binding lost	Essential part of site	82
		R→NLeu			103
Tyr 97	Highly conserved	Y→L	Increased E'_m	Influences F10 packing to Haem?	92
		Y→BrY, BrF	Decreased pK_{695m}	Increased polarity in interior?	102

Thirty residues of 104, including 13 of 21 invariant side chains, have been the object of semisynthetic modification. Suggested causes of functional effects of mutation are those proffered in the original reference, except where indicated by a question mark. Abbreviations are NVal, norvaline; NLeu, norleucine; I Tyr, iodotyrosine; O-Ac-Tyr, O-acetyltyrosine; Acim Lys, ε-acetimidyl lysine; p-FPhe, 4–fluorophenylalanine; Aba, α-aminobutyric acid; CbzK, ε-carbobenzoxylysine; o-FPhe, o-Phe, 1-fluorophenylalanine; BrY, 3-bromotyrosine; BrF, 4-bromophenylalanine; Dmp Orn, dimethylpyrimidylornithine: SeM, selenomethionine. Conserved means unchanged within a kingdom; highly conserved means complete or very nearly complete functional group conservatism.

of the stability of the heme crevice to chaotropic conditions. More detailed information can be obtained from nuclear magnetic resonance spectroscopy, which has been applied to several analogs.[107] The development of two-dimensional NMR techniques will allow detailed analysis of subtle conformational change. Equally sensitive to small changes at the surface of the molecule are monoclonal antibodies. Using panels of antibodies, differential changes in affinity can be used to map mutationally-induced conformational change.[44] Presently, the ultimate test of conformational state is X-ray crystallography. As yet, no semisynthetic analogs have been analyzed in this way, but the recent development of methods to make semisynthetic analogs of the yeast sequence (easier to crystallize than the horse protein) may soon change this. In the absence of determined structures, molecular modeling programs are being used, in our laboratory and others, to predict the consequences of mutations in sequence. A comparison of the properties of semisynthetic [Hse75] cytochrome c, the [Met75] mutant from which it was derived, and the parent yeast cytochrome c bears out the value of such predictions for point substitutions in well known structures.[30]

The kinetics of fragment reassociation in non-covalent complexes have been studied by stopped-flow techniques that monitor both absorbance and fluorescence changes.[32] Thermodynamic studies of complex stability have used both fragment exchange (with labeled and cold fragments),[31] and equilibrium gel filtration and dialysis.[118]

The interaction of semisynthetic cytochrome c with physiological partners has also been subject to scrutiny. Michaelis constants for the oxidase reaction give a sense of the relative affinities,[106] but a more direct measure of K_d and stoichiometry with oxidase can be had from the study of spectral changes accompanying binding.[120] Circular dichroism (CD) and magnetic circular dichroism (MCD) spectra can be informative about conformational changes due to the interaction.[120]

A principal attribute of cytochrome c is its redox potential. E'_m of semisynthetic cytochromes has generally been measured using the uncomplicated method of mixtures in which the absorbance difference at 550 nm is titrated with the ferricyanide-ferrocyanide couple.[88] This technique is suitable for the determination of E'_m values up to about 100 mV less than that of the native protein. Other related spectrophotometric methods have used the Fe^{III}/Fe^{II}-EDTA couple, or the ascorbate-dehydroascorbate couple, which is useful down to below -100 mV.[97] An alternative to redox titrations is the use of differential pulse polarography.[98]

The prime role of the protein is electron transfer between complexes III and IV of the respiratory chain, so measures of the electron transfer efficiency of derivatives are essential to understand structure-function relationships in cytochrome c. Generally kinetic studies have employed either cytochrome oxidase, or the succinate oxidase (cytochrome c reductase) activity of cytochrome c-depleted whole mitochondria.[17] The oxidase reaction can be followed either spectrophotometrically or polarographically.[92,120] Other kinetic assays are possible with the subsidiary redox enzyme systems of the intermembrane space; lactate dehydrogenase,[25] cytochrome c peroxidase, or sulfite oxidase.[121]

The relationship between activity in the depleted mitochondria (reductase) assay and redox potential of analogs has been discussed in Section 8.3.2. The two-fragment complexes with which it was developed did not include any modification to residues in the electron port of cytochrome c (the surface region bounded by the *docking ring* of positive charge and within which the exposed heme edge emerges) or to the residues contacting the ligand sphere of the iron.[17] When such residues are mutated, the relationship is not necessarily followed, and in these cases changes in reaction rate are clearly not simply due to changes in driving force ($\Delta E'_m$) but will indicate a role of the residue either in electron transfer mechanism, influencing the reorganizational energy term of the Marcus equation, or in protein-protein interactions. Thus, such deviation can be a useful pointer to a residue's functional role.

Another source of structural information is the susceptibility to oxidation by environmental O_2. The native protein is relatively inert, but weakening of the internal stability will permit penetration by O_2, and more rapid oxidation.[102] Alterations in conformation resulting from mutations that perturb the distribution of surface charge are often revealed by changes in elution time in cation-exchange high-performance liquid chromatography.[97]

We have been using semisynthetic analogs of the protein to examine another functional property of cytochrome c that we consider biologically significant, the strong affinity for specific anions. Such studies require measures of the number of anions bound under varying conditions. We found that equilibrium gel filtration was most suitable under conditions of low ionic strength and with limited quantities of analog,[45] while for higher ionic strength conditions, equilibrium dialysis has proved most valuable.[122] Another useful means of determining relative affinities has been competition experiments using affinity gels with an ATP ligand, to which cytochrome c binds most strongly.[122]

8.6.2 KEY RESIDUES IN FOLDING AND STABILITY

Semisynthetic analogs of the holoprotein and of noncovalent complexes have helped to identify the crucial primary structural elements responsible for establishing the folding pattern. It has long been clear that the heme group has a major role in directing folding since holo- and apoprotein structures are radically different, so that the final conformation is not established until after the heme group is enzymatically added. Although no analog structures have been determined, biological activity provides a very good indicator of whether the correct fold has been assumed, since the association with physiological partners is crucially dependent on electron port conformation. Thus, the radically different behavior of (1–38)H:(60–104) (totally inactive) from that of semisynthetic (1–38)H:(59–104) (50% activity) confirmed a functional role for the evolutionary conserved Trp59 that was attributed by the authors to its ability to stack with the heme, as in the native structure, and thus define a productive active site structure.[116]

The 695 nm charge transfer band of the ferric iron-thioether sulfur coordination bond is a very sensitive indicator of the perturbation of conformation in the heme crevice, and its absence or weakening in an analog will suggest that the substituted residue had a structural role to play. Thus, unsurprisingly, the placement of a nor-valine residue in the hydrophilic face of the perfect amphipathic helix that forms the conventional left side of the molecule, leads to complete loss of the band.[40] In the L-α-aminobutyric acid (Aba) 78 derivative,[96] the pK for the loss of this band (the alkaline transition) is shifted from 9.3 to <7. This phenomenon may explain the apparent total lack of a 695 nm band noted by ten Kortenaar et al. for the Val78 analog.[92] The crystal structure of the cytochrome shows that the almost completely conserved Thr78 residue is internal and is part of a hydrogen bond network found in the heme crevice.[123] The bond provided by threonine, at least, is important in inhibiting the methionine displacement by an alternative ligand. The only natural substitution known is Asn78, in *Chlamydomonas* cytochrome *c*.[124] We made this change in the horse sequence and found reduced structural stability.[96] Apparently, in the alga, either tolerance of the change is much greater than in any other species, or there is a compensatory change in sequence. In sharp contrast, the replacement in the same network of Tyr67 by Phe actually increases stability to alkaline and thermal denaturation. Thermodynamic calculations suggested that the entropic cost of burying the hydroxyl group in the heart of the molecule outweighed the enthalpic advantage of H-bonding, so that the role of tyrosine is not a structural one.[96]

The severe destabilization of the Aba78 analog is reflected in the lower yield experienced for the (1–65)H:(66–104) religation reaction used in its preparation; the same is true for the [Nva66] protein. Such low yields are therefore likely to be diagnostic of mutations that affect folding. Conversely, the ease with which active chimeric molecules may be formed, even from sequences as distant as horse and yeast,[76,88] demonstrates how well conserved the stabilizing interactions have been during the last billion years of evolution. The pair of residues, Asn70 and Pro71, are absolutely conserved and form a sharp right-angle bend between two α-helices. This *Asx-turn* and its role in defining the cytochrome fold have been probed by independent change at the two residues. The substitution of Pro71 by norvaline gives a quite inactive protein, as a consequence of a shift in the pK of alkaline transition (vide infra) from 9.3 to 5.2, which indicates the crucial role played by this residue.[101]

Semisynthetic derivatives have also revealed residues that are unnecessary to the assumption or maintenance of the cytochrome *c* fold. In particular, any change in the bottom loop or elimination of the loop altogether does not significantly alter the conformation of the remainder of the molecule.[73,74] This result was foreshadowed by the observation that the crystal structure of some bacterial cytochromes, which among many other sequence changes lack this very loop, exhibit the characteristic fold. The apparent tolerance by the folding pattern of elimination of the loop, even when the rest of the molecule is unchanged, provides strong support for the proposal that Ω-loops, of which the bottom loop of cytochrome *c* is one, are independent units of folding.[73] As such, they will rely on intraloop interactions for their stability: we have demonstrated the importance of residues 39 and 40, at the neck of the Ω-loop in this role.[51] The integration of the independent units into the whole, however, requires external interactions. In this context, Arg38 can be viewed as a screw fixing the loop to the rest of the box-like structure enclosing the heme.[38,115]

8.6.3   THE ALKALINE LIGAND-DISPLACEMENT REACTION

Ferricytochrome *c* shows five distinct electronic spectra with changing pH.[125] The change from the neutral form (State III) to State IV is known as the *alkaline transition* and, in the horse protein, occurs with a pK of 8.9 to 9.4, dependent on ionic strength. The spectroscopic changes, which include loss of the 695 nm Fe-S charge-transfer band, are believed to be due to displacement of methionine as sixth coordinating ligand. Since the iron remains low spin, the replacement is presumed to be a strong field ligand.

The origins of this phenomenon, which may have biological significance, have concerned workers since it was first observed, and semisynthesis has been used to address the question. If there is indeed displacement of the 6th ligand methionine, then by what and how far?

The most popular view in the past was that the displacing agent is the side chain of Lysine79.[123] The results of the first study using semisynthetic analogs were interpreted as supporting this view, in that lysine modifications caused significant change in the transition,[126] even though Bosshard[107] had been unable to detect any difference in susceptibility to chemical modification of the most likely lysine residues between neutral and alkaline states.[127] The results of specific chemical modification of lysines had in fact supported the view that Lysine 79 could not be the substitute, but did not exclude Lys72.[128] However, a normal alkaline transition for the plant and fungal cytochromes, where residue 72 is the permanently charged trimethyl lysine, implies that this residue cannot be the substitute, either.[129] NMR evidence, that only minor conformational change occurs during the transition,[130] suggests that just those lysine side chains located nearby can be considered as candidates for the substitution reaction. Apart from 72 and 79, there are other lysines that could conceivably approach the iron atom with modest structural perturbation, including Lysine 73. The recent semisynthesis of an ornithine 80 analog of the horse protein obviously has bearing on this issue.[97] We found that the product was low-spin and had identical electrostatic properties to the native protein, indicating that an uncharged amino group can successfully coordinate the heme iron at, or even below, neutral pH. The UV-visible spectrum had properties that are similar, but not identical, to those of State IV of the horse protein. However, the redox potential (−40 mV) is substantially higher than that (approximately −200 mV) measured for State IV.[131] The [Nva71] analog of the horse protein described above has no detectable 695 nm band at pH 7, and an extrapolated pK value for the transition of 5.2. The UV-visible spectrum at pH 7 is almost identical to that of native State IV, and like it, is resistant to ascorbate reduction. Increased flexibility in the 70s loop is a probable, but yet to be confirmed, consequence of the substitution. Therefore, the 4 pH-unit change in pK_{695} (which is likely to be primarily due to a change in the ease of the conformational transition, rather than the pK of proton ionization) suggests that the structures involved in the conformational transition are in, or in contact with, the 70s loop.

Even though the EPR and MCD signatures of the alkaline form strongly indicate His-Fe-amine coordination, alternatives to lysine side-chains have been considered. Histidines 26 or 33 are believed to provide the 6th ligand in low-spin heme peptides like (1–65) but have been excluded from this role in State IV of the protein by studies of chemically-modified or natural variants.[133] The possibility that methionine remained ligated, but in an alternative conformation with a shifted (and thus obscured) charge-transfer band, or that OH⁻ is the alternate, has been raised.[127,134]

The behavior of some new semisynthetic analogs[84] argues against these sugges-
tions.[97] When methionine is replaced by either alanine or norleucine, neither of
which can ligate, the resulting ferricytochromes remain low spin at neutral pH.
Thus, some alternative ligand must be provided by the protein, in all probability
the same that replaces methionine in the alkaline transition, since their spectra
closely resemble those of State IV. Another possibility to have been aired is Tyr67.[135]
Its –OH group is so close to the iron atom that little conformational change would
occur during the transition, in accordance with the NMR evidence.[130] Modifications
of this residue that affect the pKa of the phenolic hydroxyl do cause shifts in pK_{695}.[136]
Finally, elimination of the hydroxyl group in [Phe67] cytochrome c leads to a
marked increase in the pK of the loss of methionine coordination.[96] However, if the
tyrosine-OH were the coordinating group, then Phe67 cytochrome c could never
exist in the alkaline State IV and would presumably, at that higher pH, transform
directly to State V. In fact, the visible spectrum of [Phe67] analog at pH 11.6
closely resembles that of State IV, not State V (C.J.A. Wallace, unpublished data).
The resolution of at least part of the mystery has been achieved by the University
of British Columbia group. It was first clearly demonstrated by spectroscopic methods
that the alkaline form is a mixture of two alternative conformations, both lysine-
ligated, but by different residues.[137] When Lys79 was mutated to Ala, only a single
alkaline isomer was detected, clearly establishing that residue as one of the two
alternate ligands—the other is probably Lys73, or Lys72 in the horse protein. The
availability of two alternates is, of course, the source of the confusing and often
contradictory evidence described above. Confirmation of this analysis and identification of the other ligand may be aided by spectroscopic studies of the [Nva71] analog
that is in State IV at neutral pH, currently under way. Biological assays of this
mutant have shed some light on the physiological significance of the phenomenon.
It turns out that the analog is a potent inhibitor of electron transfer, rather than
merely inert in the system.[101]

    Thus, there will be strong evolutionary pressure to maintain a pK for the tran-
sition sufficiently high that the proportion of the alkaline isomer present is small
under normal circumstances. Given the necessary presence of many lysine residues
in the docking ring that could effectively compete with the intrinsically weak
methionine for Fe^{III} ligation, it is likely that the primary role of the 70s loop, and
especially the Pro[71] that initiates it, is to enforce the native ligation pattern by its
rigidity.

8.6.4 Binding to Physiological Partners

The strength and specificity of interaction with reductase, oxidase, and ancillary
partners in the mitochondrial membranes and intermembrane space have been largely
revealed by chemical modification studies.[6] Nonetheless, some semisynthetic derivatives have been made to address the issue. The analogs developed by Tesser's group
have invariably been tested with purified oxidase, so that K_m values for this reaction
are known in each case. Modifications of lysine residues in the ring of positive
charge that encloses the electron port on the molecule's surface increase K_m, most

strikingly when two or more are simultaneously changed,[106] providing confirmation of the chemical modification studies. However, they have been able to show that, while changing internal hydrophobic residues leaves K_m of the oxidase reaction unaffected,[107] residues in the hydrophobic stretch that parallels the exposed heme edge of the electron port do contribute to the strength of binding (inasmuch as it is signified by K_m). Thus, mutations at isoleucine 81 and phenylalanine 82 produce a noticeable effect on K_m.[92,108]

This data gave the first indication that forces other than electrostatic might contribute to the binding surface, a proposition now given weight by the determination of structure of the cytochrome c-cytochrome c peroxidase complexes.[13] It now appears that the interacting forces contain electrostatic, H-bonding, and hydrophobic elements.

A novel use to which semisynthetic analogs are being put is in studies of the orientation of the protein relative to the membrane surfaces with which it interacts. X-ray standing waves can be used for this purpose, if the protein contains a detectable heavy atom. The X-rays are reflected at a shallow angle from a silver mirror, and the incident angle determines the wavelength of the standing waves above its surface. By tuning the angle, the wavelength can be varied so that the X-ray fluorescence of a heavy atom monolayer will be maximized when its distance from the mirror is coincident with a standing wave maximum. Such a monolayer will be created if cytochrome molecules containing a heavy atom form a regular array of identical orientation on an artificial membrane layered on the mirror. By using the SeM80-labeled protein,[97] we showed that this is indeed the case,[137] with a monolayer distance from the membrane surface consistent with the proposed *face-down* orientation of cytochrome c on the phospholipid bilayer of the mitochondrial inner membrane.[138] To determine an absolute orientation, though, two reference points are required, so we have now constructed, and commenced study with, a set of four analogs containing both selenium and bromine.[102]

8.6.5 MAINTAINING AN APPROPRIATE REDOX POTENTIAL

All known eukaryotic cytochromes c have E'm values that fall in the range 260 ± 20 mV, and the vast majority are near the middle of this range. Since the protein forms part of an interdependent chain of electron carriers, it is unsurprising that such evolutionary conservatism occurs, and a principal function of the protein coat that enfolds the heme must be to maintain a stable and appropriate redox potential.

How it does so, and at a level strongly divergent from that of free heme, has been the subject of intense speculation. The protein factors suggested to play a major role in modulating heme redox potential are:

1. Axial ligation—the nature of the 5th and 6th ligands[139]
2. The polarity of the residues packing the heme[140]
3. The specific interaction of heme and Trp59[117]
4. The degree of heme exposure to solvent[141]
5. Surface charge and internal dielectric constant[142]
6. Special electrostatic interactions[143]

The evaluation of the relative importance of these factors in eukaryotic cytochrome c has been hampered by the simple fact that they all possess approximately the same midpoint potential, so that analysis has often involved comparison with prokaryotic cytochromes and the consequent simultaneous variation of several different factors.

Protein engineering provides a unique opportunity to vary potential within a common structural framework, and semisynthetic analogs have proved valuable in quantitating the effects of many of these parameters. Those modifications encompassing the bottom loop have demonstrated that its primary role is probably to close off the bottom of the heme from solvent and maintain a high potential.[17,73] Its elimination causes a 120 mV drop in E'_m.[73] Many analogs have caused changes in surface charge, and often the result is a modified redox potential. Replacing Glu66 by glutamine leads to an increase of 16 mV, replacement with lysine increases the potential by 24 mV.[40] Elimination of the negative charge of position 69 raises E'm by 8 mV.[40] In general, the changes induced by charge change are small (reversing the charge on all 19 lysines by citraconylation only leads to 100 mV drop),[133] although Lys39 to Ala or Phe makes a 50 mV difference.[51] This change probably more reflects a weakening of the structure of the bottom loop[51] than it does a suggested specially potent electrostatic interaction due to low dielectric constant of the protein matrix intervening between residue 39 and the iron atom.[115]

A special electrostatic interaction, proposed for Arg38,[143] has been tested by substitutions in a non-covalent complex by lysine and glutamine. Since [Lys38] cytochrome c (1–37H):(38–104) has a potential much lower than that of the Arg38 parent, and only a little higher than that of the [Glu38] analog, it was concluded that the primary role of Arg38 is to stabilize the bottom loop of the protein by H-bonding, and not to provide a counterion to the heme propionate.[115] Thus, the evidence for special electrostatic interactions as controlling influences is sketchy, and it is clear that electrostatic effects in general are of minor importance.

As yet, few semisynthesis experiments have been directed to modulating the polarity of the heme environment. Tyr48 has been made the slightly more hydrophobic iodotyrosine or O-acetyl tyrosine with a consequent E'_m rise of 10 mV in both cases,[83] but the results of modifying Tyr67 are puzzling. The drop in E'_m on introducing p-fluorophenylalanine might be reconciled with the slightly greater polarity of this side chain.[92] However, the clearly less polar phenylalanine and leucine show the same degree of change.[96,106] We have proposed that this influence is a consequence of another modulating factor, the electron distribution within the heme ring itself, which is likely to be controlled by a closely proximate polar group.[96] This view is supported by the shifts in absorbance band wavelengths shown by the Phe67 derivative and by the observation that another influence on electron distribution, the nature of the substituent groups on the heme, also modifies redox potential. *Crithidia* cytochrome c, with one vinyl group replacing a thioether, has a high E'_m.[144] However, the discovery that the Y67F mutation causes a change in the crystal structure that permits an additional water molecule to enter the heme crevice complicates interpretation.[145]

Most recently, the quantitative influence of axial ligation has been addressed. The results give general support for the view that this is a prime modulator of potential and that substitute ligands with particular affinity for ferric iron will cause a substantial drop in potential by stabilizing that state vs. the ferrous one.[139] Thus, Met80 → His or Orn causes a 200 to 300 mV drop,[97,98] while Met80 → Cys reduces E'_m by 300 or 600 mV.[97,99] Alternative thioether ligands, which like Met ligate Fe^{2+} well, maintained a high potential, although the seemingly subtle replacement of sulfur by selenium in the selenomethionine 80 analog led to a 50 mV potential drop.[97] The introduction of non-ligating residues also gives relatively high E'_ms, although the issue is confused by the spin-state change that occurs on reduction of [Ala80] cytochrome c. As well as displaying a myoglobin-like E'_m and absorbance spectrum, this analog binds O_2 in the reduced state.[97]

It is apparent that all of the suggested modulating factors are, in fact, operative within the cytochrome c molecule and that their qualitative influence, with the exception of electrostatic effects, can be very substantial.

8.6.6   Reducing the Barrier to Electron Transfer

Electrons do not readily move from an orbital of a redox center to one in another, distant center, and so it seems likely that the protein coat adopts a catalytic role to ensure a high rate of electron transfer. Some of the proposed mechanisms (applicable to transfers to and from any physiological partner) invoke the participation of residues within the electron port, and these have been the subject of semisynthetic modification for kinetic analysis. As pointed out earlier, rates are also dependent on thermodynamic driving force, represented by $\Delta E'_m$, so discrepancies between observed rates and E'_ms can pinpoint mechanistically important residues.[17]

Because of its absolute conservation and its central position in the active site, Phe82 has always been a prime target for replacement. Early experiments showed that substitution by o-fluorophenylalanine would greatly reduce activity with reductase, but not enough of the analog was available for E'_m determinations.[62] The Phe82 → Leu substitution leaves both E'_m and Vmax for the oxidase reaction unchanged, so that the observed difference, an increased K_m, suggests that the primary role of this residue may be in binding rather than catalysis.[92]

Ile81 is a functionally conserved residue (Ala or Val are evolutionarily acceptable alternatives), and a drastic modification to ε-carbobenzoxylysine diminishes electron transfer rates.[62] More subtle changes, to Ala, Val, or Leu, leave E'_m unaffected.[108] The Ala81 and Val81 proteins have unchanged Vmax (although K_m is increased), but electron transfer rate with oxidase is lower for [Leu81] cytochrome c, hinting that conformation in this region may in fact be important to electron transfer efficiency.

Residue 83 is Ala (or Val) in all animals, Gly in fungi, and Pro in higher plants. We noticed that the plant cytochrome c electron transfer rates in rat mitochondria were anomalously low, although the yeast cytochromes behave normally, and we suspected the cause to be this change.[129] We introduced the same substitution into the horse protein[96] and indeed found unchanged redox potential and a significantly lower rate of electron transfer with reductase. If this is a consequence of reduced conformational flexibility in the electron port, then the implication is that some movement occurs during the electron transfer act, consistent with one set of mechanistic proposals and the results of molecular dynamics simulations.[146,147] However, such putative flexibility could also be of great importance in the matching of hydrophobic surfaces during binding to minimize the distance between redox centers.

The replacement of threonine at position 78 by other residues also provoked a disproportionately large change in activity.[96] This finding implied either a direct mechanistic role for the residue or an indirect one via the methionine 80 to which it is H-bonded. This latter possibility was tested by the series of substitutions we later introduced at this site.[97] Some deviations from the redox potential-reductase activity relationship were noted. Diminished (Selenomethionine, Alanine, Norleucine), normal (S-methyl Cysteine, Thienylalanine), and enhanced (histidine) activities support the view that the ligating residue also makes a significant contribution to the act of electron transfer. In this case, however, it is more likely to be the effect of the change on reorganization energy, as discussed above, than on conductivity between redox centers, that is signalled by these deviations. Recent structural studies of cytochrome mutants, and comparisons of their redox states, suggest that there is a substantial energy cost in reorganizing the H-bond network that includes the Met80 sulfur. However, for the His18 ligand on the other side of the heme, no comparable reorganization is required. If, by analogy, a histidine at position 80 required no conformational change in going from ferrous to ferric state then this mutant would, as is indeed the case, display anomalously high electron transfer rates.[97]

8.6.7   REGULATION OF OVERALL RESPIRATION RATE

We reported the unusual observation that the gross modification of an absolutely conserved residue (Arg91) did not affect the major functional indicators of the resulting semisynthetic cytochrome c.[82] This required the proposal that the residue was conserved for some less obvious function, and we subsequently used these analogs and some modified at other residues to show that cytochrome c was capable of binding ATP in a strong and specific manner at a site that incorporated this residue.[95,121,122] If this binding had a functional role, as was indeed suggested by the conservation of the residue, then an obvious possibility was to mediate feedback inhibition. The effects of site occupancy on the interaction of cytochrome c with its physiological partners have been studied in the absence of free ATP (which could affect other system components) by preparing affinity labeled protein. The adducts have been shown to be labeled at the Arg91-containing site, which has little effect on the physiochemical properties of the protein.[148] However, reactivity with the respiratory chain, in particular the oxidase, is profoundly inhibited.[149] Peptide mapping of these adducts has shown that the C8 of ATP bonds primarily to Lys 86 and 87. Molecular modeling using energy minimization techniques suggests that the nucleotide can fit snugly into the crevice formed between the 60s helix and the 80s stretch,[150] and that lysines of the docking ring move to bind the phosphates, thus supporting a hypothesis of electrostatic hindrance by ATP of cytochrome c-partner interactions. We have also observed that blocking ([DHCH-Arg91] cytochrome c) or eliminating ([Nle91] cytochrome c) the guanido group changes the sensitivity to ATP of redox reactions of cytochrome c with its partners,[103] and of its interaction with phospholipid bilayers.

ACKNOWLEDGMENTS

My contribution to the work described above was made possible by the financial support of the Natural Sciences and Engineering Research Council of Canada, the Swiss National Science Foundation, and the Medical Research Council of the United Kingdom, the many collaborators whose names appear on the relevant publications, especially Dr. Amanda Proudfoot, and the support of my mentor, Professor Robin Offord. I am very grateful to them all and to Charles Bradshaw, Monique Rychner, Barbara Battistolo, and Angela Brigley for their valuable technical assistance.

REFERENCES

1. Margoliash, E. and Schejter, A. *How does a small protein become so popular? in Cytochrome c: a multidisciplinary approach,* Scott, R.A. and Mauk, A.G., Eds., University Science Books, Sausalito, CA, 1996, 3.

2. Dickerson, R.E. The structure and history of an ancient protein, *Sci. Amer.* April, 1972, 58.

3. Voet, D. and Voet J.G. *Biochemistry,* John Wiley and Sons, New York, 1990.

4. *Protein Science,* Cambridge University Press.

5. Brayer, G.D. and Murphy, M.E.P. *Structural studies of eukaryotic cytochromes c, in Cytochrome c: a multidisciplinary approach,* Scott, R.A. and Mauk, A.G., Eds., University Science Books, Sausalito, CA, 1996, 103.

6. Millet, F. and Durham, B. Chemical modification of surface residues of cytochrome c, Ibid, 1996, 573.

7. Sano, S. Chemical synthesis of the cytochrome c molecule, *In Structure and Function of Oxidation-Reduction Enzymes,* Akeson, A. and Ehrenberg, A., Eds. Pergamon, Oxford, 1972, 69.

8. Kent, S.B.H. Solid-phase peptide synthesis, *Annu. Rev. Biochem.* 57, 957, 1988.

9. Moroder, L., Borin, G., Marchiori, F. and Scoffone, E. Studies on cytochrome c VIII, *Biopolymers* 12, 477, 1973.

10. Moroder, L., Filippi, B., Borin, G. and Marchiori, F. Studies on cytochrome c IX, *Biopolymers* 14, 2061, 1975.

11. Pielak, G.J., Mauk, A.G. and Smith, M. Site-directed mutagenesis of cytochrome c shows that an invariant Phe is not essential to function, *Nature* 313, 152, 1985.

12. Keilin, D. *The history of cell respiration and cytochrome,* Cambridge University Press, 1966.

13. Pelletier, H. and Kraut, J. Crystal structure of a complex between electron transfer partners, *Science* 258, 1748, 1994.

14. Pettigrew, G.W. and Moore, G.R. Cytochrome c: *Biological Aspects.* Springer-Verlag, Berlin, 1987.

15. Moore, G.R. and Pettigrew, G.W. Cytochrome c: *Evolutionary, Structural and Physicochemical Aspects.* Springer-Verlag, Berlin, 1990.

16. Richards, F.M. On the enzyme activity of subtilisin-modified ribonuclease, *Proc. Natl. Acad. Sci. U.S.A.* 44, 162, 1958.

17. Wallace, C.J.A. and Proudfoot, A.E.I. On the relationship between oxidation-reduction potential and biological activity in cytochrome c analogues, *Biochem. J.* 245, 773, 1987.

18. Corradin, G. and Harbury, H.A. Reconstitution of horse heart cytochrome *c*. Interaction of the components obtained upon cleavage of the peptide bond following Met residue 65, *Proc. Natl. Acad. Sci. U.S.A.* 68, 3036, 1971.

19. Corradin, G. and Harbury, H.A. Reconstitution of horse heart cytochrome *c*. Reformation of the peptide bond linking residues 65 and 66. *Biochem. Biophys. Res. Commun.* 61, 4100, 1974.

20. Fisher, W.R., Taniuchi, H. and Anfinsen, C.B. On the role of heme in the formation of the structure of cytochrome *c*. *J. Biol. Chem.* 248, 3188, 1973.

21. Wallace, C.J.A. and Harris, D.E. The preparation of fully N-ε-acetimidylated cytochrome *c*, *Biochem. J.* 217, 589, 1984.

22. Harris, D.E. and Offord, R.E. A functioning complex between tryptic fragments of cytochrome *c*. *Biochem. J.* 161, 21, 1977.

23. Westerhuis, L.W., Tesser G.I. and Nivard, R.J.F. Formation of a biologically active complex from two complementary fragments of horse heart cytochrome *c*, *Rec. Trav. Chim. Pays-Bas* 98, 109, 1979.

24. Wilgus, H., Ranweiler, J.S., Wilson, G.S. and Stellwagen, E. Spectral and electrochemical studies of cytochrome *c* peptide complexes, *J. Biol. Chem.* 253, 3265, 1978.

25. Hantgan, R.R. and Taniuchi, H. Formation of a biologically active, ordered complex from two overlapping fragments of cytochrome *c*, *J. Biol. Chem.* 252, 1367, 1977.

26. Parr, G.R., Hantgan, R.R. and Taniuchi, H. Formation of two alternative complementary structures from a cytochrome *c* heme fragment (residues 1 to 38) and the apoprotein, *J. Biol. Chem.* 253, 5381, 1980.

27. Juillerat, M., Parr, G.R. and Taniuchi, H. A biologically active, three-fragment complex of horse heart cytochrome *c*, *J. Biol. Chem.* 255, 845, 1980.

28. Wallace, C.J.A. and Campbell, L.A. Activating peptide fragments for conformationally catalysed resynthesis of proteins, *Proc. Amer. Pept. Symp.* 11, 1043, 1990.

29. Fontana, A., Zambonin, M., DeFilippis, V., Bosco, M. and Polverino de Laureto, P. Limited proteolysis of cytochrome *c* in trifluoroethanol, *FEBS Lett.* 362, 266, 1995.

30. Woods, A.C., Guillemette, J.G., Parrish, J.C., Smith, M. and Wallace, C.J.A. Synergy in protein engineering, *J. Biol. Chem* 271, 32008, 1996.

31. Hantgan, R.R. and Taniuchi, H. Conformational dynamics in cytochrome *c*, *J. Biol. Chem.* 253, 5373, 1978.

32. Parr, G.R. and Taniuchi, H. A kinetic study of the formation of ordered complexes of ferric cytochrome *c* fragments, *J. Biol. Chem.* 254, 4836, 1979.

33. Parr, G.R. and Taniuchi, H. Kinetic intermediates in the formation of ordered complexes from cytochrome *c* fragments, *J. Biol. Chem.* 255, 8914, 1980.

34. Parr, G.R. and Taniuchi, H. A thermodynamic study of ordered complexes of cytochrome *c* fragments, *J. Biol. Chem.* 257, 10103, 1982.

35. Juillerat, M. and Taniuchi, H. Conformational dynamics of a biologically active three-fragment complex of horse cytochrome *c*, *Proc. Natl. Acad. Sci. U.S.A.* 79, 1825, 1982.

36. Fisher, A. and Taniuchi, H. A study of core domains, and the core domain-domain interaction of cytochrome *c* fragment complex, *Arch. Biochem. Biophys.* 296, 1, 1992.

37. Taniuchi, H., Parr, G.R. and Juillerat, M.A. Complementation in folding and fragment exchange, *Methods Enzymol.* 131, 185, 1986.

38. Proudfoot, A.E.I., Wallace, C.J.A., Harris, D.E. and Offord, R.E. A new non-covalent complex of semisynthetically-modified tryptic fragments of cytochrome *c*, *Biochem J.* 239, 333, 1986.

39. Wallace, C.J.A. and Tanaka, Y. Improving cytochrome *c* function by protein engineering? *J. Biochem. (Tokyo)* 115, 693, 1994.

40. Wallace, C.J.A. and Corthesy, B.E. Protein engineering of cytochrome c by semisynthesis: substitutions at glutamic acid 66, *Protein Eng.* 1, 23, 1986.
41. Wallace, C.J.A., Corradin, G., Marchiori, F. and Borin, G. Cytochrome c chimerae from natural and synthetic fragments, *Biopolymers* 24, 2121, 1986.
42. Marcus, R.A. and Sutin, N. Electron transfers in chemistry and biology, *Biochim. Biophys. Acta* 811, 265, 1985.
43. Antalis, T.M. and Palmer, E. Kinetic characterization of the interaction between cytochrome c oxidase and cytochrome c, *J. Biol. Chem.* 257, 6194, 1982.
44. Collawn, J.F., Wallace, C.J.A., Proudfoot, A.E.I. and Paterson, Y. Monoclonal antibodies as probes of conformational change in protein-engineered cytochrome c, *J. Biol. Chem.* 263, 8625, 1988.
45. Corthesy, B.E. and Wallace, C.J.A. The oxidation-state dependent ATP-binding site of cytochrome c: a possible physiological significance, *Biochem. J.* 236, 359, 1986.
46. Dykes, D.F., Creighton, T. and Sheppard, R.C. Spontaneous reformation of a broken peptide chain, *Nature* 247, 202, 1974.
47. Wallace, C.J.A. Chemical studies on cytochrome c, *D. Phil. Thesis,* University of Oxford. University Microfilms Inc., 1976.
48. Harbury, H.A. Cytochrome c: reconstitution, formation of hybrids, and semisynthesis, in *Semisynthetic peptides and proteins,* Offord, R.E. and Di Bello, C., Eds. Academic Press, London 73, 1978.
49. Galpin, I.J. and Hoyland, D.A. Semisynthesis III-the homoserine 12, 105 analogue of hen egg-white lysozyme, *Tetrahedron* 41, 907, 1985.
50. Sheppard, R.C. Selective chain cleavage and combination in protein partial synthesis, *Proc. Amer. Pept. Symp.* 6, 577, 1980.
51. Proudfoot, A.E.I., Rose, K. and Wallace, C.J.A. Conformation-directed recombination of enzyme-activated peptide fragments, *J. Biol. Chem.* 254, 8764, 1989.
52. Kullman, W. *Enzymatic peptide synthesis,* CRC Press, Boca Raton 87, 1987.
53. Juillerat, M. and Homandberg, G.A. Clostripain-catalysed reformation of a peptide bond in a cytochrome c fragment complex, *Int. J. Peptide Protein Res.* 18, 335, 1981.
54. Proudfoot, A.E.I., Offord, R.E., Rose, K., Schmidt, M. and Wallace, C.J.A. A case of spurious product formation during attempted resynthesis of proteins by reverse proteolysis, *Biochem. J.* 221, 325, 1984.
55. Harris, D.E. D. Phil. Thesis, University of Oxford, University Microfilms Inc., 1977.
56. Sano, S. and Tanaka, K. Recombination of protoporphyrinogen with cytochrome c apoprotein, *J. Biol. Chem.* 239, pc3109, 1964.
57. Basile, G., Di Bello, C. and Taniuchi, H. Formation of an iso-1-cytochrome c-like species containing a covalently bonded heme group from the apoprotein by a yeast cell-free system in the presence of hemin, *J. Biol. Chem.* 255, 7181, 1980.
58. Taniuchi, H., Basile, G., Taniuchi, M. and Veloso, D. Evidence for formation of two thioether bonds to link heme to apocytochrome c by partially purified cytochrome c synthetase, *J. Biol. Chem.* 258, 10963, 1983.
59. Veloso, D., Basile, E., and Taniuchi, H. Formation of a cytochrome c-like species from horse apoprotein and hemin catalysed by yeast mitochondrial cytochrome c synthetase, *J. Biol. Chem.* 256, 8646, 1981.
60. Veloso, D., Juillerat, M. and Taniuchi, H. Synthesis of a heme fragment of horse cytochrome c which forms a productive complex with a native apo fragment, *J. Biol. Chem.* 259, 6067, 1984.
61. DiBello, C., Vita, C. and Gozzini, L. Total synthesis of horse heart cytochrome c, *Biochem. Biophys. Res. Commun.* 183, 258, 1992.

62. Wallace, C.J.A. The semisynthesis of some structural analogs of cytochrome *c*, *Proc. Amer. Pept. Symp.* 6, 609, 1979.

63. Boon, P.J. and Tesser, G.I. Protection and deprotection of horse cytochrome *c*, *Int. J. Peptide Protein Res.* 25, 510, 1988.

64. Ledden, D.J., Nix, P.T. and Warme, P.K. Protected natural peptides as intermediates for preparing semisynthetic peptides and protein analogs, *Biochem. Biophys. Acta* 578, 401, 1979.

65. Shaw, W.V. Protein engineering, *Biochem J.* 246, 1, 1987.

66. Wallace, C.J.A. The effect of complete or partial specific acetimidylation on the biological properties of cytochrome *c* and cytochrome *c*-T, *Biochem. J.* 217, 595, 1984.

67. Nix, P.T. and Warme, P.K. Semisynthetic analogs of cytochrome *c* constructed from natural and synthetic peptides, *Biochem. Biophys. Acta* 578, 413, 1979.

68. Offord, R.E. The possible use of cyanogen bromide fragments in the semisynthesis of proteins and polypeptides, *Biochem. J.* 129, 499, 1972.

69. Wallace, C.J.A. and Offord, R.E. The semisynthesis of fragments corresponding to residues 66-104 of horse-heart cytochrome *c*, *Biochem. J.* 179, 169, 1979.

70. Boon, P.J., Tesser, G.I. and Nivard, R.J.F. *Semisynthesis of [Hse65] cytochrome c, in Semisynthetic peptides and proteins,* Offord, R.E. and Di Bello, C., Eds. Academic Press, London, 115, 1978.

71. Rose, K., Herrero, C., Proudfoot, A.E.I., Wallace, C.J.A. and Offord, R.E. Enzyme-assisted semisynthesis of polypeptide active esters for subsequent spontaneous coupling, *Proc. Eur. Pept. Symp.* 19, 219, 1987.

72. Rose, K., Herrero, C., Proudfoot, A.E.I., Offord, R.E. and Wallace, C.J.A. Enzyme-assisted semisynthesis of polypeptide active esters and their use, *Biochem. J.* 249, 83, 1988.

73. Wallace, C.J.A. Functional consequences of the excision of an Ω-loop, residues 40-55, from mitochondrial cytochrome *c*, *J. Biol. Chem.* 262, 16767, 1987.

74. Proudfoot, A.E.I., Rose, K. and Wallace C.J.A. Conformation-directed recombination of enzyme-activated peptide fragments, *J. Biol. Chem.* 254, 8764, 1989.

75. Wallace, C.J.A. Developing a general method for protease-promoted fragment condensation semisynthesis, *Proc. Eur. Pept. Symp.* 21, 260, 1991.

76. Wallace, C.J.A., Guillemette, J.G., Hibiya, Y. and Smith, M. Enhancing protein engineering capabilities by combining mutagenesis and semisynthesis, *J. Biol. Chem.* 266, 21355, 1991.

77. Wallace, C.J.A., Guillemette, J.G., Smith, M. and Hibiya, Y. *Synergy of semisynthesis and site-directed mutagenesis, in Techniques in Protein Chemistry III,* Angeletti, R.H., Ed. Academic Press, New York, 209, 1992.

78. Shoelson, S.E., Polonsky, K.S., Zeidler, A., Rubenstein, A.H. and Tager, A.S. Human insulin B24 (Phe→Ser). Secretion and metabolic clearance of the abnormal insulin in man and in a dog model, *J. Clin. Invest.* 73, 1351, 1984.

79. Markussen, J. and Schaumburg, K. Reaction mechanism in trypsin catalyzed synthesis of human insulin studied by ^{17}O-NMR spectroscopy, *Proc. Eur. Pept. Symp.* 17, 387, 1983.

80. Corradin, G. and Harbury, H.A. Reconstitution of horse heart cytochrome *c*, *Fed. Proc.* 33, 1302, 1974.

81. Harbury, H.A. *Cytochrome c; reconstitution, formation of hybrids and semisynthesis, in Semisynthetic Peptides and Proteins,* Offord, R.E. and Di Bello, C., Eds. Academic Press, London, 73, 1978.

82. Wallace, C.J.A. and Rose, K. The semisynthesis of analogues of cytochrome c: modifications of arginines 38 and 91, *Biochem. J.* 215, 651, 1983.

83. Wallace, C.J.A. The semisynthesis of cytochrome c analogues modified at the invariant tyrosine residue, in *Proceedings of the first Forum Peptides,* Castro, B. and Martinez, J., Eds. Centre de Pharmacologie-Endocrinologie, Montpellier, 434, 1986.

84. Wilgus, H. and Stellwagen, E. Alkaline isomerization of ferricytochrome c: identification of the lysine ligands, *Proc. Natl. Acad. Sci. U.S.A.* 71, 2892, 1974.

85. Barstow L.E., Young, R.S., Yakali, E., Sharp, J.J., O'Brien, J.C., Berman, P.W. and Harbury, H.A. Semisynthetic cytochrome c, *Proc. Natl. Acad. Sci. U.S.A.* 74, 4248, 1977.

86. Atherton, E., Wooley, V. and Sheppard, R.C. Total synthesis of a 39-residue fragment of horse cytochrome c, *J. Chem. Soc. Chem. Commun.,* 970, 1980.

87. Borin, G., Corradin, G., Calderan, A., Marchiori, F. and Wallace, C.J.A. Synthesis of fragments by classical solution methods for use in cytochrome c semisynthesis, *Biopolymers* 25, 2269, 1986.

88. Wallace, C.J.A., Corradin, G., Marchiori, F. and Borin, G. Cytochrome c chimerae from natural and synthetic fragments, *Biopolymers* 24, 2121, 1986.

89. Ten Kortenaar, P.B.W., Tesser, G.I. and Nivard, R.J.F. Synthesis of cytochrome c – (66-104) nonatriacontapeptide and analogues, *Proc. Eur. Pept. Symp.* 17, 349, 1983.

90. Ten Kortenaar, P.B.W., Tesser, G.I. and Nivard, R.J.F. Semisynthesis and properties of some cytochrome c analogues, *Proc. Amer. Pept. Symp.* 7, 211, 1983.

91. Tesser, G.I., Ten Kortenaar, P.B.W., Boots, H.A. Semisynthesis of cytochrome c analogues, *Proc. Eur. Pept. Symp.* 18, 221, 1985.

92. Ten Kortenaar, P.B.W., Adams, P.J.H.M. and Tesser, G.I. Semisynthesis of horse cytochrome c from two or three fragments, *Proc. Natl. Acad. Sci. U.S.A.* 82, 8279, 1985.

93. Di Bello, C., Tonellato, M., Lucciari, A., Buso, O. and Gozzini, I. Conformationally-driven covalent semisynthesis of apocytochrome c, in *Peptide Chemistry 1987,* T. Shiba and S. Sakakibara, Eds. Protein Res. Found., Osaka 409, 1988.

94. Di Bello, C., Tonellato, M., Lucchiari, A., Buso, O., Gozzini, L. and Vita, C. Solid-phase synthesis of the native sequence of horse heart cytochrome c – (66-104) non-atriacontapeptide and of a C-terminal carboxyamide analog selectively modified at Met-80, *Int. J. Protein Peptide Res.* 35, 336, 1990.

95. Wallace, C.J.A., Proudfoot, A.E.I., Mascagni, P. and Kent, S.B.H. Protein engineering of cytochrome c: substitutions of Tyr67, Thr78, and Ala83 of the horse protein by semisynthesis, *Proc. Eur. Pept. Symp.* 20, 283, 1989.

96. Wallace, C.J.A., Mascagni, P., Chait, B.T., Collawn, J.F., Paterson, Y., Proudfoot, A.E.I. and Kent, S.B.H. Substitutions engineered by chemical synthesis at three conserved sites in mitochondrial cytochrome c, *J. Biol. Chem.* 264, 15199, 1989.

97. Wallace, C.J.A. and Clark-Lewis, I. Functional role of heme ligation in cytochrome c, *J. Biol. Chem.* 267, 3852, 1992.

98. Raphael, A.L. and Gray, H.B. Axial ligand replacement in horse heart cytochrome c by semisynthesis, *Proteins: structure, function, genetics* 6, 338, 1989.

99. Raphael, A.L. and Gray, H.B. Semisynthesis of axial ligand (position 80) mutants of cytochrome c, *J. Amer. Chem. Soc.* 113, 1038, 1991.

100. Frauenhoff, M.M. and Scott, R.A. The role of tyrosine 67 in the cytochrome c heme crevice structure studied by synthetic site-67 substitution. *Proteins* 14, 202, 1992.

101. Wallace, C.J.A. and Clark-Lewis, I. A rationale for the absolute conservation of Asn70 and Pro71 in mitochondrial cytochrome c suggested by protein engineering, *Biochemistry* 36, 14732, 1997.

102. Wallace, C.J.A. and Clark-Lewis, I. Using semisynthesis to insert heavy-atom labels in functional proteins, *In Techniques in Protein Chemistry VII,* Marshak, D.R., Ed: Academic Press, San Diego, 499, 1996.

103. Tuominen, E.K.J., Wallace, C.J.A., and Kinnunen, P.K.J. The invariant Arg91 is required for the rupture of liposomes by cytochrome *c, Biochem. Biophys. Res. Commun.,* 238, 140, 1997.

104. Tesser, G.I., Boon, P.J. and Nivard, R.J.F. Semisynthesis of a cytochrome *c* analog *Proc. Eur. Pept. Symp.* 15, 671, 1979.

105. Boon, P.J., Tesser, G.I. and Nivard, R.J.F. Semisynthetic horse heart [65-homoserine] cytochrome *c* from three fragments, *Proc. Natl. Acad. Sci. U.S.A.* 76, 61, 1979.

106. Boon, P.J., van Raay, A.J.M., Tesser, G.I. and Nivard, R.J.F. Semisynthesis, conformation and cytochrome *c* oxidase activity of eight cytochrome *c* analogues, *FEBS Lett.* 108, 131, 1979.

107. Boon, P.J., Tesser,G.J., Brinkhof, H.H.K. and Nivard, R.J.F. Semisynthetic horse heart cytochrome *c* analogues: Hse65, Leu67 and Hse65, Leu74 cytochrome *c,* physicochemical and biochemical evaluation, *Proc. Eur. Pept. Symp.* 16, 301, 1981.

108. Boots, H.A. and Tesser, G.I. Synthesis and properties of [Hse65, Ala81] and [Hse65, Val81] cytochrome *c, Proc. Eur. Pept. Symp.* 19, 211, 1987.

109. Koul, A.K., Wasserman, G.F. and Warme, P.K. Semi-synthetic analogs of cytochrome *c* at positions 67 and 74, *Biochem. Biophys. Res. Commun.* 89, 1253, 1979.

110. Wallace, C.J.A. Semisynthetic analogues of cytochrome *c*: modifications to fragment (66-104), in *Semisynthetic Peptides and Proteins,* Offord, R.E. and Di Bello, C., Eds. Academic Press, London, 101, 1978.

111. Gozzini, L., Taniuchi, H and Di Bello, C. Complexation which facilitates rejoining of horse cytochrome *c* apo fragment [Hse65] (1-65) or [Hse65] (23-65) to apo fragment (66-104), *Int. J. Peptide Protein Res.* 37, 293, 1991.

112. Vita, C., Gozzini, L. and DiBello, C. Total synthesis of horse heart apocytochrome *c* by conformation-assisted condensation of two chemically synthesised fragments, *Eur. J. Biochem.* 204, 631, 1992.

113. Harris, D.E. Cytochrome *c-T, in Semisynthetic Peptides and Proteins,* Offord, R.E. and Di Bello, C. Eds. Academic Press, London, 127, 1978.

114. Harris, D.E. Semisynthetic analogs of cytochrome *c* prepared by the noncovalent association of modified tryptic fragments, *Proc. Amer. Pept. Symp.* 6, 613, 1979.

115. Proudfoot, A.E.I. and Wallace, C.J.A. Semisynthesis of cytochrome *c* analogues. The effect of modifying the conserved residues 38 and 39, *Biochem. J.* 248, 965, 1987.

116. Westerhuis L.W., Tesser, G.I. and Nivard, R.J.F. A functioning complex of two cytochrome *c* fragments with deletion of the (39-58) eicosapeptide. *Int. J. Peptide Protein Res.* 19, 290, 1982.

117. Myer, Y.P., Saturno, A.F., Verma, B.C. and Pande, A. Horse-heart cytochrome *c*: the redox potential and protein structures, *J.Biol. Chem.* 254, 11202, 1979.

118. Juillerat, M.A. and Taniuchi, H. A study of the role of evolutionary invariant leucine 32 of cytochrome *c, J. Biol. Chem.* 261, 2697, 1986.

119. Poerio, G., Parr, G.R. and Tanuichi, H. A study of roles of evolutionary invariant Proline 30 and glycine 34 of cytochrome *c, J. Biol. Chem.* 261, 10976, 1986.

120. Michel, B., Proudfoot, A.E.I., Wallace, C.J.A. and Bosshard, H.R. The cytochrome *c* oxidase-cytochrome *c* complex, *Biochemistry* 28, 456, 1989.

121. Craig, D.B. and Wallace, C.J.A. Studies of 8-azido ATP adducts reveal two mechanisms by which ATP binding to cytochrome *c* could inhibit respiration, *Biochemistry* 34, 2686, 1995.

122. Craig, D.B. and Wallace, C.J.A. The specificity and Kd at physiological ionic strength of an ATP-binding site on cytochrome *c* suit it to a regulatory role, *Biochem J.* 279, 781, 1991.

123. Takano, T. and Dickerson, R.E. Conformation change of cytochrome *c*, *J. Mol. Biol.* 153, 95, 1981.

124. Amati, B.B., Goldschmidt-Clermont, M., Wallace, C.J.A. and Rochaix, J-D. cDNA and deduced amino acid sequences of cytochrome *c* from Chlamydomonas *J. Mol. Evol.* 28, 151, 1988.

125. Theorell, H. and Akesson, A. Studies on cytochrome *c*, *J. Amer. Chem. Soc.* 63, 1804, 1941.

126. Wallace, C.J.A. Modulation of the alkaline transition in cytochrome *c* and cytochrome *c* – T by full or specific partial acetimidylation, *Biochem. J.* 217, 601, 1984.

127. Bosshard, H.R. Alkaline isomerisation of ferricytochrome *c*, *J. Mol. Biol.* 153, 1125, 1981.

128. Smith, H.T. and Millet, F. Involvement of lysines 72 and 79 in the alkaline isomerization of horse heart ferricytochrome *c*, *Biochemistry* 19, 1117, 1980.

129. Wallace, C.J.A. and Boulter, D. Spectroscopic and redox properties of plant mitochondrial cytochromes c, *Phytochemistry* 27, 1947, 1988.

130. Boswell, A.P., Moore, G.R., Williams, R.J.P., Harris, D.E., Wallace, C.J.A., Bocieck, S and Welti, D. Ionisation of tyrosine and lysine residues in nature and modified horse cytochrome *c*, *Biochem. J.* 213, 679, 1983.

131. Barker, P.D. and Mauk, A.G. pH-linked conformational regulation of a metalloprotein oxidation-reduction equilibrium, *J. Amer. Chem. Soc.* 114, 3619, 1992.

132. Gadsby, P.M.A., Peterson, J., Foote, N., Greenwood, C. and Thomson, A.J. Identification of the ligand-exchange process in the alkaline transition of horse-heart cytochrome *c*, *Biochem. J.* 246, 43, 1987.

133. Wallace, C.J.A. and Corthesy, B.E. Alkylamine derivatives of cytochrome *c*, *Eur. J. Biochem.* 170, 293, 1987.

134. Pettigrew, G.W., Aviram, I and Schejter, A. The role of the lysines in the alkaline heme-linked ionization of ferric cytochrome *c*, *Biochem. Biophys. Res. Commun.* 68, 807, 1976.

135. Salemme, F.R., Kraut, J. and Kamen, M.D. Structural basis for function in cytochrome *c*, *J. Biol. Chem.* 248, 7701, 1973.

136. Skov, K. and Williams, G.R. Correlations between ORD and other spectroscopic properties of ferricytochrome *c*, *Can. J. Biochem.* 49, 441, 1971.

137. Wang, J., Wallace, C.J.A., Clark-Lewis, I. and Caffrey, M. Structure characterization of membrane bound and surface adsorbed protein, *J. Mol. Biol.* 237, 1, 1994.

138. Salemme, F.R. Structure and function of cytochromes *c*, *Annu. Rev. Biochem.* 46, 299, 1977.

139. Moore, G.R. and Williams, R.J.P. Structural basis for the variation in redox potential of cytochromes, *FEBS Lett.* 79, 229, 1977.

140. Kassner, R.J. A theoretical model for the effects of local non polar heme environments on the redox potential of heme complexes, *J. Amer. Chem. Soc.* 95, 2674, 1973.

141. Stellwagen, E. Heme exposure as the determinant of oxidation-reduction potential of proteins, *Nature* 275, 73, 1978.

142. Rees, D.C. Experimental evaluation of the effective dielectric constant of proteins, *J. Mol. Biol.* 141, 323, 1980.

143. Moore, G.R. Control of redox properties of cytochrome *c* by special electrostatic interactions, *FEBS Lett* 161, 171, 1983.

144. Moore, G.R., Harris, D.E., Leitch, F.A. and Pettigrew, G.W. Characterization of ionisations that influence the redox potential of mitochondrial cytochrome *c*, *Biochem. Biophys. Acta* 764, 331, 1984.

145. Berghuis, A.M., Guillemette, J.G., Smith, M. and Brayer, G.D. Mutation of Tyr[67] to Phe in cytochrome *c* significantly alters the local heme environment, *J. Mol. Biol.* 235, 1326, 1994.

146. Poulos, T.L. and Kraut, J. A hypothetical model of the cytochrome *c* peroxidase: cytochrome *c* electron transfer complex, *J. Biol. Chem.* 255, 10322, 1980.

147. Wendeloski, J.J., Mathew, J.B., Weber, P.C. and Salemme, F.R. Molecular dynamics of a cytochrome *c*-cytochrome b$_5$ electron transfer complex, *Science* 238, 794, 1987.

148. Craig, D.B. and Wallace, C.J.A. ATP binding to cytochrome *c* diminishes electron flow in the mitochondrial respiratory pathway, *Protein Science* 2, 966, 1993.

149. Craig, D.B. and Wallace, C.J.A. Studies of 8-azido-ATP adducts reveal two mechanisms by which ATP binding to cytochrome *c* could inhibit respiration, *Biochemistry* 34, 2686, 1995.

150. McIntosh, D.B., Parrish, J.C. and Wallace, C.J.A. Definition of a nucleotide binding site on cytochrome *c* by photoaffinity labeling, *J. Biol. Chem.*, 271, 18379, 1996.

151. Pollock, W.B.R., Rosell, F.I., Twitchett, M.B., Dumont, M.E. and Mauk, A.G., Bacterial expression of a mitochondrial cytochrome c. Trimethylation of Lys72 in yeast iso-1-cytochrome c and the alkaline conformal transition, *Biochemistry,* 37, 6124, 1998.

Index